LONDON MATHEMATICAL SOCIETY STUDEN

T0296336

London Mathematical Society Student Texts. 14

Combinatorial Group Theory: a topological approach

DANIEL E. COHEN

Reader in Pure Mathematics, Queen Mary College, University of London

The right of the University of Cambridge to print and sell all manner of books was granted by Henry VIII in 1534. The University has printed and published continuously since 1584.

CAMBRIDGE UNIVERSITY PRESS

Cambridge

New York Port Chester Melbourne Sydney

CAMBRIDGE UNIVERSITY PRESS
Cambridge, New York, Melbourne, Madrid, Cape Town, Singapore, São Paulo, Delhi

Cambridge University Press
The Edinburgh Building, Cambridge CB2 8RU, UK

Published in the United States of America by Cambridge University Press, New York

www.cambridge.org
Information on this title: www.cambridge.org/9780521341332

First published 1989
Re-issued in this digitally printed version 2008

A catalogue record for this publication is available from the British Library

ISBN 978-0-521-34133-2 hardback
ISBN 978-0-521-34936-9 paperback

To the memories of Ralph Fox, from whom I first learned of the connections between group theory and topology, and of Hanna Neumann and Roger Lyndon, whose works extended my interest in group theory.

INTRODUCTION

Combinatorial group theory can be regarded as that branch of group theory which considers groups given by generators and relations. Some of its basic results involve manipulation with words; that is, products of the generators and their inverses. It is this aspect to which the word "combinatorial" refers; it is not connected with that branch of mathematics known as combinatorics.

From its earliest stages this theory has been closely connected with topology. To any topological space there is an associated group, called the *fundamental group* of the space. In trying to investigate properties of certain spaces we are led to problems in combinatorial group theory. Conversely, some problems in combinatorial group theory are best solved by geometric and topological discussions of suitable fundamental groups.

It is this interplay between group theory and topology which is the theme of this book. The texts on combinatorial group theory by Magnus, Karrass, and Solitar (1966) and by Lyndon and Schupp (1977) go much deeper into the group theory, but have little to say about the topology, while the texts by Massey (1967) and Stilwell (1980) concentrate on the topology rather than the group theory.

Chapter 1 contains the main constructions of combinatorial group theory; free groups, presentations, free products, amalgamated free products, and the HNN extension. The various Normal Form Theorems are proved, with several different proofs, and applications of the constructions are made (for instance, to show the existence of a finitely generated infinite simple group).

Chapter 2 has some topological preliminaries; ways of building new spaces from old ones, and a discussion of paths in spaces.

From looking at paths in spaces, we are led to consider objects very like groups, but for which the multiplication is not defined everywhere. These objects are called *groupoids*, and their theory is developed in Chapter 3.

Chapter 4 contains the main properties of fundamental groups, including van Kampen's theorem which gives the fundamental group of the union of two spaces in terms of the fundamental groups of the spaces and their intersection. The proof of this theorem is simplified by the use of the fundamental groupoid. There follows a brief account of covering spaces, the deeper properties being left to a later chapter. It is shown how these properties can be used to obtain results about the complex plane; in particular, to give a proof of the Fundamental Theorem of Algebra. The chapter concludes with a discussion of some surprising situations which can occur when the spaces considered are not nice enough.

In Chapter 5 we look at graphs and complexes. Part of their theory parallels that of Chapter 4, but without the topological complications that may occur; occasionally, however, there is some cost in avoiding these technicalities. Among the material in this chapter is a discussion of the fixed subgroup of an automorphism of a free group. This uses the very easy approach of Goldstein and Turner, and its recent extension by M. Cohen and Lustig.

In Chapter 6 we discuss coverings of spaces and complexes. This is put to use in group theory in Chapter 7. In that chapter we prove Schreier's theorem on subgroups of a free group, and Kurosh's theorem on subgroups of a free product. These theorems are proved both using coverings and by purely algebraic means, and the connection between these two approaches is investigated in detail. The material in this chapter is perhaps the heart of the interplay between topology and group theory.

The long Chapter 8 is devoted to the theory of groups acting on trees, due to Bass and Serre. To some extent, this can be described as a one-dimensional rewriting of the theory of coverings, and this one-dimensional aspect makes it easier to work with than the other approaches, once the main theorems have been proved. We begin with a discussion of free actions of groups on trees, which leads to yet another proof of Schreier's theorem. We then look at Nielsen's approach to subgroups of free groups. After that we develop the general results of Bass-Serre theory. These results are then applied, in the first place to give a simple proof of the Kurosh Subgroup Theorem and related theorems about the subgroups of amalgamated free products and HNN extensions. Further applications are to Grushko's theorem on free products, and to the theorems of M. Hall and of Howson about finitely generated subgroups of free groups. These theorems find their best versions, including various generalisations, in this context. The chapter concludes with a discussion of general ways of constructing trees, including a statement of the important new theorem due to Dicks and Dunwoody, showing how this theorem can be applied to give the structure of groups with a free subgroup of finite index.

Chapter 9 is about decision problems for groups. It begins with a discussion about the nature of decision problems. Some easy decision problems are then looked at, together with an introduction to the word problem. We then show that there is a group with unsolvable word problem, and we prove Higman's Embedding Theorem, in both cases using the method of modular machines, which provides short proofs of these results. Various other related problems are then looked at, including the unsolvability of the isomorphism problem. The chapter concludes with a modern proof of Magnus's Freiheitssatz for one-relator groups and his solution of the word problem for these groups.

Chapter 10 is a quick account, without proofs, of some other topics in the theory, including small cancellation theory and Whitehead's results on automorphisms.

This book is an expanded version of a book produced in the Queen Mary College Lecture Notes series some ten years ago. The latter was an account of a two-term course of postgraduate lectures given in 1976-7.

In my original two-term course I covered most of the material in Chapters 1 to 7. For a one-semester course the best choice is to concentrate either on the group theory or the topology. I am currently giving a one-semester course, in which I cover Chapter 1, the first three sections of Chapter 5, and the part of Chapter 7 dealing with Schreier's Theorem using graphs but not coverings. Alternatively, Chapters 2,4, possibly parts of 5, 6, and some of 7 could form a course on the fundamental group, while Chapter 2 and parts of Chapter 4 (especially the early part and the section on the circle and the complex plane) could even form part of an undergraduate topology course.

Readers will notice that I have used the phrase "it is easy to see that..." very often. I believe that, whenever I have used this phrase, all the results are genuinely easy to prove. Readers should regard such results as additional exercises.

I am grateful to Ian Chiswell for reading the early parts of this book and for pointing out obscurities and inaccuracies, which I have endeavoured to correct.

The typist of a complex mathematical work like this deserves credit, and I am happy to acknowledge all my hard work in typing this book. I would also like to thank Franz Schmerbeck, the creator of the Signum word processor (running on the Atari ST), without whose valuable work the production of this book would have been much more difficult.

TABLE OF CONTENTS

1 COMBINATORIAL GROUP THEORY

1.1 *FREE GROUPS*

Let X be a generating subset of a group G. Certain products of members of X and their inverses will be 1 whatever X and G are; for instance, $xyyz^{-1}zy^{-1}y^{-1}x^{-1}$. Other products, such as xyz or xx, will be 1 for some choices of X and G but not for other choices. Those pairs G and X for which a product of elements in $X \cup X^{-1}$ is 1 only when the properties holding in all groups require it to be 1 are obviously of interest.

They are called *free groups;* a more formal definition will be given later. If G is such a group, any function f from X to a group H can be extended uniquely to a homomorphism from G to H. For any $g \in G$ can be written as $x_{i_1}^{\varepsilon_1} \dots x_{i_n}^{\varepsilon_n}$ where $\varepsilon_r = \pm 1$ and $x_{i_r} \in X$ for $r = 1, \dots, n$. Now suppose that g can also be written as $x_{j_1}^{\delta_1} \dots x_{j_m}^{\delta_m}$, where $\delta_s = \pm 1$ and $x_{j_s} \in X$ for $s = 1, \dots, m$. Then

$$x_{i_1}^{\varepsilon_1} \dots x_{i_n}^{\varepsilon_n} x_{j_m}^{-\delta_m} \dots x_{j_1}^{-\delta_1} = 1$$

and our assumption on G and X then tells us we must have

$$(x_{i_1}f)^{\varepsilon_1} \dots (x_{i_n}f)^{\varepsilon_n} (x_{j_m}f)^{-\delta_m} \dots (x_{j_1}f)^{-\delta_1} = 1.$$

Hence the element of H given by $(x_{i_1}f)^{\varepsilon_1} \dots (x_{i_n}f)^{\varepsilon_n}$ depends only on g and not on how g is written as a product of elements of $X \cup X^{-1}$. It follows that we can define a function $\varphi: G \to H$ by requiring $g\varphi$ to be this element. It is easy to check that φ is a homomorphism and that $x\varphi = xf$ for all $x \in X$. Since every element of G is a product of elements of X and their inverses, there can only be one homomorphism with specified values on X. Conversely, suppose that X is a subset of a group G such that any function from X into an arbitrary group H extends uniquely to a homomorphism from G to H. Let $x_{i_1}^{\varepsilon_1} \dots x_{i_n}^{\varepsilon_n}$ equal 1 in G, and let f be a function from X into a group H. Since f extends to a homomorphism from G to H, it follows that $(x_{i_1}f)^{\varepsilon_1} \dots (x_{i_n}f)^{\varepsilon_n}$ equals 1 in H.

Since this holds for all H and f, this amounts to saying that the product of elements of $X \cup X^{-1}$ equals 1 in G only if the corresponding products in all groups are also 1. It can be shown that this holds iff we can derive the fact that the product equals 1 from the group laws. These observations lead to the formal definition of free groups.

Definition Let X be a set, G a group, and $i:X \to G$ a function. The pair (G,i) is called *free on X* if for every group H and function $f:X \to H$ there is a unique homomorphism $\varphi:G \to H$ such that $f = i\varphi$.

In particular, the trivial group is free on the empty set, and the infinite cyclic group $\mathbb{Z}$ is free on any one element set $\{x\}$, where xf is the integer 1.

The following three questions are natural ones to ask. Do free groups on an arbitrary set exist? Are free groups on a given set unique? If (G,i) is free on X is i injective? We leave the first question till later. The answer to the second question and a partial answer to the third are of an abstract nature, applying in many situations in abstract algebra and category theory.

It will be useful at times to use the language of category theory, which frequently sheds light on general situations. (Barry Mitchell once wrote "The purpose of category theory is to show that what is trivial is trivially trivial.") For instance, in asking whether free groups on an arbitrary set exist we are asking whether the forgetful functor from groups to sets has an adjoint, and the fact that free groups on a given set are essentially unique is just a uniqueness property of adjoints (or of initial objects in categories). Readers who have no knowledge of category theory can safely ignore any remarks about it; such remarks are made only to give extra insight to those with the relevant knowledge.

Plainly, if (G,i) is free on X and $\varphi:G \to H$ is an isomorphism then $(H,i\varphi)$ is also free on X. Our first proposition is the converse of this.

Proposition 1 *Let (G_1,i_1) and (G_2,i_2) be free on X. Then there is an isomorphism $\varphi:G_1 \to G_2$ such that $i_1\varphi = i_2$.*

Proof Since (G_1,i_1) is free on X, there is, by definition, a homomorphism $\varphi:G_1 \to G_2$ such that $i_1\varphi = i_2$. Similarly, there is a homomorphism $\psi:G_2 \to G_1$ such that $i_2\psi = i_1$. We then have $i_1\varphi\psi = i_1 = i_1 I_1$, where I_1 is the identity function on G_1. The uniqueness property in the definition now requires that

$\varphi\psi = I_1$, and, similarly, $\psi\varphi = I_2$, the identity function on G_2. Hence φ is an isomorphism.//

Proposition 2 *Let (F,i) be free on X.*

(i) If there is a group G with an injective function from X to G then i is injective.

(ii) There is a group into which X maps injectively; for instance. the set $\mathbb{Z}^X$ of all functions from X to $\mathbb{Z}$, where $\alpha + \beta$ is defined by $x(\alpha+\beta) = x\alpha + x\beta$ for all x.

(iii) i is injective.

Proof (i) Let $f:X \to G$ be an injection. Then there is a homomorphism $\varphi:F \to G$ such that $f = i\varphi$. It then follows that i is injective. Notice that this part of the argument is of an abstract nature, holding in many situations, whereas (ii) requires a specific example.

(ii) Plainly $\mathbb{Z}^X$ is a group. If we define α_x by $x\alpha_x = 1$ and $y\alpha_x = 0$ for $y \neq x$ then the function sending x to α_x is plainly injective. Part (iii) is now immediate.//

We now proceed to construct a group free on X, beginning with an auxiliary construction. Let X be any set, and let $M(X)$ denote the set of all finite sequences $(x_{i_1}, \ldots, x_{i_n})$ of elements of X, where $n \geq 0$ (the case $n = 0$ corresponds to the empty sequence). Define a multiplication on $M(X)$ by

$$(x_{i_1}, \ldots, x_{i_n})(x_{j_1}, \ldots, x_{j_m}) = (x_{i_1}, \ldots, x_{i_n}, x_{j_1}, \ldots, x_{j_m}).$$

This multiplication is obviously associative, with an identity element which we call 1 (namely, the empty sequence). Also, $x \to (x)$ is obviously one-one, and, if we identify x with (x), every element of $M(X)$ can be uniquely written as a product $x_{i_1} \ldots x_{i_n}$ for some n; we shall always use this notation. We call $M(X)$ the *free monoid* on X.

By a *segment* of $x_{i_1} \ldots x_{i_n}$ we mean an element $x_{i_r} x_{i_{r+1}} \ldots x_{i_s}$, where $1 \leq r \leq s \leq n$; this is called an *initial segment if* $r = 1$, and a *final segment* (or *terminal segment*) if $s = n$, and it is a *proper segment* unless $r = 1$ and $s = n$.

One technical point has been glossed over in this construction. Since X can be any set, it is possible that some element of X is itself a finite sequence of other elements of X. We would then want to distinguish between this sequence as an element of X and as an element of $M(X)$. The simplest way of dealing with this problem is to replace X by the set X' which is defined to be the set of all $\{x\}$ for $x \in X$, and to define the free monoid on X to be $M(X')$. However, we shall ignore this technicality in future.

We now proceed to construct the free group on X. Take a set $\bar{X}$ bijective with X under a bijection which sends x to $\bar{x}$, and such that $X \cap \bar{X} = \emptyset$. We usually denote $\bar{x}$ by x^{-1}, and we may also write x^1 instead of x. The elements of $M(X \cup \bar{X})$ are called *words on X*. Let w be the word $x_{i_1}^{\varepsilon_1} \ldots x_{i_n}^{\varepsilon_n}$. Then n is called the *length* of w, written $|w|$ or $l(w)$, and we call the elements $x_{i_r}^{\varepsilon_r}$ the *letters* of w.

The word w is called *reduced* if, for $1 \leq r \leq n-1$, either $i_{r+1} \neq i_r$ or $i_{r+1} = i_r$ but $\varepsilon_{r+1} \neq -\varepsilon_r$; the empty word is also called reduced. Suppose that w is not reduced, and choose r such that $i_{r+1} = i_r$ and $\varepsilon_{r+1} = -\varepsilon_r$. Let w' be the word obtained from w by deleting the adjacent pair of letters $x_{i_r}^{\varepsilon_r}$ and $x_{i_{r+1}}^{\varepsilon_{r+1}}$. We say that w' comes from w by an *elementary reduction*. If w'' is obtained from w by a sequence of elementary reductions we say that w'' comes from w by *reduction*. It is usually convenient to allow w'' to be w in this definition (corresponding to the empty sequence of elementary reductions); readers will be left to decide for themselves on each occasion whether this case is permitted or not.

Examples Let w be the word $zxx^{-1}zy^{-1}y$. Then both $zzy^{-1}y$ and $zxx^{-1}z$ come from w by elementary reductions, and zz comes from w by reduction.

Let w be $zxx^{-1}xy$. Then zxy comes from w by elementary reduction. Two elementary reductions may be applied to w, both giving the same result; the first deletes the pair of letters xx^{-1}, while the second deletes the pair $x^{-1}x$.

Write, for the moment, $w \approx w'$ iff either w is identical to w' or there is a sequence of words $w_1, \ldots, w_k$ for some k such that w_1 is w and w_k is w' and, for each $j < k$, one of w_{j+1} and w_j comes from the other by elementary reduction. Plainly $\approx$ is an equivalence relation. We denote the set of equivalence classes by $F(X)$. Whenever we have an equivalence relation on a set, the equivalence class of the element w will be denoted by $[w]$. This notation will be used without further comment (but occasionally the notation will have a different meaning; for instance, $[a,b]$ is used for a closed interval of real numbers).

It is easy to see that if u, v, w, and w' are words such that $w \approx w'$ then $uwv \approx uw'v$. It follows, since $\approx$ is an equivalence relation, that if $u \approx u'$ and $w \approx w'$ then $uw \approx u'w'$ (as both are equivalent to uw'). This enables us to define a multiplication on $F(X)$, by requirng $[u][w]$ to be $[uw]$. This multiplication is plainly associative, and has an identity (the class of the empty sequence) which we denote by 1 (just as the empty sequence itself is also denoted by 1). When w is the word $x_{i_1}^{\varepsilon_1}\dots x_{i_n}^{\varepsilon_n}$ we let w^{-1} denote the word $x_{i_n}^{-\varepsilon_n}\dots x_{i_1}^{-\varepsilon_1}$. Then $ww^{-1}\approx 1$, as we see easily. Hence $F(X)$ is a group. Define a function $i{:}X\to F(X)$ by $xi = [x]$. Plainly $F(X)$ is generated by Xi.

Theorem 3 *$(F(X),i)$ is free on X.*

Proof Let G be a group and $f{:}X\to G$ a function. Then f extends to a function from $M(X\cup\bar{X})$ to G sending the word $x_{i_1}^{\varepsilon_1}\dots x_{i_n}^{\varepsilon_n}$ to the element $(x_{i_1}f)^{\varepsilon_1}\dots(x_{i_n}f)^{\varepsilon_n}$. It is immediate that if w' comes from w by an elementary reduction then w and w' have the same image. It follows that if $w\approx w''$ then w and w'' have the same image. Hence we may define a function φ from $F(X)$ to G by requiring $[w]\varphi$ to be the image of w. It is obvious that φ is a homomorphism and that $xf = xi\varphi$. Further, since Xi generates $F(X)$, there can be at most one homomorphism from $F(X)$ to G with specified values on Xi.//

Clearly, because elementary reduction decreases length, when we start with any word and apply elementary reductions one after another in an arbitrary way until no more can be applied, we will ultimately reach a reduced word. In particular, every equivalence class contains at least one reduced word.

Theorem 4 (Normal form theorem for free groups) *There is exactly one reduced word in each equivalence class.*

Remarks As a special case of this theorem, we find that there is exactly one reduced word which can be obtained by reduction from a given word.

Because this theorem is so important, we give several proofs. Similar proofs apply in many situations.

We have seen that every equivalence class contains at least one reduced word. The difficult property is that each class contains at most one reduced word, and we now give proofs of this fact.

Proofs (I) Canonical reduction.

Define a function $\lambda{:}M(X\cup\bar{X})\to M(X\cup\bar{X})$ inductively by

$1\lambda = 1$, $x^{\varepsilon}\lambda = x^{\varepsilon}$, $(ux^{\varepsilon})\lambda = (u\lambda)x^{\varepsilon}$ if $u\lambda$ does not end in $x^{-\varepsilon}$, and $(ux^{\varepsilon})\lambda = v$ if $u\lambda$ is $vx^{-\varepsilon}$.

It is easy to check, by induction on the number of letters in w, that $w\lambda$ is reduced for any word w, that $w\lambda$ is w if w is reduced, and that $w\lambda$ comes from w by reduction. Also, because $u\lambda$ is reduced, $(ux^{\varepsilon}x^{-\varepsilon})\lambda = u\lambda$ for any u. It then follows, by induction on the length of v, that $(ux^{\varepsilon}x^{-\varepsilon}v)\lambda = (uv)\lambda$ for all v; that is, $w\lambda = w'\lambda$ if w' comes from w by elementary reduction. From this we see that $w\lambda = w''\lambda$ if $w \approx w''$. In particular, if w and w'' are reduced words with $w \approx w''$ then $w = w\lambda = w''\lambda = w''$, as required.

(II)The Diamond Lemma.

We first show that if both w' and w'' come from w by elementary reduction then either w' is the same as w'' or there is a word $w*$ which comes from both w' and w'' by elementary reductions.

If w' and w'' come from w by different elementary reductions then there are two possibilities (interchanging w' and w'' if necessary), both of which were illustrated in the examples earlier. The first is that w is $ux^{\varepsilon}x^{-\varepsilon}x^{\varepsilon}v$ for some (possibly empty) words u and v, and that w' is obtained by deleting $x^{\varepsilon}x^{-\varepsilon}$ and w'' is obtained by deleting $x^{-\varepsilon}x^{\varepsilon}$. In this case w'' is the same as w'. The second possibility is that w is $ux^{\varepsilon}x^{-\varepsilon}ty^{\delta}y^{-\delta}v$ for some (possibly empty) words t, u, and v, and that w' is $uty^{\delta}y^{-\delta}v$ and w'' is $ux^{\varepsilon}x^{-\varepsilon}tv$. In this case the word utv is obtained by elementary reduction from both w' and w''.

Now take two equivalent words w and w' and consider a sequence $w_1, \ldots, w_n$ such that w_1 is w, w_n is w', and, for each i, one of w_{i+1} and w_i comes by elementary reduction from the other. Suppose that there is some r such that w_{i+1} comes by elementary reduction from w_i for all $i < r$ and w_i comes from w_{i+1} by elementary reduction for all $i \geq r$. Then w_r comes from both w and w' by reduction.

If there is no such r then there must be some k such that both w_{k-1} and w_{k+1} come from w_k by elementary reduction. In this case, the argument above tells us that we may obtain a new sequence which also shows the equivalence of w and w', either by deleting w_k and w_{k+1} or by replacing w_k by a word $w*$ such that both w_{k-1} and w_{k+1} reduce to $w*$. Since the sum of the lengths of the members of the sequence decreases when we make this change, it follows by induction that there must be some word which comes from both w and w' by reduction.

In particular, if w and w' are equivalent reduced words then they must be the same.

A slightly more complicated version of this proof is given in the exercises, which has the advantage that it can be applied in more general situations.

(III) van der Waerden's method.

This method is one we shall use in several later theorems.

Let S be the set of all reduced words, and let G be the group of all permutations of S. We shall define a homomorphism $\varphi: F(X) \to G$ such that $[w]\varphi$ acting on the empty sequence ( ) give the sequence w whenever w is a reduced word. In particular, if w and w' are reduced words with $w \approx w'$ then $[w] = [w']$ and so w and w', being the results of acting on ( ) by $[w]\varphi$ and $[w']\varphi$, must be the same.

Since we have already shown that $F(X)$ is free on X, we can obtain φ by defining a function $f: X \to G$. We define xf as the permutation sending w to wx if w does not end in x^{-1} and sending w to u if w is ux^{-1}. It is easy to check that this is a permutation of S, whose inverse sends w to wx^{-1} if w does not end in x and sends w to v if w is vx. This holds because a reduced word cannot end in xx^{-1} or $x^{-1}x$.

If w is the reduced word $x_{i_1}^{\varepsilon_1} \dots x_{i_n}^{\varepsilon_n}$ then $[w]\varphi$ is the product of the permutations $(x_{i_r} f)^{\varepsilon_r}$ and, inductively, the reuslt of acting on ( ) by $[w]\varphi$ is just w itself, as needed.

(IV) For the final method, we show that if w is a reduced word distinct from 1 then there is a homomorphism φ from $F(X)$ to a finite group such that $[w]\varphi$ is not the identity. Since $[1]\varphi$ is the identity, we cannot have $w \approx 1$.

Suppose that this has been shown, and let u and v be reduced words such that $u \approx v$; then $uv^{-1} \approx 1$. The word uv^{-1} need not be reduced, but if u and v are different it is easy to see that by reduction from uv^{-1} we will obtain a reduced word w different from 1. Further, $w \approx uv^{-1} \approx 1$, which is prohibited by the previous paragraph.

So let w be the reduced word $x_{i_1}^{\varepsilon_1} \dots x_{i_n}^{\varepsilon_n}$. We will define a homomorphism φ into the group S_{n+1} of permutations of $\{1, \dots, n+1\}$ such that $[w]\varphi$ sends 1 into $n+1$, whence $[w]\varphi$ is not the identity. To do this we require a function $f: X \to S_{n+1}$ such that $(x_{i_r})^{\varepsilon_r}$ sends r into $r+1$ for all $r \leq n$.

So we need xf to send r to $r+1$ if x_{i_r} is x and $\varepsilon_r = 1$ and to send $r+1$ to r if x_{i_r} is x and $\varepsilon_r = -1$. If these conditions define a one-one function from some subset of $\{1, \dots, n+1\}$ to another subset we can extend this function to a permutation of $\{1, \dots, n+1\}$, and then define xf to be this permutation.

Can this definition require us to send r to two different numbers? This could only occur if x_{i_r} were x with $\varepsilon_r = 1$ and $x_{i_{r-1}}$ were also x with $\varepsilon_{r-1} = -1$. This cannot happen since w is a reduced word. Also this function could only map two different numbers to s if $x_{i_{s-1}}$ were x and $\varepsilon_{s-1} = 1$ and x_{i_s} were also x with $\varepsilon_s = -1$, which is again impossible as w is reduced.//

We can now obtain a new proof of Proposition 2(iii).

Corollary $i: X \to F(X)$ *is injective.*

Proof If x and y are distinct elements of X then they are distinct reduced words. Hence they lie in different equivalence classes.//

Exercise 1 Prove the claimed results about canonical reduction. In particular, show that $w\lambda$ is always reduced, that $w\lambda$ is w if w is reduced, and that $w\lambda$ comes from w by reduction. Show also that $(ux^{\varepsilon}x^{-\varepsilon})\lambda$ is the same as $u\lambda$, and, by induction on the number of letters in v, that $(ux^{\varepsilon}x^{-\varepsilon}v)\lambda$ is the same as $(uv)\lambda$.

Exercise 2 Let w be $ux^{\varepsilon}x^{-\varepsilon}v$. If ux^{ε} is reduced we say that uv comes from w by leftmost reduction. Show that $w\lambda$ comes from w by repeating leftmost reduction until we get a reduced word.

We may prove the Normal Form Theorem, starting from the first result in the Diamond Lemma approach, by methods which apply to very general reduction relations where the previously given induction on length cannot be used. This is done in the exercises below.

Exercise 3 Let u and v be obtained by reduction from w. Show that there is a word w^* which is obtained from both u and v by reduction. (An indication of the argument is given in Figure 1. Here the top diamond exists by the Diamond Lemma approach, the left-hand diamond exists inductively, and then the right-hand diamond also exists inductively.)

Exercise 4 Let $w \approx w'$, and let $w_1, \ldots, w_n$ be a sequence with w_1 being w and w_n being w' such that, for all i, one of w_{i+1} and w_i is an elementary reduction of the other. If w_{k-1} and w_{k+1} are elementary reductions of w_k we say k is a peak. Use Exercise 3 to show that if the sequence has a peak then there is another sequence joining w to w' with fewer peaks. (Figure 2 provides a sketch of the argument.) Deduce the Normal Form Theorem.

Exercise 5 It is possible to give a proof of the Normal Form Theorem by the Diamond Lemma approach without using induction on the number of peaks. Consider the relation between words in which u is related to v iff there is a word w such that both u and v reduce to w. Using Exercise 3, show that this is an equivalence relation. Deduce that this relation coincides with the relation $\approx$.

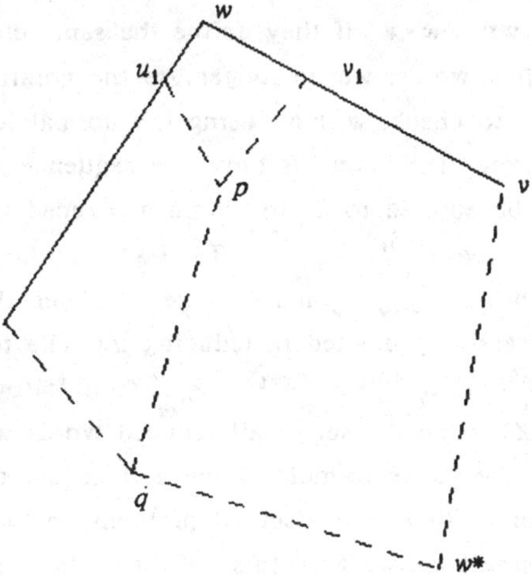

Figure 1

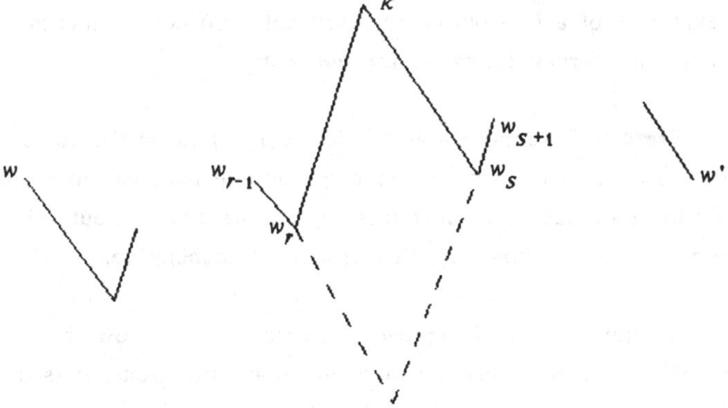

Figure 2

Exercise 6 Show that the function from the set of reduced words to itself which sends w to wx if w does not end in x^{-1} and which sends vx^{-1} to v is a permutation of the set of reduced words.

We usually regard X as a subset of $F(X)$ with i being inclusion; consequently we shall usually omit mention of i in future. We frequently identify elements of $F(X)$ with the corresponding reduced words. At times we need to regard words as elements of $M(X \cup \bar{X})$ and at other times we want to regard them as giving elements of $F(X)$. We write $w \equiv w'$ when w and w' are the same word, while we write $w = w'$ if they define the same element of $F(X)$ (that is, if they are equivalent words; we no longer use the notation $\approx$).

It is easy to check, without using the normal form theorem, that if u and v are reduced words then there is only one sequence of elementary reductions which can be applied to uv to obtain a reduced word. Let u be $x_{i_1}^{\varepsilon_1} \dots x_{i_n}^{\varepsilon_n}$ and let v be $x_{j_1}^{\delta_1} \dots x_{j_m}^{\delta_m}$. Take $s \geq 0$ as large as possible such that for all $r \leq s$ we have $i_{n+1-r} = j_r$ and $\varepsilon_{n+1-r} = -\varepsilon_r$. Then the pairs $x_{i_{n+1-r}}^{\varepsilon_{n+1-r}} x_{j_r}^{\varepsilon_r}$ for $r = 1, \dots, s$ are successively deleted in reducing uv. The reduced word we obtain is therefore $x_{i_1}^{\varepsilon_1} \dots x_{i_{n-s}}^{\varepsilon_{n-s}} x_{j_{s+1}}^{\delta_{s+1}} \dots x_{j_m}^{\delta_m}$. It follows that it would be possible to define $F(X)$ to be the set of all reduced words with the product of u and v being given by the above formula. Some authors use this approach, but it is not really satisfactory. There is a practical problem, in that the proof of associativity is surprisingly messy with this definition (see Exercise 7) There is a more important theoretical objection to this approach, as it confuses the question of the existence of a free group and the question of the nature of normal forms for the elements. We frequently find in algebraic situations that it is easy to prove the existence of a free object but extremely difficult (sometimes impossible) to find normal forms for the elements.

Exercise 7 Suppose that we define $F(X)$ to be the set of reduced words, with the product of u and v being the unique reduced word obtained from uv by reduction. (We have seen that this word is unique without using the Normal Form Theorem.) Show that this product is associative.

A totally different approach may be used to show the existence of free groups. This approach works in other algebraic situations. It is of a categorical nature, amounting to the fact that we are looking for an adjoint to the forgetful functor. Unfortunately the proof runs into some set-theoretical difficulties. To make the proof easier, I first give the argument in the (incorrect!) form ignoring these difficulties, and then show how to resolve them.

Consider all pairs (f, G_f) where G_f is a group and f is a function from X to G_f. Take the cartesian product over all f of the groups G_f, which we call K. The set-theoretic problem arises here, since in formal set theory this product cannot be constructed because there are too many functions f; this problem will be ignored for the moment. There is a function $i: X \to K$ such that the component of xi in the factor G_f is xf. Let F be the subgroup of K generated by Xi. Let f be any function from X to a group. Then this group is

G_f, and the projection of K on this factor provides the needed homomorphism φ such that $f = i\varphi$. Because F is generated by Xi, there cannot be more than one such homomorphism.

We now deal with the problem. To begin with, observe that we need only consider those f such that Xf generates G_f. For let f be a function that does not satisfy this property, and let H_f be the subgroup of G_f generated by Xf. As above, we have the corresponding homomorphism from F to H_f, and this can be regarded as a homomorphism into G_f.

It follows that we need only look at those pairs (f,G_f) for which G_f has cardinality at most $\max(|X|,|\mathbb{N}|)$. This holds because G_f, being generated by Xf, has cardinality at most that of the set of finite sequences on $(Xf) \cup (Xf)^{-1}$, and hence at most that of the set of finite sequences on $X \cup X^{-1}$, which has the stated cardinality (using the Axiom of Choice). However the set-theoretic difficulties have not yet been resolved.

Next we note that if we have two pairs (f_1,G_1) and (f_2,G_2) and an isomorphism $j:G_1 \to G_2$ such that $f_1 j = f_2$ then there is no need to include both groups in the product (though we may do so if we wish). For if G_1 is included then we have $\varphi_1 : F \to G_1$ with $f_1 = i\varphi_1$ and we need only define $\varphi_2 : F \to G_2$ by $\varphi_2 = \varphi_1 j$.

Now take a fixed set S of cardinality $\max(|X|,|\mathbb{N}|)$. Let (f,G) be a pair such that G has at most this cardinality. Then there is a bijection ϑ from G to a subset H of S. We can then make H into a group by defining the product of h_1 and h_2 to be $(h_1\vartheta^{-1}.h_2\vartheta^{-1})\vartheta$, and we may use the pair $(f\vartheta,H)$ instead of (f,G). It follows that in our cartesian product we need only consider functions into those groups whose underlying set is a subset of S. Those readers with some experience of formal set theory should have no difficulty in showing that the collection of such f forms a set, and consequently the cartesian product exists; readers with no knowledge of formal set theory will have to be content with my assurance that the originally suggested construction cannot be made, while this modified construction is permitted.

Proposition 5 *Free groups are residually finite. That is, if F is free and $1 \neq w \in F$ then there is a normal subgroup N of F with $w \notin N$ and F/N finite.*

Proof Let w be given by a non-trivial reduced word. Method (IV) for proving Theorem 4 then gives a homomorphism φ from F into the finite group S_{n+1} for some n such that $w\varphi$ is not the identity.//

Corollary *Finitely generated free groups are hopfian. That is, any onto endomorphism is an automorphism.*

Proof This holds for any finitely generated residually finite group G.

Let α be an endomorphism onto G, and let $N \triangleleft G$ with G/N finite. Then $G/N\alpha^{-n}$ is isomorphic to G/N for any n. Now there are only finitely many homomorphisms from G to G/N, since any homomorphism is determined by its values on the finite generating set of G and there are only finitely many elements of G/N to which a given generator can map.

Since $N\alpha^{-n}$ is the kernel of such a homomorphism for every n, it follows that there must be m and n with $m > n$ such that $N\alpha^{-n} = N\alpha^{-m}$. Hence $N = N\alpha^{-(m-n)}$, and so $N \supseteq \ker\alpha$. Since G is residually finite, the intersection of all such N is $\{1\}$, and so $\{1\} \supseteq \ker\alpha$, as required.//

If X is infinite, $F(X)$ cannot be hopfian. For there will be a function f from a proper subset Y of X onto X, and we can define an endomorphism of $F(X)$ onto itself with non-trivial kernel as the homomorphism obtained from the function sending $y \in Y$ to yf and $x \in X - Y$ to 1.

Proposition 6 *$F(X)$ is isomorphic to $F(Y)$ iff $|X| = |Y|$.*

Proof Let f be a bijection from X to Y. Then f extends to a homomorphism φ from $F(X)$ to $F(Y)$, while f^{-1} extends to a homomorphism ψ from $F(Y)$ to $F(X)$. Now both $\varphi\psi$ and the identity function on $F(X)$ extend the identity function on X, so the uniqueness in the definition of a free group requires that $\varphi\psi$ is the identity function. Similarly $\psi\varphi$ is the identity function, and so φ is an isomorphism.

Now suppose that $F(X)$ is isomorphic to $F(Y)$. The number of homomorphisms from $F(X)$ to $\mathbb{Z}_2$, the cyclic group of order 2, is the same as the number of functions from X to $\mathbb{Z}_2$, and so equals $2^{|X|}$. Hence $2^{|X|} = 2^{|Y|}$. It follows at once that if one of X and Y is finite then so is the other and they are bijective.

If both X and Y are infinite we use a different approach requiring the Axiom of Choice. (To obtain the result from the previous paragraph would require some such set-theoretic principle as the Generalised Continuuum Hypothesis, and this implies the Axiom of Choice.) It can be shown, using the Axiom of Choice, that $|M(X \cup \bar{X})| = |X \cup \bar{X}| = |X|$, and as $F(X)$ is a set of equivalence classes of $M(X \cup \bar{X})$ it follows that $|F(X)| \leq |X|$. Since X embeds in $F(X)$, we see that $|F(X)| = |X|$. Hence, as required, $|X| = |Y|$ if $F(X)$ is isomorphic to $F(Y)$. Note that, without using the Axiom of Choice, we can show that $F(X)$ is countable iff X is either finite or countably infinite.//

The same argument shows that if A generates $F(X)$ then $|A| \geq |X|$.

A group G is called a ***free group*** if it is isomorphic to $F(X)$ for some X. If $i: F(X) \to G$ is an isomorphism then Xi is called a ***basis*** of G; we also say that G is ***free on*** Xi. Plainly, if α is a basis of G and β is any automorphism of G then $A\beta$ is also a basis of G. Conversely, if both A and B are bases of G then $|A| = |B|$, by the previous proposition, and then any bijection from A to B extends to an automorphism of G. The cardinality of a basis of a free group is called its ***rank***.

Further, if α is any automorphism of $F(X)$ then αi is also an isomorphism from $F(X)$ to G, so that $X\alpha i$ is also a basis of G. In particular, $X\alpha$ is a basis of $F(X)$. Any automorphism α comes from a function $f: X \to F(X)$, and any such function f gives rise to an endomorphism α of $F(X)$. This will be an automorphism iff there is another function $f': X \to F(X)$, and corresponding endomorphism α' such that both $f\alpha'$ and $f'\alpha$ are the identity on X. By the corollary to Proposition 5, when X is finite α will be an automorphism if it maps onto $F(X)$. This holds if for every x in X there is a word w_x such that $x = w_x\alpha$.

Examples Let F be $F(x,y)$. Then $\{x^{-1}, x^2y\}$ is a basis of F. So also are $\{xy, xy^2\}$ and $\{xy, yxyxy\}$. In each case, we have an endomorphism of F sending x and y to the two corresponding elements. This endomorphism is onto, since it is easy to find an element which maps onto y, and, using this, to find another element which maps onto x.

The set $\{yx, xy^2\}$ is not a basis of F. For if it were then there would be a reduced word w such that $w\alpha = x$, where α is the endomorphism sending x to yx and y to xy^2. But $w\alpha = w'$, where w' is obtained from w by replacing x by yx, y by xy^2, x^{-1} by $x^{-1}y^{-1}$, and y^{-1} by $y^{-2}x^{-1}$. It is easy to check that w' is reduced, because w is reduced, and also easy to see, because of the way w' is obtained from w by substitution, that w' has length ≥ 2 unless w is the identity. However, the next proposition shows that the subgroup of F generated by $\{yx, xy^2\}$ has this pair of elements as a basis, because w' is reduced whenever w is reduced.

Also $\{xy, x^2y^3\}$ is not a basis of F. For if it were then the set $\{xy, x^2y^3(xy)^{-1}\}$ would also be a basis. That is, $\{xy, x^2y^2x^{-1}\}$ would be a basis. If we conjugate by x we would get another basis. But the result of conjugation by x is $\{yx, xy^2\}$, which we know is not a basis.

Let X be any set, $Y \subseteq X$, and let $x \in X - Y$. Then the map sending y to $yx^{\pm 1}$ for $y \in Y$ and z to z for $z \notin Y$ induces an automorphism of $F(X)$, as does the map sending y to $x^{\pm 1}y$ and z to z. Any permutation of X also induces an automorphism of $F(X)$, as does the map sending y to y^{-1} for $y \in Y$ and fixing the other elements of X. When X is finite, the latter two types of automorphism, together with the first type with Y consisting of a single element, generate Aut $F(X)$; this will be proved in section 8.2.

Proposition 8 *Let X be a subset of the group G. Then the following are equivalent: (i) G is free with basis X, (ii) every element of G can be uniquely written as $x_{i_1}^{\varepsilon_1} \ldots x_{i_n}^{\varepsilon_n}$ for some $n \geq 0$, $x_{i_r} \in X$, $\varepsilon_r = \pm 1$, where $\varepsilon_{r+1} \neq -\varepsilon_r$ if $i_{r+1} = i_r$, (iii) X generates G, and 1 is not equal to any product $x_{i_1}^{\varepsilon_1} \ldots x_{i_n}^{\varepsilon_n}$ with $n > 0$, $x_{i_r} \in X$, $\varepsilon_r = \pm 1$, and $\varepsilon_{r+1} \neq -\varepsilon_r$ if $i_{r+1} = i_r$.*

Proof If (ii) holds then (iii) obviously holds. If (iii) holds then any element of G can be written as in (ii), since X generates G. Also, if an element could be written in two different ways then, multiplying one by the inverse of the other and performing all possible cancellations, we would obtain a way of writing 1 as a non-trivial product of the kind considered in (iii).

If G is free on X then G has these properties, since $F(X)$ does. If G has properties (ii) and (iii) consider the homomorphism from $F(X)$ to G which is induced by the identity map on X. It maps onto G, since G is generated by X. It is one-one, since (iii) tells us that a non-trivial reduced word on X does not map to 1 in G.//

Corollary 1 *Let X generate G, and let $\varphi: G \to H$ be a homomorphism which is one-one on X and such that $G\varphi$ is free with basis $X\varphi$. Then G is free with basis X.*

Proof Condition (iii) holds for X and G because it holds for $X\varphi$ and $G\varphi$.//

Corollary 2 *Let G be free with basis X, and let $Y \subset X$. Then the subgroup $\langle Y \rangle$ of G is free with basis Y.*

Proof Condition (iii) obviously holds for $\langle Y \rangle$ and Y because it holds for G and X.//

Corollary 3 *Let F be free with basis $\{x,y\}$. Let $\varphi: F \to \mathbb{Z}$ send x to 1 and y to 0. Then $\ker\varphi$ is free with basis $\{x^{-i}yx^{i}$, all $i \in \mathbb{Z}\}$.*

Proof Plainly all these elements are in $\ker\varphi$. Also it is easy to check that any element of F can be written as ux^n for some n and some u in the stated subgroup. Since we must have $n = 0$ for an element of the kernel, we see that $\ker\varphi$ is the subgroup generated by these elements .

Now let x_i be $x^{-i}yx^{i}$, and consider any reduced word $x_{i_1}^{\varepsilon_1} \dots x_{i_n}^{\varepsilon_n}$. This word equals $x^{j_1}y^{\varepsilon_1}x^{j_2}y^{\varepsilon_2} \dots y^{\varepsilon_n}x^{j_{n+1}}$, where $j_r = i_{r-1} - i_r$, with $i_0 = 0 = i_{n+1}$. Such a product is not 1, as F is free on $\{x,y\}$.//

Let g be an element of $F(X)$. We regard it as a reduced word As such, it has a length, which we write as $|g|$. If h is another element of $F(X)$, which we also regard as a reduced word, the word gh need not be reduced (but the product gh in $F(X)$ is the equivalence class of this word). If the word gh is reduced we say the product gh is *reduced as written*. More generally, if $g_1, \dots, g_k$ are reduced words and the word $g_1 \dots g_k$ is a reduced word, we say the product is reduced as written.

It is easy to check the following properties. (i) $|gh| \le |g| + |h|$, with equality iff gh is reduced as written. (ii) Either the first letter of gh is the first letter of g or g cancels completely in the product gh; this latter holds iff $h \equiv g^{-1}k$ for some reduced word k. Similarly. either gh ends with the last letter of h or h cancels completely in the product gh; this latter holds iff $g \equiv kh^{-1}$ for some reduced word k.

Definition Let $g \equiv x_{i_1}^{\varepsilon_1} \dots x_{i_n}^{\varepsilon_n}$ be a reduced word. Then g is *cyclically reduced* if either $i_n \ne i_1$ or $i_n = i_1$ but $\varepsilon_n \ne -\varepsilon_1$. Also 1 is cyclically reduced.

Clearly g is cyclically reduced iff gg is reduced as written. If g is cyclically reduced then, inductively, g^n is cyclically reduced as written, and $|g^n| = n|g|$.

If $g \equiv u^{-1}vu$, reduced as written, and v is cyclically reduced, then $g^n = u^{-1}v^n u$, reduced as written.

By a *cyclic permutation* of $x_{i_1}^{\varepsilon_1} \dots x_{i_n}^{\varepsilon_n}$ is meant any word $x_{i_r}^{\varepsilon_r} \dots x_{i_n}^{\varepsilon_n} x_{i_1}^{\varepsilon_1} \dots x_{i_{r-1}}^{\varepsilon_{r-1}}$.

Proposition 9 (*i*) *Any element of $F(X)$ is conjugate to a cyclically reduced word.*

(*ii*) *Any cyclic permutation of a cyclically reduced word is cyclically reduced.*

(*iii*) *Two cyclically reduced words are conjugate iff they are cyclic permutations of each other.*

Proof (i) Let $g \equiv x_{i_1}^{\varepsilon_1} \dots x_{i_n}^{\varepsilon_n}$ be reduced but not cyclically reduced. Then $g \equiv x_{i_1}^{\varepsilon_1} g' x_{i_1}^{-\varepsilon_1}$, where g' is the word $x_{i_2}^{\varepsilon_2} \dots x_{i_{n-1}}^{\varepsilon_{n-1}}$. The result follows by induction.

(ii) The word $x_{i_n}^{\varepsilon_n} x_{i_1}^{\varepsilon_1} \dots x_{i_{n-1}}^{\varepsilon_{n-1}}$ is cyclically reduced since, by assumption, either $i_n \neq i_1$ or $\varepsilon_n \neq -\varepsilon_1$. The result follows inductively.

(iii) Any cyclic permutation of the cyclically reduced word g is a conjugate of g.

Conversely, take any conjugate $u^{-1}gu$ of g. If $u^{-1}gu$ is reduced as written, with $u \neq 1$, then it is not cyclically reduced, since it ends with the last letter of u and begins with its inverse, the first letter of u^{-1}. If $u^{-1}gu$ is not reduced as written then the first letter of u is either $x_{i_1}^{\varepsilon_1}$ or $x_{i_n}^{-\varepsilon_n}$. In either case $u^{-1}gu = v^{-1}hv$, where $|v| < |u|$ and h is a cyclic permutation of g. The result follows by induction on the length.//

Proposition 10 *Free groups are torsion-free.*

Proof Let $g \in F$, where F is free and $g \neq 1$. Taking conjugates if necessary, we may assume that g is cyclically reduced. Then as already remarked, $|g^n| = n|g| \neq 0$, so $g^n \neq 1$.//

Proposition 11 *Let F be free, and g and h in F. If $g^k = h^k$ for some $k \neq 0$ then $g = h$.*

Proof We may assume that $k \geq 0$. Taking conjugates if necessary, we may assume that g is cyclically reduced. Then g^k is cyclically reduced as written. If h is not cyclically reduced then h^k is not cyclically reduced, and so $g^k \neq h^k$. If h is cyclically reduced then h^k is cyclically reduced as written. Hence g^k consists of g repeated k times, and similarly for h^k, so $g^k = h^k$ requires $g = h$.//

Exercise 8 Let u and v be elements of a free group, and let m and n be non-zero integers. Prove the following results. (i) If $uv^n = v^n u$ then $uv = vu$. (ii) If $u^m v^n = v^n u^m$ then $uv = vu$.

Proposition 12 *Let F be free, and take g and h in F. If $gh = hg$ then $\langle g,h\rangle$ is cyclic; that is, $\exists u \in F$ and integers r and s with $g = u^r$ and $h = u^s$.*

Proof By induction on $|g| + |h|$, the case $|g| + |h| \leq 2$ being obvious. Taking conjugates if necessary, we can assume (without increasing $|g| + |h|$) that g is cyclically reduced.

If gh is reduced as written then $|gh| = |g| + |h|$, and so $|hg| = |h| + |g|$. Hence hg is also reduced as written. Then g is an initial segment of gh and h is an initial segment of hg. As hg is the same as gh, one of g and h must be an initial segment of the other. That is, interchanging g and h if necessary, we have $g = hw$, reduced as written. We then have $hw = wh$, and the result follows by induction.

If gh is not reduced as written we have $g = vx^{\varepsilon}$ and $h = x^{-\varepsilon}t$, both reduced as written. Now hg begins with the first letter of h, which is $x^{-\varepsilon}$, unless h cancels completely in the product. Also gh begins with the first letter of g unless g cancels completely in the product. Since g is cyclically reduced and ends in x^{ε} it cannot begin with $x^{-\varepsilon}$. So gh and hg are different unless g cancels completely in gh or h cancels completely in hg. Hence either $h = g^{-1}w$ or $g = h^{-1}w$, reduced as written in either case. The result now follows by induction.//

1.2 *GENERATORS AND RELATORS*

Proposition 13 *Any group G is a quotient of some free group.*

Proof The identity map from G to G extends to a homomorphism from $F(G)$ to G, which is plainly onto G.//

Definitions Let G be a group, X a set, and $\varphi: F(X) \to G$ an epimorphism. Then X is called a set of *generating symbols* for G (*under* φ), and the family $\{x\varphi; x \in X\}$ is called a family of *generators* of G. (Plainly $G = \langle G\varphi\rangle$. Since we can have $x\varphi = y\varphi$ with x different from y, we refer to a family rather than a set.) We also refer to X simply as a set of generators of G; the reader then has to decide on each occasion whether or not the generators are to be regarded as elements of G.

We call $\ker\varphi$ the set of *relators* of G (under φ). If $u \equiv x_{i_1}^{\varepsilon_1} \ldots x_{i_n}^{\varepsilon_n}$ and $v \equiv x_{j_1}^{\delta_1} \ldots x_{j_m}^{\delta_m}$ are (not necessarily reduced) words with uv^{-1} representing an element of $\ker\varphi$, and $x_i\varphi$ is a_i, we say that $a_{i_1}^{\varepsilon_1} \ldots a_{i_n}^{\varepsilon_n} = a_{j_1}^{\delta_1} \ldots a_{j_m}^{\delta_m}$ is a *relation* in G. In particular, if u represents an element of $\ker\varphi$ then the corresponding relation in G is $a_{i_1}^{\varepsilon_1} \ldots a_{i_n}^{\varepsilon_n} = 1$.

For any subset S of a group H, the normal closure $\langle S\rangle^H$ in H of the subgroup $\langle S\rangle$ is called the set of *consequences* of S in H, or simply the *normal subgroup of H generated by S*.

If $\ker\varphi$ is the set of consequences of some subset R of $F(X)$ we call R a set of *defining relators of G* (*under* φ). We have a corresponding set of defining relations. We shall also say that a relation $u = v$ is a consequence of a set of defining relators (or defining relations) if the corresponding relator uv^{-1} is a consequence of the defining relators.

A *presentation* $\langle X;R\rangle^{\varphi}$ of G consists of a set X, an epimorphism φ from $F(X)$ to G, and a set R of defining relators of G under φ. Frequently we omit mention of φ, especially when φ is the natural map from $F(X)$ to $F(X)/\langle R\rangle^{F(X)}$ or when φ is one-one on X (in which case we regard X as a subset of G). We write $G = \langle X;R\rangle^{\varphi}$ when $\langle X;R\rangle^{\varphi}$ is a presentation of G. At times it is more convenient to replace each relator r by the corresponding relation $r = 1$, or, more generally, by a relation $u = v$ where r is uv^{-1} (or $v^{-1}u$). If convenient, we can mix relators and relations. For instance, if we know that a group G generated by a set X is abelian, it is easier to write relations $xy = yx$, where x and y are in X, rather than the corresponding relator $x^{-1}y^{-1}xy$, whether or not we replace the other relators by relations. Examples of presentations of specific groups will have to wait until after the next theorem.

If both X and R are finite, we refer to a *finite presentation* and to a *finitely presented group*.

Examples The free group on X has a presentation $\langle X;R\rangle$ with R empty.

The presentation $\langle x, y;\, xy^2 = y^3x,\ yx^2 = x^3y\rangle$ presents the trivial group. To show this we have to obtain a considerable number of consequences of the defining relations.

First observe that $xy^4x^{-1} = (xy^2x^{-1})^2 = y^6$, and so $x^2y^4x^{-2} = xy^6x^{-1} = (xy^2x^{-1})^3 = y^9$. It follows that $x^2y^4x^{-2} = yx^2y^4x^{-2}y^{-1} = yx^2y^{-1}y^4yx^{-2}y^{-1} = x^3y^4x^{-3}$, and so $xy^4 = y^4x$. But then $y^6 = x^2y^4x^{-2} = y^4$. Hence $y^2 = 1$. From this and $xy^2 = y^3x$ we deduce that $y = 1$, and the second relation gives $x = 1$.

It would be significantly harder to show in $F(x,y)$ that x and y are in the subgroup of consequences of the corresponding two relators. In general, the advantage of using relations rather than relators is that we are working in the quotient group $F(X)/\langle R\rangle^{F(X)}$ rather than looking in $F(X)$ at congruence modulo the normasl subgroup $\langle R\rangle^{F(X)}$.

Theorem 14 (von Dyck's Theorem) *Let G be $\langle X;R\rangle^{\varphi}$, let $f{:}X \to H$ be a function into some group H, and let $\vartheta{:}F(X) \to H$ be the corresponding homomorphism. Then there is a homomorphism $\psi{:}G \to H$ such that $xf = x\varphi\psi$ for all $x \in X$ if $r\vartheta = 1$ for all $r \in R$. Further, φ is an epimorphism if Xf generates H.*

Proof We are told that $R \subseteq \ker\vartheta$, and, since $\ker\varphi$ is, by definition, the normal subgroup generated by R, we see that $\ker\varphi \subseteq \ker\vartheta$. It follows, as is well-known, that the required homomorphism ψ may be defined by requiring $g\psi$ to be $w\vartheta$ for any w such that $w\varphi = g$. Plainly $G\varphi = \langle Xf\rangle$, and so $G\varphi = H$ if Xf generates H.//

As a particular case of the theorem, the inclusion of X in $X \cup Y$ induces a homomorphism from $\langle X;R\rangle$ to $\langle X \cup Y; R \cup S\rangle$ for any subset S of $F(X \cup Y)$. This homomorphism will be used in future without explicitly stating how it is constructed.

By the same argument as in the theorem, we can prove the following result, which we will also refer to as von Dyck's Theorem (more strictly, it is a case of one of the homomorphism theorems). *Let R be a subset of a group A, and let $\vartheta{:}A \to H$ be a homomorphism such that $R\vartheta = \{1\}$. Then there is a homomorphism $\psi{:}A/\langle R\rangle^{A} \to H$ such that $\vartheta = \pi\psi$, where π is the homomorphism from A onto $A/\langle R\rangle^{A}$.*

When we are looking for a presentation for a group H the following procedure is often useful. Begin by finding a set X of generators of H. Then find a set R of relators of H with generators X, chosen so that one has reason to believe (or just guesses) that R is a set of defining relators. By von Dyck's theorem, there is an epimorphism from $G = \langle X;R\rangle$ to H. It may be that the relators R enable us to write the elements of G in a simple form from which we can see that the map is one-one. Alternatively, if H is finite we may be able to show that G is also finite and that $|G| \leq |H|$, which will again show that the map is one-one. There is a procedure due to Todd and Coxeter (computer implementations of which are efficient in practice, though not in theory) which, given any presentation $\langle X;R\rangle$ with X and R finite, will give a bound to the order of the group (and will even give the multiplication table if required) provided the group is finite; however the procedure will continue for ever if the group is infinite, and so cannot be used to tell whether or not the group is finite.

Examples The cyclic group $\mathbb{Z}_n$ of order n has presentation $\langle x; x^n\rangle$, since the group with this presentation certainly has an epimorphism to $\mathbb{Z}_n$ and is easily seen to have at most n elements.

Similarly, the dihedral group D_n of symmetries of a regular n-gon has presentation $\langle x,y; x^n, y^2, (xy)^2\rangle$. For it is clear that D_n is generated by a rotation through $2\pi/n$ and a reflection, and that the stated relators hold between these generators. We can easily see that D_n has exactly $2n$ elements. On the other hand, the group G with the stated presentation is easily seen to have at most $2n$ elements, namely the elements x^i and x^iy where $0 \leq i < n$. It then follows that the homomorphism from G onto D_n must be one-one, and that G has exactly $2n$ elements. It was not immediately obvious from the presentation for G that no equalities between the elements x^i and x^iy for $1 \leq i \leq n$ follow from the defining relators.

Further, the group D_∞ of distance-preserving bijections of $\mathbb{Z}$ has presentation $\langle x,y; y^2,(xy)^2\rangle$. For D_∞ is seen easily to be generated by the functions $n \to n+1$ and $n \to -n$, which satisfy these relators. The group with the stated presentation has an epimorphism to D_∞, and it is easily checked that its elements are either of the form x^i or of the form x^iy for some integer i. Since the groups are infinite, we cannot use a counting argument to show that the map is one-one. However, we can easily find the images in D_∞ of these elements, and check that only the identity element maps to the identity in D_∞.

Any group G has a multiplication table presentation, which is finite when G is finite. The generators of this presentation form a set X bijective with G; the element corresponding to g is written x_g. The relators are the elements $x_g x_h x_k^{-1}$ for all g, h, and k with $gh = k$. By von Dyck's Theorem, the group H with this presentation has a homomorphism onto G. As a consequence of the relators, any element of H equals in H some element x_g, and so the homomorphism is one-one.

Suppose that $G = \langle X;R\rangle^\varphi$ and that $H = \langle Y;S\rangle^\psi$, where $X \cap Y = \emptyset$. Let K be a group for which there is an epimorphism $\pi:K \to H$ with kernel G. Define $\vartheta:F(X \cup Y) \to K$ by requiring that $x\vartheta = x\varphi$ for all $x \in X$ and, for all $y \in Y$, choosing

$y\vartheta$ to be some element of K such that $y\vartheta\pi = y\psi$. Then, for all $s \in S$, we have $s\vartheta\pi = s\psi = 1$, so that $s\vartheta \in \ker\pi = G$. Consequently we can find an element u_s of F(X) such that $s\vartheta = u_s\varphi = u_s\vartheta$. Further, as $G \triangleleft K$, for any $x \in X$ and $y \in Y$ we have $(y^{-1}xy)\vartheta \in G$, and so we can find elements v_{xy} and $v*_{xy}$ in F(X) such that $(y^{-1}xy)\vartheta = v_{xy}\varphi = v_{xy}\vartheta$ and $(yxy^{-1})\vartheta = v*_{xy}\varphi = v*_{xy}\vartheta$.

We can then show that K has the presentation $\langle X,Y;\ R,\ \{su_s^{-1},\ s \in S\},\ \{y^{-1}xyv_{xy}^{-1},\ yxy^{-1}v*_{xy}^{-1},\ \text{for all } x \in X \text{ and } y \in Y\}\rangle^{\vartheta}$.

First observe that for any $k \in K$ we can find $w \in F(Y)$ such that $k\pi = w\psi = w\vartheta\pi$. Hence $k(w\vartheta)^{-1}$ is in $\ker\pi$, and so equals $w_1\vartheta$ for some $w_1 \in$ F(X). Hence K is generated by $(X \cup Y)\vartheta$. The stated relators plainly hold in K.

Let M be the group with the stated presentation, so that ϑ can be regarded as an epimorphism from M to K. Take $w \in F(X \cup Y)$ with $w\vartheta = 1$. Because of the relations $y^{-1}xy = v_{xy}$ and $yxy^{-1} = v*_{xy}$ there are words $w_1 \in F(X)$ and $w_2 \in F(Y)$ such that $w = w_1w_2$ in M. Then $1 = w\vartheta\pi = w_1\varphi\pi.w_2\psi = w_2\psi$. Hence, in $F(Y)$, w_2 equals the product of conjugates of s or s^{-1} for $s \in S$. Now $s = u_s$ in M, and any conjugate of u_s equals in M some element of $F(X)$ (because u_s is in $F(X)$), so there must be an element w_3 in $F(X)$ such that $w_2 = w_3$ in M. So $w = w_1w_3$ in M. Then $1 = w\vartheta = (w_1w_3)\vartheta = (w_1w_3)\varphi$. This shows that w_1w_3 is, in $F(X)$, a product of conjugates of elements of R and their inverses, and so $w_1w_3 = 1$ in M, as required.

This presentation of K can be simplified. Since $X\varphi$ generates G, so does $\{v_{xy}\varphi\}$, for each $y \in Y$, since conjugation by $y\vartheta$ preserves G. Hence, for each $x \in X$ and $y \in Y$, there is some product of the elements $v_{zy}\varphi$ for $z \in X$ and their inverses which equals $x\varphi$ in G. Let $v*_{xy}$ in $F(X)$ denote the product of the corresponding elements z and their inverses, and let w denote the product of the corresponding v_{zy} and their inverses.. Then the relation $y^{-1}v*_{xy}y = w$ is a consequence of the relations $y^{-1}zy = v_{zy}$. Also $w\varphi = x\varphi$, so that $w = x$ is a consequence of the relators R. Hence the relation $y^{-1}v*_{xy}y = x$ is a consequence of the other relations; equivalently, the relators $yxy^{-1}(v*_{xy})^{-1}$ for all y are consequences of the other relators. As remarked below, it follows that these relators can be deleted from the presentation.

Plainly a group G can have many presentations, even for given X and φ. We now look at how different presentations of the same group compare with each other.

Let $\langle X;R\rangle^{\varphi}$ be a presentation of G. Then so is $\langle X;R \cup S\rangle^{\varphi}$ for any set S contained in $\langle R\rangle^{F(X)}$. We say that $\langle X;R \cup S\rangle^{\varphi}$ comes from $\langle X;R\rangle^{\varphi}$ by a *general Tietze transformation* of type I, and that $\langle X;R\rangle^{\varphi}$ comes from $\langle X;R \cup S\rangle^{\varphi}$ by a general Tietze transformation of type I'. If $|S| = 1$, we refer to *simple Tietze transformations*.

Let Y be a set such that $X \cap Y = \emptyset$, and let u_y be an element of $F(X)$ for each $y \in Y$. Then $\langle X \cup Y;\ R \cup \{yu_y^{-1},\ \text{all } y \in Y\}\rangle^{\psi}$ also presents G, where $x\psi = x\varphi$ and $y\psi = u_y\varphi$. For let N be the normal subgroup of $F(X \cup Y)$ generated by $R \cup \{yu_y^{-1}\}$. Then ψ induces an epimorphism $\pi: F(X \cup Y)/N \to G$, since $(R \cup \{yu_y^{-1}\})\psi = 1$. But, by von Dyck's Theorem, there is also an epimorphism $\vartheta: G \to F(X \cup Y)/N$ with $(x\varphi)\vartheta = xN$. Plainly $\vartheta\pi$ is the identity map. Also, since $(yN)\pi\vartheta = y\psi\vartheta = u_y\psi\vartheta = u_y\varphi\vartheta = u_yN = yN$, $\pi\vartheta$ is the identity.

We say that $\langle X \cup Y;\ R \cup \{yu_y^{-1},\ \text{all } y\}\rangle^{\psi}$ comes from $\langle X;R\rangle^{\varphi}$ by a *general Tietze transformation* of type II, and that $\langle X;R\rangle^{\varphi}$ comes from $\langle X \cup Y;\ R \cup \{yu_y^{-1},\ \text{all } y\}\rangle^{\psi}$ by a general Tietze transformation of type II'. If $|Y| = 1$ we refer to *simple* Tietze transformations.

Plainly a general Tietze transformation with $|S|$ or $|Y|$ finite can be obtained by performing a finite number of simple Tietze transformations.

Theorem 15 *Any two presentations of the same group can be obtained from each other by a sequence of general Tietze transformations. If both presentations are finite then each can be obtained from the other by a sequence of simple Tietze transformations.*

Proof Let $\langle X;R\rangle^{\varphi}$ and $\langle Y;S\rangle^{\psi}$ both present G. First assume that $X \cap Y = \emptyset$. For each $y \in Y$ choose $u_y \in F(X)$ with $y\psi = u_y\varphi$, and for each $x \in X$ choose $v_x \in F(Y)$ with $x\varphi = v_x\psi$.

Defining ϑ by $x\vartheta = x\varphi$ and $y\vartheta = u_y\varphi$, we get a presentation $\langle X \cup Y;\ R \cup \{yu_y^{-1},\ \text{all } y\}\rangle^{\vartheta}$ of G, obtained from the presentation $\langle X;R\rangle^{\varphi}$ by a general Tietze transformation of type II. Now $y\vartheta = u_y\varphi$, and this equals $y\psi$ by definition. It follows that $w\vartheta = w\psi$ for any $w \in F(Y)$. In particular, $s\vartheta = s\psi = 1$ and $v_x\vartheta = v_x\psi$, which, by definition, equals $x\varphi$, which in turn is $x\vartheta$.

We find from this that the presentation $\langle X \cup Y;\ R, S,\ \{yu_y^{-1}\}, \{xv_x^{-1}\}\rangle^{\vartheta}$ comes from the previous presentation by a general Tietze transformation of type I. By symmetry, this presentation also comes from $\langle Y;S\rangle^{\psi}$ by a general Tietze transformation of type II followed by one of type I. Hence $\langle Y;S\rangle^{\psi}$ comes from this presentation by a general Tietze transformation of type I' followed by one of type II'.

Now suppose that X meets Y. We can find a set $X*$ bijective with X and not meeting X or Y. Plainly G is presented by $\langle X*;R*\rangle^{\varphi*}$, where $R*$ comes from R by replacing each x by the corresponding $x*$, and similarly for $\varphi*$. The previous discussion shows that this presentation comes from each of the other two by a sequence of general Tietze transformations, as required.

Finally, if X, Y, R, and S are all finite each general Tietze transformation used can be replaced by a finite sequence of simple Tietze transformations.//

Proposition 16 *Let G be finitely generated, and let $\langle Y;S\rangle^{\psi}$ be a presentation of G. Then there is a finite subset of Y which generates G.*

Proof Let X be a finite subset of G with $G = \langle X\rangle$. Each $x \in X$ is the product of a finite number of elements of $Y\psi$ and their inverses. Hence there is a finite subset Y_1 of Y such that $X \subseteq \langle Y_1\psi\rangle$. It follows that Y_1 generates G.//

Notice, however, that the set of defining relators for this generating set is usually larger than $S \cap F(Y_1)$.

Proposition 17 *Let G have presentations $\langle X;R\rangle^{\varphi}$ and $\langle Y;S\rangle^{\psi}$. If X, R, and Y are all finite then there is a finite subset S_1 of S such that G is $\langle Y;S_1\rangle^{\psi}$.*

Proof We use the notation of Theorem 15. Then G has the presentation $\langle X \cup Y;\ T,\ \{xv_x^{-1}\}\rangle^{\vartheta}$, where $T = R \cup \{yu_y^{-1}\}$. Let t^* come from t by replacing each element x by the corresponding v_x, and let T^* be $\{t^*;\ \text{all } t \in T\}$. Let N be $\langle xv_x^{-1}\rangle^{F(X \cup Y)}$, and let $\pi: F(X \cup Y) \to F(X \cup Y)/N$ be the natural map. Since $v_x\pi = x\pi$, we see that $t^*\pi = t\pi$. Thus t^* is a consequence of t and the relators xv_x^{-1}. It follows that a Tietze transformation of type I gives us a presentation $\langle X \cup Y;\ T,\ T^*,\ \{xv_x^{-1}\}\rangle^{\vartheta}$, and then a Tietze transformation of type I' gives the presentation $\langle X \cup Y;\ T^*,\ \{xv_x^{-1}\}\rangle^{\vartheta}$. Now v_x is in $F(Y)$, and so $T^* \subseteq F(Y)$. So we may apply a Tietze transformation of type II' to get the presentation $\langle Y;T^*\rangle^{\psi}$.

Since X, Y, and R are all finite, T^* is finite. Since $T^* \subseteq \langle S\rangle^{F(Y)}$, each element of T^* is the product of finitely many conjugates of elements of S and their inverses, and we choose one such expression for each t^*. Let S_1 consist of those elements of S which occur in some such expression. Then S_1 is finite, and as $\ker\psi$ equals $\langle S\rangle^{F(Y)}$ and also equals $\langle T^*\rangle^{F(Y)}$ it must also equal $\langle S_1\rangle^{F(Y)}$, as required.//

Exercise 9 Show that $\mathbb{Z}_6$ can be presented as $\langle y,z;\ y^3,\ z^2,\ yz = zy\rangle$.

Exercise 10 In $F(x,y)$, write x as a product of conjugates of $xy^2x^{-1}y^{-3}$ and $yx^2y^{-1}x^{-3}$ and their inverses.

Exercise 11 Show that the following are presentations of the trivial group.

(i) $\langle a,b;\ ab^na^{-1} = b^{n+1}, ba^nb^{-1} = a^{n+1}\rangle$.

(ii) $\langle a,b;\ a^n = b^{n+1},\ aba = bab\rangle$.

(iii) $\langle a,b,c;\ aba^{-1} = b^2,\ bcb^{-1} = c^2,\ cac^{-1} = a^2\rangle$.

(iv) The group with generators x, y, and z, with defining relators $(xyx^{-1}y^{-2})(yzy^{-1}z^{-2})(xyx^{-1}y^{-2}) = (yzy^{-1}z^{-2})^2$ and the two other relators obtained by permuting x, y, and z cyclically.

Exercise 12 Find presentations of the symmetric group S_n and the alternating group A_n for $n \leq 5$.

1.3 *FREE PRODUCTS*

Definition Let $\{G_\alpha\}$ be a family of groups, G a group, and let $i_\alpha: G_\alpha \to G$ be homomorphisms. Then $(G, \{i_\alpha\})$ is called a *free product* of the groups G_α if for every group H and homomorphisms $f_\alpha: G_\alpha \to H$ there is a unique homomorphism $f: G \to H$ such that $f_\alpha = i_\alpha f$ fo all α.

As with free groups, we ask if free products exist, if they are unique, and if the maps i_α are monomorphisms. The proofs of Propositions 1 and 2(i) generalise immediately. Because of the uniqueness, we usually refer to "the free product" of the groups G_α, rather than to "a free product".

Proposition 18 *If* $(G,\{i_\alpha\})$ *and* $(H,\{j_\alpha\})$ *are both free products of the groups* G_α *then there is a (unique) isomorphism* $f: G \to H$ *such that* $i_\alpha f = j_\alpha$ *for all* α.//

Proof There is a homomorphism $f: G \to H$ such that $i_\alpha f = j_\alpha$ for all α, and also a homomorphism $f': H \to G$ such that $j_\alpha f' = i_\alpha$ for all α. Since $i_\alpha ff' = i_\alpha$ for all α, the uniqueness of homomorphisms in the definition ensures that ff' is the identity map on G, and similarly $f'f$ is the identity on H.//

Proposition 19 *Let* $(G,\{i_\alpha\})$ *be the free product of the groups* G_α. *Then (i)* i_α *is a monomorphism iff there is some group* H *and homomorphisms* $f_\beta: G_\beta \to H$ *for all* β *such that* f_α *is a monomorphism; (ii)* i_α *is a monomorphism for all* α.

Proof (ii) follows from (i) by taking H to be G_α with f_α being the identity function and all other f_β being trivial, while (i) holds because there is a homomorphism f with $f_\alpha = i_\alpha f$//

Theorem 20 *Any family of groups G_α has a free product.*

Proof Let G_α have presentation $\langle X_\alpha; R_\alpha\rangle^{\varphi_\alpha}$. We may assume (replacing each X_α by a set bijective with it, if necessary) that $X_\alpha \cap X_\beta = \emptyset$ for $\alpha \neq \beta$. Let G be the group $F(\cup X_\alpha)/\langle \cup R_\alpha\rangle^{F(\cup X_\alpha)}$, so that G has presentation $\langle \cup X_\alpha; \cup R_\alpha\rangle^{\varphi}$, where φ is the natural map. By von Dyck's Theorem, the inclusion of X_α in $\cup X_\alpha$ induces a homomorphism i_α from G_α to G. We shall show that $(G, \{i_\alpha\})$ is the free product.

So let $f_\alpha : G_\alpha \to H$ be homomorphisms. Now f_α induces a homomorphism ψ_α from $F(X_\alpha)$ to H such that $R_\alpha \psi_\alpha = 1$. There is then a homomorphism ψ from $F(\cup X_\alpha)$ to H such that ψ is ψ_α on X_α. Since $(\cup R_\alpha)\psi = 1$, ψ induces a homomorphism $f: G \to H$ with $\varphi f = \psi$. Also, for every $x_\alpha \in X_\alpha$, $(x_\alpha \varphi_\alpha) i_\alpha f = x_\alpha \varphi f = x_\alpha \psi$, by definition of f and i_α, and $x_\alpha \psi = x_\alpha \psi_\alpha = x_\alpha \varphi_\alpha f_\alpha$, by definition of ψ and ψ_α. Since $X_\alpha \varphi_\alpha$ generates G_α, we see that $f_\alpha = i_\alpha f$. Also f is unique, since G is generated by $\cup X_\alpha \varphi_\alpha i_\alpha$.//

The free product of the family of groups G_α is denoted by $* G_\alpha$, while the free product of two groups is usually written as $G_1 * G_2$; more generally, the free product of n groups G_i is written $G_1 * \ldots * G_n$. If we regard the free product as being given by the presentation of the previous theorem, we see that $G_2 * G_1$ is the same as $G_1 * G_2$, while $G_1 * G_2 * G_3$ is the same as $(G_1 * G_2) * G_3$ and as $G_1 * (G_2 * G_3)$. These identifications, and other more complicated ones, will be made without further comment. Further, since each i_α is a monomorphism, we usually regard G_α as a subgroup of $*G_\beta$ with i_α being inclusion, so that we do not need to mention it.

Examples The free group $F(X)$ is the free product of the groups C_x for all $x \in X$, where C_x is infinite cyclic with generator x.

The free product of two cyclic groups of order 2 has presentation $\langle a,b;\ a^2, b^2\rangle$. Applying Tietze transformations, this group may also be written as $\langle a,b,c;\ a^2, b^2, c^{-1}ab\rangle$, as $\langle a,c;\ a^2, (c^{-1}a)^{-2}\rangle$, as $\langle a,d;\ a^2, (da)^2$, or $\langle a,d;\ a^2, (ad)^2\rangle$, or as $\langle a,c;\ a^2, a^{-1}cac\rangle$, which shows that this group is the infinite dihedral group.

Let $\{G_\alpha\}$ and $\{H_\alpha\}$ be collections of groups, and let $\varphi_\alpha: G_\alpha \to H_\alpha$ be homomorphisms. Let G be $*G_\alpha$, and let H be $*H_\alpha$. Regarding H_α as a subgroup of H, the homomorphisms φ_α from G_α to H give rise to a unique homomorphism from G to H, which we denote by $*\varphi_\alpha$.

Theorem 21 (Normal Form Theorem) *Let $(G,\{i_\alpha\})$ be the free product of the groups G_α. Then (i) each i_α is a monomorphism; (ii) regarding i_α as inclusion, any element of G can be uniquely written as $g_1 \dots g_n$, where $n \geq 0$, $g_i \in G_{\alpha_i}$ for some α_i, $g_i \neq 1$, and $\alpha_r \neq \alpha_{r+1}$ for $r < n$.*

Proof Temporarily denote $g_\alpha i_\alpha$ by $\bar{g}_\alpha$ for $g_\alpha \in G_\alpha$. We prove both parts (the first part has already been proved by another method) by showing that any $u \in G$ can be uniquely written as $\bar{g}_1 \dots \bar{g}_n$, with $n \geq 0$, $g_i \in G_{\alpha_i}$, $g_i \neq 1$, and $\alpha_r \neq \alpha_{r+1}$.

Since our construction of G shows that $\cup G_\alpha i_\alpha$ generates G, any u can certainly be written as $\bar{g}_1 \dots \bar{g}_n$, with $n \geq 0$, $g_i \in G_{\alpha_i}$, $g_i \neq 1$, where $\alpha_r = \alpha_{r+1}$ is permitted. If $\alpha_r = \alpha_{r+1}$ and $g_{r+1} \neq g_r^{-1}$ then we may write u as $\bar{g}_1 \dots \bar{g}_{r-1} \bar{h} \bar{g}_{r+2} \dots \bar{g}_n$, where $1 \neq h = g_r g_{r+1} \in G_{\alpha_r}$, while if $\alpha_r = \alpha_{r+1}$ and $g_{r+1} = g_r^{-1}$ we may write u as $\bar{g}_1 \dots \bar{g}_{r-1} \bar{g}_{r+2} \dots \bar{g}_n$. By induction on n, there is at least one way of writing u in the required form.

To prove uniqueness we follow van der Waerden's method (method III of Theorem 4). Let S be the set of all sequences $(g_1, \dots, g_n)$ with $n \geq 0$, $g_i \in G_{\alpha_i}$, $g_i \neq 1$, and $\alpha_r \neq \alpha_{r+1}$; in particular, the empty sequence ( ) is in S.

Take $g_\alpha \in G_\alpha - \{1\}$. A map from S to S is defined by mapping

$(g_1, \dots, g_n)$ to $(g_1, \dots, g_n, g_\alpha)$ if $\alpha \neq \alpha_n$,

$(g_1, \dots, g_n)$ to $(g_1, \dots, g_{n-1}, g_n g_\alpha)$ if $\alpha = \alpha_n$ and $g_n g_\alpha \neq 1$,

$(g_1, \dots, g_n)$ to $(g_1, \dots, g_{n-1})$ if $\alpha = \alpha_n$ and $g_n g_\alpha = 1$.

This defines a function φ_α from G_α to the set of maps from S to itself (defining $1\varphi_\alpha$ to be the identity). It is easy to check that φ_α preserves multiplication. It follows that φ_α maps into Sym$S$, the group of permutations of S, since $g_\alpha \varphi_\alpha$ has inverse $g_\alpha^{-1} \varphi_\alpha$.

Let $\varphi: G \to \text{Sym}S$ be the homomorphism such that $\varphi_\alpha = i_\alpha \varphi$ for all α. Take $u \in G$ and write it as $h_1 \dots h_m$, with $h_i \neq 1$, $h_i \in G_{\alpha_i}$, and $\alpha_i \neq \alpha_{i+1}$. It is easy to see (by induction) that $u\varphi$ applied to the empty sequence ( ) gives the sequence $(h_1, \dots, h_m)$, and so $h_1, \dots, h_m$ are uniquely determined by u.//

An alternative proof of uniqueness may be given along the lines of method IV of Theorem 4, and this enables us to obtain results about residual finiteness of free products. Take some particular u expressed as in the previous paragraph. Let S_m consist of those elements of S with $n \leq m$. We can define a map ψ_α from G_α to the set of functions from S_m to itself by using the previous definition except in the case where $n = m$ and $\alpha \neq \alpha_m$; in this case g_α is required to send $(g_1, \ldots, g_m)$ to $(g_1, \ldots, g_m)$. As before, we get a homomorphism ψ from G to $\mathrm{Sym}S_m$ such that $u\psi$ is not the identity.

The advantage of this method is that S_m is finite if there are only finitely many G_α all of which are finite. Thus we have proved the following result.

Proposition 22 *The free product of finitely many finite groups is residually finite.//*

In fact a stronger result is true. If we take the free product G of an arbitrary collection of residually finite groups G_α, and an element $u \in G - \{1\}$, there is a homomorphism φ from G onto the free product of finite groups $\overline{G}_\alpha$, all but finitely many of which are trivial, such that $u\varphi \neq 1$ (see Exercise 17). Applying the previous argument to $u\varphi$, we get the following.

Corollary *The free product of residually finite groups is residually finite.//*

We have an analogue of Proposition 8.

Proposition 23 *Let G_α be subgroups of a group G. Then the following are equivalent. (i) G is the free product of the subgroups G_α; (ii) every element of G can be uniquely written as $g_1 \ldots g_n$, with $n \geq 0$, $g_i \in G_{\alpha_i}$, $g_i \neq 1$, and $\alpha_i \neq \alpha_{i+1}$; (iii) G is generated by the subgroups G_α and 1 cannot be written as a product $g_1 \ldots g_n$ with $n > 0$, $g_i \in G_{\alpha_i}$, $g_i \neq 1$, and $\alpha_i \neq \alpha_{i+1}$.*

Proof (ii) and (iii) are equivalent, as in Proposition 8. If (i) holds so does (ii), by the Normal Form Theorem. Suppose that (iii) holds. We know that there is a homomorphism from $*G_\alpha$ to G which is the inclusion on each G_α, and (iii) tells us that this homomorphism is onto and has trivial kernel.//

Let u in $*G_\alpha$ be $g_1 \ldots g_n$, with $g_i \in G_{\alpha_i}$, $g_i \neq 1$, and $\alpha_i \neq \alpha_{i+1}$. We say that u has *length* n, and that the product $g_1 \ldots g_n$ is *reduced*. If v is written as a reduced product $h_1 \ldots h_m$, we say that uv is *reduced as written* if g_n and h_1 are not in the same G_α. If uv is not reduced as written, we say there is *cancellation* if $g_nh_1 = 1$, while we say there is *coalescence* (or *amalgamation;* but this latter phrase is in danger of being confused with another concept) if $g_nh_1 \neq 1$. Plainly $|uv| = |u| + |v|$ if uv is reduced as written, while $|uv| = |u| + |v| - 1$ if there is coalescence. If there is cancellation then $|uv| < |u| + |v|$.

With u written as above, we say u is *cyclically reduced* if $n = 1$ or $\alpha_n \neq \alpha_1$. As with free groups, every element is conjugate to a cyclically reduced element. If v is cyclically reduced and $|v| > 1$ then $|v^n| = n|v|$. It follows that any element of finite order in a free product must be in a conjugate of one of the free factors.

The exercises indicate an alternative way of constructing free products, similar to that used for free groups.

Exercise 13 Show that the function φ_α defined in the proof of the Normal Form Theorem, and the function ψ_α used in proving Proposition 22 preserve multiplication.

Exercise 14 Let $\{G_\alpha\}$ be a collection of groups such that $G_\alpha \cap G_\beta = 1$ for $\alpha \neq \beta$. Let M be the free monoid $M(\cup G_\alpha)$. The word w' is said to come from w by *elementary reduction* if w is $g_1 \ldots g_n$ and either w' is $g_1 \ldots g_{i-1}hg_{i+2} \ldots g_n$ where g_i and g_{i+1} are in the same factor G_α and h is g_ig_{i+1} or else w' is $g_1 \ldots g_{i-1}g_{i+2} \ldots g_n$ with g_ig_{i+1} being 1 (note that the second type of reduction is obtained by two applications of the first, except for the reduction of the sequence 1 to the empty sequence). Obtain an equivalence relation from this as in the construction of free groups. Show that the set of equivalence classes is the free product.

Exercise 15 Using the construction of the previous exercise, prove the Normal Form Theorem both by canonical reduction and by the Diamond Lemma.

Exercise 16 Let G_A be the free product of the groups G_α for $\alpha \in A$, and let G_B be the free product of the groups G_α for $\alpha \in B$, where $B \subset A$. Show that G_B is a subgroup of G_A, and that there is a homomorphism from G_A to G_B which is the identity on G_B.

Exercise 17 Let G be a residually finite group, and let S be a finite subset of G. Show that there is a normal subgroup N of finite index in G such that $N \cap S = \emptyset$. Use this to show that if the groups G_α are residually finite then for any $u \neq 1$ in their free product there is a homomorphism φ from the free product to a free product of finitely many finite groups such that $u\varphi \neq 1$. Deduce the corollary to Proposition 22.

Exercise 18 Let $G = A * B$, where $A \neq \{1\} \neq B$. Let $g \in G$ be such that $A \cap g^{-1}Ag \neq \{1\}$. Show that $g \in A$. Deduce that A contains no non-trivial normal subgroup of G.

1.4 *PUSH-OUTS AND AMALGAMATED FREE PRODUCTS*

Definition Let G_0, G_1, and G_2 be groups, and let $i_1:G_0 \to G_1$ and $i_2:G_0 \to G_2$ be homomorphisms. Let G be a group, and let $j_1:G_1 \to G$ and $j_2:G_2 \to G$ be homomorphisms. We call (G, j_1, j_2) the *push-out* of (i_1, i_2) if

(1) $i_1 j_1 = i_2 j_2$,

(2) for any group H and homomorphisms $\varphi_r:G_r \to H$ $(r = 1,2)$ with $i_1\varphi_1 = i_2\varphi_2$ there is a unique homomorphism $\varphi:G \to H$ such that $\varphi_r = j_r\varphi$ $(r = 1,2)$.

We then say that we have a *push-out square*

$$\begin{array}{ccc} G_0 & \xrightarrow{i_1} & G_1 \\ {\scriptstyle i_2}\downarrow & & \downarrow{\scriptstyle j_1} \\ G_2 & \xrightarrow[j_2]{} & G \end{array} .$$

As usual, the push-out is unique up to isomorphism.

Theorem 24 *Any pair* (i_1, i_2) *has a push-out.*

Proof For $r = 1,2$, let G_r have presentation $\langle X_r; R_r \rangle^{\vartheta_r}$, where $X_1 \cap X_2 = \emptyset$. Let Y generate G_0, and choose, for all $y \in Y$, $w_{yr} \in F(X_r)$ such that $yi_r = w_{yr}\vartheta_r$. Let G be the group with presentation $\langle X_1 \cup X_2; R_1, R_2, \{w_{y1}^{-1}w_{y2}\}\rangle$. Thus we have the natural maps j_r from G_r into G, induced by the inclusions of X_r into $X_1 \cup X_2$, and G is generated by $G_1j_1 \cup G_2j_2$. Hence there can be at most one homomorphism from G with specified values on $G_1j_1 \cup G_2j_2$.

Suppose that we have homomorphisms φ_r from G_r to a group H such that $i_1\varphi_1 = i_2\varphi_2$. Then φ_r defines a homomorphism ψ_r from $F(X_r)$ to H which is trivial on R_r. By von Dyck's Theorem, the homomorphism from $F(X_1 \cup X_2)$ which is ψ_r on X_r defines a homomorphism φ from G to H, which, by construction, satisifes $j_r\varphi = \varphi_r$.//

Let G_2 be trivial. By using the above proof, or by arguing directly from the definition, we find that the pushout is the quotient group $G_1/\langle G_0 i_1 \rangle^{G_1}$.

The maps j_1 and j_2 need not be one-one, even if one of i_1 and i_2 is one-one. For instance, suppose that G_1 is simple, and that i_1 is one-one and i_2 is onto but not one-one. Take $w \in G_0$ with $w \neq 1$ but $wi_2 = 1$. Then $wi_1 \neq 1$ but $wi_1j_1 = 1$. As G_1 is simple, this ensures that j_1 is trivial. Since i_2 is onto and $i_2j_2 = i_1j_1$ is trivial, it follows that j_2 is trivial. Hence the push-out is trivial.

When both i_1 and i_2 are one-one, the pushout G is called the *amalgamated free product* of G_1 and G_2 with G_0 amalgamated. In this case we usually regard G_0 as a subgroup of G_1 and G_2, and regard i_1 and i_2 as inclusions. The usual notation for this situation is $G_1 *_{G_0} G_2$. Sometimes it is mor convenient to use the notation $G_1\ {}_{G_0}{*}_{H_0}\ G_2$ where $G_0 \subseteq G_1$, $H_0 \subseteq G_2$, i_1 is the inclusion, and i_2 is an isomorphism between G_0 and H_0. For more precision, we could mention the specific isomorphism from G_0 to H_0.

For an amalgamated free product we shall see that j_1 and j_2 are one-one, and we regard them as inclusions. We shall also see that, with this convention, $G_1j_1 \cap G_2j_2 = G_0$.

Let G be $A *_C B$. Let S and T be left transversals of C in A and B with $1 \in S \cap T$ (that is, S contains one member of each coset aC). We shall obtain a normal form theorem using S and T.

As in Propositions 2 and 19, in order to show that the maps j_A and j_B from A and B to G are monomorphisms it is enough to find some group H and monomorphisms $\varphi: A \to H$ and $\psi: B \to H$ such that $\varphi = \psi$ on C. We can take H to be the group of permutations of $S \times T \times C$. Let $a\varphi$ be the function sending (s,t,c) to (s_1,t,c_1), where $s_1c_1 = sca$. It is easy to check that $a\varphi$ is a permutation (we have a bijection of $S \times C$ with A, and so a bijection of $S \times T \times C$ with $T \times A$, under which $a\varphi$ corresponds right multiplication by a), and that φ is a homomorphism. We define ψ similarly. It is easy to check that both φ and ψ are one-one, and that $\varphi = \psi$ on C, as required.

If we take B to be the same as A then, provided $C \neq A$, we obtain two distinct homomorphisms from A to H which agree on C. Also H will be finite if A is finite.

In an arbitrary category a morphism $\eta: X \to Y$ is defined to be an epimorphism if, for any Z and morphisms φ and ψ from Y to Z, the condition $\eta\varphi = \eta\psi$ implies that $\varphi = \psi$. In the category of rings with identity, the inclusion of $\mathbf{Z}$ in $\mathbf{Q}$ is easily seen to be an epimorphism which is not an onto map. From the preceding paragraph, it is easy to check that in the category of groups (or of finite groups) epimorphisms are the same as onto homomorphisms.

Similarly the morphism η is a monomorphism if for any morphisms φ and ψ from Z to X such that $\varphi\eta = \psi\eta$ we have $\varphi = \psi$. It is rather easier to prove that monomorphisms are the same as one-one homomorphisms in the category of groups (or the category of finite groups). We need only observe that if $H \triangleleft G$ then the trivial map and the inclusion map from H to G give the same result when composed with the map from G to G/H.

Theorem 25 (Normal Form Theorem) *Let G be $A *_C B$. With the above notation*

(*i*) j_A *and* j_B *are monomorphisms,*

(*ii*) $Aj_A \cap Bj_B = Cj_A = Cj_B$,

(*iii*) *regarding* j_A *and* j_B *as inclusions, any element of G can be uniquely written as* $u_1 \ldots u_n c$, *where* $n \geq 0$, $c \in C$, *and* $u_1, \ldots, u_n$ *come alternately from* $S - \{1\}$ *and* $T - \{1\}$.

Remark By the definition of an amalgamated free product, we know that $j_A = J_B$ on C. Also, regarding J_A and J_B as inclusions, (ii) can be written as $A \cap B = C$.

Proof (ii) follows at once from uniqueness in (iii).

We will temporarily denote aj_A and bj_B by $\bar{a}$ and $\bar{b}$. As in the normal form theorem for free products, to prove both (i) and (iii) it is enough to show that for any $g \in A *_C B$ there are unique $u_1, \ldots, u_n$, and c with $n \geq 0$, $c \in C$, and $u_1, \ldots, u_n$ alternately from $S - \{1\}$ and $T - \{1\}$ such that $g = \bar{u}_1 \ldots \bar{u}_n \bar{c}$.

Now any g can be written as $\bar{g}_1 \ldots \bar{g}_k$, where $g_i \in A \cup B$, for some k. If g_i and g_{i+1} are both in A or both in B we may write g as $\bar{g}_1 \ldots \bar{g}_{i-1} \bar{h} \bar{g}_{i+2} \ldots \bar{g}_k$, where $h = g_i g_{i+1}$. Continuing like this, we see that g can be written either as $\bar{c}$ or as $\bar{g}_1 \ldots \bar{g}_n$, where $g_1, \ldots, g_n$ are alternately from $A - C$ and $B - C$.

Inductively, write $\bar{g}_1 \ldots \bar{g}_{n-1}$ as $\bar{u}_1 \ldots \bar{u}_{n-1} \bar{c}$, where $u_i \in (S \cup T) - \{1\}$ and $u_1 \in g_1 C$ and $u_i \in C g_i C$ for each $i > 1$, so that u_i comes alternately from $S - \{1\}$ and $T - \{1\}$. Then

$$g = \bar{u}_1 \ldots \bar{u}_{n-1} \bar{c} \bar{g}_n = \bar{u}_1 \ldots \bar{u}_{n-1} \bar{h}, \text{ where } h = c g_n,$$
which equals $\bar{u}_1 \ldots \bar{u}_{n-1} \bar{u}_n \bar{d}$, where $h = u_n d$ with $u_n \in S \cup T$.

Since $u_n d = c g_n$, we have $u_n \in C g_n C$, as required. Also $u_n \neq 1$, since $g_n \notin C$. So we have shown that every element of G can be written in the required form.

We prove uniqueness by van der Waerden's method (another proof is given in the next section). Let X be the set of all sequences $(u_1, \ldots, u_n, c)$

with $n \geq 0$, $c \in C$, and $u_1, \ldots, u_n$ alternately in $S - \{1\}$ and $T - \{1\}$. We define a homomorphism φ from G to Sym$X$ such that, when g is written as $\bar{u}_1 \ldots \bar{u}_n \bar{c}$ of the required form, the action of $g\varphi$ on the sequence (1) is the sequence $(u_1, \ldots, u_n, c)$. This will prove uniqueness of the representation of g.

To construct φ it is enough, by definition of the amalgamated free product, to define φ on $A \cup B$ so that it maps into the set of functions from X to X and is multiplication-preserving on A and on B (then it automatically maps A into Sym$X$, since $a^{-1}\varphi$ will be the inverse of $a\varphi$). We define φ on A by requiring the action of $a\varphi$ on $(u_1, \ldots, u_n, c)$ to be

$(u_1, \ldots, u_n, v, d)$ if $u_n \notin A$, $ca = vd$ with $v \in S - \{1\}$ and $d \in C$,
$(u_1, \ldots, u_n, d)$ if $u_n \notin A$ and $ca = d$ with $d \in C$ (that is, if $a \in C$)
$(u_1, \ldots, u_{n-1}, v, d)$ if $u_n \in A$ and $u_n ca = vd$ with $v \in S - \{1\}$ and $d \in C$,
$(u_1, \ldots, u_{n-1}, d)$ if $u_n \in A$ and $u_n ca = d$ with $d \in C$.

Readers are left to check for themselves that $(aa')\varphi$ equals $(a\varphi)(a'\varphi)$. We define φ on B by similar formulae. It is easy to check that the definitions of φ on A and on B agree on C.//

There is an alternative version of the Normal Form Theorem, using right transversals instead of left ones, which may be proved symmetrically. Because of the inconvenience of working with transversals (and the fact that they can be arbitrary transversals) the weaker theorem below is often more useful than the Normal Form Theorem.

Theorem 26 (Reduced Form Theorem) *With the above notation, and regarding j_A and j_B as inclusions,*

(i) any $w \in G - C$ can be written as $g_1 \ldots g_n$, where $n \geq 1$, and the g_i are alternately from $A - C$ and $B - C$,

(ii) if we can also write w as $h_1 \ldots h_m$ with the h_j alternately from $A - C$ and $B - C$ then $m = n$ and $h_1 \in g_1 C$, $h_n \in C g_n$, and $h_i \in C g_i C$ for all other i,

(iii) if $n > 1$ then $w \notin A \cup B$,

(iv) such a product cannot be in C.

Proof In the proof of the Normal Form Theorem we proved (i). We also showed that if w is expressed as in (i) then its normal form is $u_1 \ldots u_n c$ where $u_1 \in g_1 C$, $u_n \in C g_n$, and $u_i \in C g_i C$ for all other i. Then (ii) follows at once from the uniqueness of the normal form. The Normal Form Theorem also tells us that such a product cannot be in C, and can only be in $A \cup B$ if $n = 1$.//

It is easy to see that the Normal Form Theorem could be deduced from the Reduced Form Theorem (and the fact that j_A and j_B are one-one), if we had an alternative proof of the Reduced Form Theorem. Readers may wonder if one of these theorems (and the residual finiteness of suitable amalgamated free products) can be proved along the lines of Proposition 22 . Such proofs do exist, but the only one I know is notationally complicated. We shall give another approach in the exercises to the next section.

If w is written as $g_1 \ldots g_n$ with the g_i alternately from $A - C$ and $B - C$ we refer to this expression as a *reduced form* of w. Then the integer n (which we have just seen depends only on w and not on how it is written as such a product) is called the *length* of w. Thus w has length 1 iff it is in $(A - C) \cup (B - C)$. An element of C is said to have length 0. We have results similar to those for free groups and free products about the length of products. Also any element not in C is easily seen to be conjugate to an element $g_1 \ldots g_n$ in reduced form with g_1 and g_n in diifferent factors if $n \neq 1$. It follows that w has finite order only if it is conjugate to an element of $A \cup B$.

Proposition 27 *Let A and B be subgroups of a group G, and let C be $A \cap B$. Then $G = A *_C B$ iff every element of $G - C$ can be written as a product $g_1 \ldots g_n$ with the g_i alternately from $A - C$ and $B - C$, and no such product is* 1.

Proof We know that $A *_C B$ has this property. Let G have the property. The inclusions of A and B into G give rise to a homomorphism from $A *_C B$ into G. The conditions tell us that this homomorphism is both one-one and onto.//

Proposition 28 *Let G be $A *_C B$. Let $A_1 \subseteq A$ and $B_1 \subseteq B$ be such that $A_1 \cap C = B_1 \cap C = C_1$, say. Then the subgroup $\langle A_1, B_1 \rangle$ of G is $A_1 *_{C_1} B_1$. Also $\langle A_1, B_1 \rangle \cap A = A_1$ and $\langle A_1, B_1 \rangle \cap B = B_1$.*

Proof The first part is immediate, by applying Proposition 27 to $\langle A_1, B_1 \rangle$. The second part then follows from part (iii) of the Reduced Form Theorem, as this tells us that the product of elements alternately from $A_1 - C_1$ and $B_1 - C_1$ cannot be in A if its length is >1.

Proposition 29 *Let G be $A *_C B$. If G and C are finitely generated then A and B are finitely generated.*

Proof Write each element of G as the product of finitely many elements of $A \cup B$. Let A_1 be the subgroup of A generated by C and the finitely many elements of A which occur in the chosen expressions for each generator of G, and similarly for B_1. Then $G = \langle A_1, B_1 \rangle$, by construction. Thus Proposition 28 tells us that $A = A_1$, which gives the result.//

Remark More generally, given groups G_α and monomorphisms $i_\alpha : C \to G_\alpha$ we can define the amalgamated free product $*_C G_\alpha$, with similar properties. In particular, if $C = \{1\}$, this group is just the free product of the G_α. The general case is perhaps best regarded as a special case of the *tree products* of section 8.3

Exercise 19 Suppose that we have a pushout square with G_2 trivial. Show directly from the definition that G is $G_1 / \langle G_0 i_1 \rangle^{G_1}$.

Exercise 20 Suppose that we have three groups G_1, G_2, and G_3, and subgroups H_r of G_r for $r = 1, 2, 3$, together with monomorphisms i_r from H_r into G_{r+1} (where G_4 is G_1) such that $H_r i_r \cap H_{r+1} = \{1\}$ for all r. Give an example to show that if we take the quotient of $G_1 * G_2 * G_3$ by the normal subgroup generated by $\{(h_r \varphi_r) h_r^{-1}$, all r, all $h_r \in H_r\}$ then the resulting group may be trivial (one of the earlier exercises gives an example of this situation). Show that if we make the corresponding construction using $k > 3$ groups then the natural map from each G_r to the quotient group is a monomorphism.

1.5 *HNN EXTENSIONS*

In 1949 Higman, Neumann, and Neumann studied a construction related to amalgamated free products. This construction is therefore now called an HNN extension. Later Britton (1963) used this construction in his proof of the unsolvability of the word problem for finitely presented groups (see Chapter 9). In recent years it has become clear that HNN extensions (originally obtained as subgroups of certain amalgamated free products) should be treated in their own right as one of the basic constructions of combinatorial group theory.

This time we shall not use a universal property in the definition, but will obtain this property later. Let G and A be groups, and let i_0 and i_1 be monomorphisms from A to G. Let P be an infinite cyclic group with generator p. Let N be the normal subgroup of $G*P$ generated by the set $\{p^{-1}(ai_0)p(ai_1)^{-1}\}$ where a runs over A (or just over a set of generators of A, since this provides the same subgroup N). Let H denote $(G*P)/N$. Then H is called the *HNN extension of the base group G with stable letter p and associated subgroups* Ai_0 *and* Ai_1. We usually take A as a subgroup of G with i_0 being inclusion, and we write H as $\langle G,p;\ p^{-1}Ap = B\rangle$, where B is Ai_1. This notation, however, does not make i_1 explicit, and we may want to be more precise by writing H as $\langle G,p;\ p^{-1}Ap = Ai_1\rangle$.

Let G have presentation $\langle X;R\rangle^{\varphi}$ with $p \notin X$, and let Y be a subset of $F(X)$ and let ϑ be a one-one map from Y into $F(X)$. Let $Y\varphi$ generate a subgroup A and let $Y\vartheta\varphi$ generate a subgroup B. If ϑ induces an isomorphism from A to B then the HNN extension H has presentation $\langle X,p;\ R,\ p^{-1}yp = y\vartheta\rangle$. Conversely, a presentation of this form is an HNN extension provided that ϑ induces an isomorphism from $\langle Y\varphi\rangle$ to $\langle Y\vartheta\varphi\rangle$.

Let g_0 and g_1 be in G and define j_0 and j_1 from A to G by $aj_r = (ai_r)^{g_r}$ for $r = 0,1$. Then the HNN extensions $\langle G,p;\ (ai_0)^p = ai_1\rangle$ and $\langle G,q;\ (aj_0)^q = aj_1\rangle$ are easily seen to be isomorphic, with the isomorphism sending g to g and p to $g_0qg_1^{-1}$.

More generally, we can take a family of groups A_α and monomorphisms $i_{0\alpha}$ and $i_{1\alpha}$ from A_α to G. Let P be free on $\{p_\alpha\}$, and let the normal subgroup of $G*P$ generated by $\{p_\alpha^{-1}(a_\alpha i_{0\alpha})p_\alpha(ai_{1\alpha})^{-1};\ \text{all } \alpha, \text{ all } a_\alpha \in A_\alpha\}$ be N. Then $H = (G*P)/N$ is called the HNN extension of the base group G with stable letters $\{p_\alpha\}$ and associated pairs of subgroups $A_\alpha i_{0\alpha}$ and $A_\alpha i_{1\alpha}$. The notations and remarks of the previous two paragraphs can be extended to this more general case. It is occasionally useful to consider the same construction when the maps $i_{0\alpha}$ and $i_{1\alpha}$ are not monomorphisms; in this case we refer to a *pseudo-HNN extension*.

Examples The free group with basis $\{p_\alpha\}$ is the HNN extension of the trivial group with stable letters p_α.

Let G be any group, andf let A be a subgroup of G. Then we can form the HNN extension $\langle G,k;\ k^{-1}ak = a \text{ for all } a \in A\rangle$.

The group with presentation $\langle a,b;\ a^{-1}ba = b^2\rangle$ is the HNN extension of the infinite cyclic group $\langle b\rangle$ with stable letter a and associated subgroups $\langle b\rangle$ and $\langle b^2\rangle$.

The inclusion of G in $G*P$ induces a homomorphism $j;G \to H$, which we shall see later is a monomorphism. The universal property of HNN extensions is given in the proposition below.

Proposition 30 *Let φ be a homomorphism from G to a group K. Suppose that K has elements k_α such that $k_\alpha^{-1}(a_\alpha i_{0\alpha}\varphi)k_\alpha = a_\alpha i_{1\alpha}\varphi$ for all α and all $a_\alpha \in A_\alpha$. Let H be the HNN extension of G with stable letters p_α and associated subgroups $A_\alpha i_{0\alpha}$ and $A_\alpha i_{1\alpha}$. Then there is a unique homomorphism $\psi: H \to K$ such that $j\psi = \varphi$ and $p_\alpha\psi = k_\alpha$ for all α.*

Proof Since H is generated by Gj and $\{p_\alpha\}$ there can be at most one such homomorphism. There is certainly a homomorphism from $G*P$ to K which is φ on G and which sends p_α to k_α, and this will send the elements $p_\alpha^{-1}(a_\alpha i_{0\alpha})p_\alpha(a_\alpha i_{1\alpha})^{-1}$ to 1. Since the normal subgroup of $G*P$ generated by these elements is the kernel of the natural map from $G*P$ to H, this homomorphism will induce the required ψ.//

As usual, there can be only one group (up to isomorphism) with the property of this proposition; hence, if we construct a group with this property we know it must be the HNN extension.

In particular, there is a homomorphism from H to P sending each p_α to itself and mapping G trivially. It follows that the subgroup of H generated by the image $\{p_\alpha N\}$ of $\{p_\alpha\}$ is free with these elements as basis. Because of this, we regard P as a subgroup of H.

Let H be the HNN extension $\langle G,p_\alpha;\ p_\alpha^{-1}A_\alpha p_\alpha = A_{-\alpha}\rangle$, with an isomorphism $i_{1\alpha}$ from A_α to $A_{-\alpha}$. Let S_α and $S_{-\alpha}$ be left transversals of A_α and $A_{-\alpha}$ in G, all containing 1. We have the following Normal Form and Reduced Form theorems.

Theorem 31 (Normal Form Theorem) (*i*) *The homomorphism j is a monomorphism.* (*ii*) *Regarding G as a subgroup of H, any $h \in H$ can be uniquely written as*

$$h = g_0 p_{\alpha_0}^{\varepsilon_0} g_1 \ldots g_{n-1} p_{\alpha_{n-1}}^{\varepsilon_{n-1}} g_n$$

where $n \geq 0$, $\varepsilon_i = \pm 1$ (for $n = 0$ the expression is just g_0) and

$g_n \in G$,
$g_i \in S_{\alpha_i}$ *if* $\varepsilon_i = 1$ *and* $g_i \in S_{-\alpha_i}$ *if* $\varepsilon_i = -1$,
if $\alpha_{i-1} = \alpha_i$ *and* $\varepsilon_{i-1} = -\varepsilon_i$ *then* $g_i \neq 1$.

We say that an element $g_0 p_{\alpha_0}^{\varepsilon_0} g_1 \dots g_{n-1} p_{\alpha_{n-1}}^{\varepsilon_{n-1}} g_n$ of $G * P$, where $n \geq 0$, $\varepsilon_i = \pm 1$, $g_i \in G$ has a *pinch* if there is some r such that $\alpha_r = \alpha_{r-1}$ and either $\varepsilon_r = -\varepsilon_{r-1} = 1$ and $g_r \in A_r$ or $\varepsilon_r = -\varepsilon_{r-1} = -1$ and $g_r \in A_{-r}$. When G has presentation $\langle X;R \rangle$, it is also convenient to say that an element of $F(X \cup \{p_\alpha$, all $\alpha\})$ has a pinch if the corresponding element of $G * P$ has a pinch.

Theorem 32 (Reduced Form Theorem, or Britton's Lemma)
(*i*) *Regarding G as a subgroup of H, any* $h \in H$ *can be written as*

$$h = g_0 p_{\alpha_0}^{\varepsilon_0} g_1 \dots g_{n-1} p_{\alpha_{n-1}}^{\varepsilon_{n-1}} g_n$$

where $n \geq 0$, $\varepsilon_i = \pm 1$, $g_i \in G$ *and the corresponding element of* $G * P$ *has no pinch.*

(*ii*) *If h can also be written as*

$$v_0 p_{\beta_0}^{\eta_0} \dots p_{\beta_{m-1}}^{\eta_{m-1}} v_m$$

with similar conditions on the v_j *then* $m = n$ *and, for all i,* $\alpha_i = \beta_i$ *and* $\varepsilon_i = \eta_i$.

(*iii*) *If h has an expression as above with no pinch and* $n > 0$ *then* $h \notin G$.

(*iv*) *If the product* $g_0 p_{\alpha_0}^{\varepsilon_0} g_1 \dots g_{n-1} p_{\alpha_{n-1}}^{\varepsilon_{n-1}} g_n$, *with* $g_i \in G$, $\varepsilon_i = \pm 1$, *and* $n > 0$, *equals* 1 *then the corresponding element of* $G * P$ *has a pinch.*

(*v*) *If a product* $g_0 p_{\alpha_0}^{\varepsilon_0} g_1 \dots g_{n-1} p_{\alpha_{n-1}}^{\varepsilon_{n-1}} g_n$, *with* $g_i \in G$, $\varepsilon_i = \pm 1$, *and* $n > 0$, *lies in G then the corresponding element of* $G * P$ *has a pinch.*

Remark In both theorems, the case $n = 0$ gives an element g_0 of G; the product $g_0 p_{\alpha_0}^{\varepsilon_0}$ corresponds to $n = 1$ with $g_1 = 1$.

Proofs Parts (iii) and (iv) of Theorem 32 are immediate from (ii). Part (v) follows from part (iv) since $g_0 \dots g_n = g$ can be written as $g_0 \dots (g_n g^{-1}) = 1$.

We use van der Waerden's method to prove that j is a monomorphism and to prove that an element of H has at most one representation of the form in Theorem 31(ii). So we consider the set W of all sequences $(g_0, \alpha_0, \varepsilon_0, g_1, \dots, g_{n-1}, \alpha_{n-1}, \varepsilon_{n-1}, g_n)$ satisfying the conditions of Theorem 31(ii).

We can obviously define a monomorphism $\varphi : G \to \mathrm{Sym}W$ by $(g_0, \ldots, \varepsilon_{n-1}, g_n).g\varphi = (g_0, \ldots, \varepsilon_{n-1}, g_n g)$. We can also define a function $\pi_\alpha : W \to W$ by

$$(g_0, \ldots, \varepsilon_{n-1}, g_n)\pi_\alpha = (g_0, \ldots, \varepsilon_{n-1}, s, \alpha, 1, b)$$

unless $\alpha_{n-1} = \alpha$, $\varepsilon_{n-1} = -1$, and $g_n \in A_\alpha$, where we write g_n as sa for $s \in S_\alpha$ and $a \in A_\alpha$ and $b = ai_{1\alpha}$, while

$$(g_0, \ldots, \varepsilon_{n-1}, g_n)\pi_\alpha = (g_0, \ldots, \varepsilon_{n-2}, g_{n-1}b)$$

when $n > 1$, $\alpha_{n-1} = \alpha$, $\varepsilon_{n-1} = -1$, $g_n \in A_\alpha$ and $b = g_n i_{1\alpha}$, and, finally, $(g_0, \alpha, -1, g_1)\pi_\alpha = g_0 b$ when $g_1 \in A_\alpha$ and $b = g_1 i_{1\alpha}$. (Note that the case $n = 0$ is covered by the first situation).

It is easy to check that π_α is a permutation of W and that $(a_\alpha\varphi)\pi_\alpha = \pi_\alpha(a_\alpha i_{1\alpha}\varphi)$ for all a_α in A_α.

Hence, by Proposition 30, there is a homomorphism $\psi : H \to \mathrm{Sym}W$ such that $j\psi = \varphi$ and and also $p_\alpha\psi = \pi_\alpha$. Since φ is a monomorphism so is j, proving (i) of the Normal Form Theorem.

Now suppose that $h \in H$ can be written as in (ii) of the Normal Form Theorem. It is easy to check, inductively, that $h\psi$ acting on (1) gives the element $(g_0, \alpha_0, \varepsilon_0, \ldots, \varepsilon_{n-1}, g_n)$ of W. It follows that the expression for h is unique, since it is determined by the action of $h\psi$ on (1).

Take any $h \in H$. It can certainly be written as

$$h = g_0 p_{\alpha_0}^{\varepsilon_0} g_1 \ldots g_{n-1} p_{\alpha_{n-1}}^{\varepsilon_{n-1}} g_n$$

where $n \geq 0$, $\varepsilon_i = \pm 1$, and $g_i \in G$. Suppose that $\alpha_{i-1} = \alpha_i$, $\varepsilon_i = 1 = -\varepsilon_{i-1}$, and $g_i \in A_{\alpha_i}$. Let $b \in A_{-\alpha_i}$ be the image of g_i. Then h can also be written as

$$h = \ldots p_{\alpha_{i-2}}^{\varepsilon_{i-2}} (g_{i-1} b g_{i+1}) p_{\alpha_{i+1}}^{\varepsilon_{i+1}} \ldots$$

with a similar expression if $\alpha_{i-1} = \alpha_i$, $\varepsilon_i = -1 = -\varepsilon_{i-1}$, and $g_i \in A_{-\alpha_i}$.

It follows (by induction on n, for instance) that any $h \in H$ can be written as in (i) of Britton's Lemma.

Now let h be written in this form. It is easy to see that, for any $r < n$, we have another expression for h in which g_r is replaced by $g_r a^{-1}$ and g_{r+1} is replaced by bg_{r+1}, the remainder of the expression being unchanged, where a

is in A_{α_r} if $\varepsilon_r = 1$ and in $A_{-\alpha_r}$ if $\varepsilon_r = -1$, and b is the corresponding element of $A_{-\alpha_r}$ or A_{α_r}, respectively. In particular, we can replace g_r by an element of the relevant transversal, altering g_{r+1} but no other term of the product.

Applying this change successively to $r = 0,1,\ldots$ gives an expression for h in normal form.

Now suppose that we have two reduced forms for h. Applying the above process to each of these gives two normal forms for h, which we have shown must be the same. Since the process of obtaining the normal form from a reduced form does not affect the p_α, part (ii) of Theorem 32 follows at once.//

Conversely, uniqueness of normal forms, and also part (ii) of Theorem 32, follows from part (iv) of Theorem 32 together with the fact that the natural map from G to H is a monomorphism, For suppose h has both

$$g_0 p_{\alpha_0}^{\varepsilon_0} g_1 \ldots g_{n-1} p_{\alpha_{n-1}}^{\varepsilon_{n-1}} g_n \text{ and } v_0 p_{\beta_0}^{\eta_0} \ldots p_{\beta_{m-1}}^{\eta_{m-1}} v_m$$

as reduced forms. Then

$$1 = v_m^{-1} \ldots p_{\beta_0}^{-\eta_0} v_0^{-1} g_0 p_{\alpha_0}^{\varepsilon_0} \ldots .$$

Because the two expression for h are in reduced form, if Theorem 32(iv) holds then $\alpha_0 = \beta_0$, $\varepsilon_0 = \eta_0$, and $v_0^{-1} g_0$ is in A_{α_0} if $\varepsilon_0 = 1$ and in $A_{-\alpha_0}$ if $\varepsilon_0 = -1$.

If both expressions for h are normal forms this requires that $v_0 = g_0$. Then both $g_1 p_{\alpha_1}^{\varepsilon_1} \ldots$ and $v_1 p_{\beta_1}^{\eta_1} \ldots$ are normal forms for $p_{\alpha_0}^{-\varepsilon_0} g_0^{-1} h$. Inductively we see that the two normal forms must coincide.

When we simply have reduced forms we have already seen (because of the condition on $v_0^{-1} g_0$) that the second reduced form can be replaced by a third reduced form, changing v_0 to g_0, changing v_1 to another element, and not changing any other term. We will then get two shorter reduced forms for $p_{\alpha_0}^{-\varepsilon_0} g_0^{-1} h$, and Theorem 32(ii) follows inductively.

We can give yet another proof of Britton's Lemma, using methods similar to method (IV) for Theorem 4 and Proposition 22, which will also give a result on residual finiteness. We have already seen that the key part of Britton's Lemma is part (iv), together with the fact that G embeds in H, the remainder of the theorem following from this.

First we reduce to the case where there are only finitely many stable letters. For in the general case if a product is 1 in H the corresponding product in $G * P$ is the product of conjugates of elements $p_\alpha^{-1}(a_\alpha i_{0\alpha}) p_\alpha (a_\alpha i_{1\alpha})$

and their inverses. The set Q of those p_β which occur in these conjugates is finite, and our given product will then also be 1 in the HNN extension $\langle G, p_\beta;\ p_\beta^{-1}(a_\beta i_{0\beta})p_\beta(a_\beta i_{1\beta}),$ all β with p_β in $Q\rangle$. This argument is a version of the direct limit argument of the examples in section 3.2.

Thus we may assume that there are n stable letters for some finite n, and use induction on n. Let H be the HNN extension of G with stable letters p_r and associated subgroups A_r and A_{-r} for $r \leq n$, and let K be the HNN extension of G with stable letters p_r for $r < n$, with the same associated subgroups as before. Inductively, we can regard G as a subgroup of K, and so we can form the HNN extension $\langle K, p_n;\ p_n^{-1}A_np_n = A_{-n}\rangle$. A comparison of presentations shows that this HNN extension is just H. As K embeds in H (by the initial case of the induction, which we prove later), we see that G embeds in H.

Now look at a product of elements of G and $p_1^{\pm 1},\ldots,p_n^{\pm 1}$. Regarding H as an HNN extension of K just amounts to bracketing together those terms between two successive occurrences of $p_n^{\pm 1}$. By the case for a single stable letter, such a product can only be 1 if it contains a portion $p_n^{-1}kp_n$ with k in A_n or a portion $p_nkp_n^{-1}$ with k in A_{-n}. If k does not involve any p_r with $r < n$ we immediately have the situation we want. If k does involve some p_r then, inductively by Theorem 32(iii), the product k must have a portion $p_r^{-1}gp_r$ with g in A_r or a portion $p_rgp_r^{-1}$ with g in A_{-r}. Thus we have the situation we require. We still have to prove the result when there is only one stable letter.

So we now look at an HNN extension with a single stable letter, say $H = \langle G, p;\ p^{-1}Cp = D\rangle$. Let G/C denote the set of left cosets of C. For the moment we will suppose that $|G/C| = |G/D|$. This certainly holds if G is finite, since $|C| = |D|$, but need not be true if G is infinite (for instance, if G is an infinite dimensional vector space over $\mathbb{Z}_2$ and C and D are subspaces whose complements have different finite dimensions).

As we still have to prove that j is a monomorphism, we shall temporarily denote by g the image in H of g. We have seen that any $h \in H$ is either g or has a reduced form $g_0p^{\varepsilon_0}\ldots p^{\varepsilon_{n-1}}g_n$. In the latter case we will obtain a homomorphism ψ from H to $\mathrm{Sym}(G\times N)$, where $N = \{0,\ldots,n\}$, such that $h\psi \neq 1$. Also ψ will be such that $g\psi \neq 1$ for all $g \in G - \{1\}$, showing that G embeds in H.

Let $\varphi: G \to \mathrm{Sym}(G\times N)$ be given by $(a,r).g\varphi = (ag,r)$ for all $x \in X$, a and g in G, and $r \leq n$. If π is any permutation of $G\times N$ satisfying the condition

(1) for any $c \in C$, $g \in G$, and $r \leq n$, if $(g,r)\pi = (g',s)$ then $(gc,r)\pi = (g'd,s)$, where d is the image in D of c under the given isomorphism $i: C \to D$,

then, by Proposition 30, there is a homomorphism $\psi: H \to \mathrm{Sym}(G \times N)$ sending p to π and with $j\psi = \varphi$.

Plainly any bijection from $G/C \times N$ to $G/D \times N$ can be lifted to a permutation of $G \times N$ satisfying (1). Further, if for each r and each coset U of C we choose an element u_r in U we can require $(u_r, r)\pi$ to be the element (v,s), for any v such that the bijection sends (U,r) to (vD,s).

We shall define a relation between $G/C \times N$ and $G/D \times N$, and will show that it defines a one-one function from a finite subset of $G/C \times N$ to $G/D \times N$. As we are assuming that $|G/C| = |G/D|$, this function can be extended to a bijection from $G/C \times N$ to $G/D \times N$.

We define the relation as follows. If $\varepsilon_r = 1$ then $(g_0 \dots g_r C, r)$ is related to $(g_0 \dots g_r D, r+1)$, while if $\varepsilon_r = -1$ then $(g_0 \dots g_r C, r+1)$ is related to $(g_0 \dots g_r D, r)$. Now (gC,r) occurs as the first member of a pair of this relation iff either $\varepsilon_r = 1$ and $gC = g_0 \dots g_r C$ or $\varepsilon_{r-1} = -1$ and $gC = g_0 \dots g_{r-1} C$. Since our word is reduced, if $\varepsilon_r = 1 = -\varepsilon_{r-1}$ we cannot have $g_r \in C$. Hence (gC,r) occurs as the first member of at most one pair, so the relation is a function.

Similarly (gD,r) occurs as the second member of some pair iff either $\varepsilon_r = -1$ and $gD = g_0 \dots g_r D$ or $\varepsilon_{r-1} = 1$ and $gD = g_0 \dots g_{r-1} D$. Again, at most one of these can hold since we have a reduced word. Hence the function is one-one.

As remarked, this function can be extended to a bijection of the whole sets, which can then be lifted to a permutation π of $G \times N$. Further, since the relevant members of $G/C \times N$ and $G/D \times N$ correspond, π can be chosen so that if $\varepsilon_r = 1$ then $(g_0 \dots g_r, r)\pi = (g_0 \dots g_r, r+1)$ while if $\varepsilon_r = -1$ then $(g_0 \dots g_r, r+1)\pi = (g_0 \dots g_r, r)$.

By the construction of ψ, $h\psi$ sends $(1,0)$ to $(g_0 \dots g_n, n)$. It follows that $h\psi \neq 1$, as required.

We still have to consider the case where $|G/C| \neq |G/D|$. We take any set X such that $|X \times G/C| = |X \times G/D|$. (Properties of cardinal numbers using the Axiom of Choice enable us to take X to be G. Readers who feel that the Axiom of Choice should not be needed are reminded that in the proof of Theorem 31 we used transversals, and the existence of a transversal in general requires the Axiom of Choice. For a proof not involving the Axiom of Choice, see Exercise 29). Fix some element $x_0 \in X$.

Now let $\varphi: G \to \mathrm{Sym}(X \times G \times N)$ be given by $(x,a,r)g\varphi = (x,ag,r)$. If π is a permutation of $X \times G \times N$ satisfying a modified version of (1) then there is a homomorphism $\psi: H \to \mathrm{Sym}(X \times G \times N)$ sending p to π and such that $j\varphi = \psi$. As before, π can be obtained by taking a bijection of $X \times G/C \times N$ to $X \times G/D \times N$ and lifting it, and this bijection can be obtained by extending a one-one function from a finite subset of $X \times G/C \times N$ into $X \times G/D \times N$. This function is obtained from the earlier function by modifying it to map a subset of $\{x_0\} \times G/C \times N$ to $\{x_0\} \times G/D \times N$. We now find that $h\psi \neq 1$, because $h\psi$ maps $(x_0,1,0)$ to $(x_0,g_0 \ldots g_n,n)$. The details are left to the reader.

If G is finite so is $\mathrm{Sym}(G \times N)$. So the discussion above, in addition to proving Britton's Lemma, proves the following result.

Proposition 33 *If G is finite then the HNN extension* $\langle G,p;\ p^{-1}Cp = D\rangle$ *is residually finite.//*

We can also obtain Britton's Lemma from the Reduced Form Theorem for amalgamated free products. Let Y be free with basis $\{y_\alpha\}$ and let Z be free with basis $\{z_\alpha\}$. Let L be the subgroup $\langle G, y_\alpha^{-1}A_\alpha y_\alpha\rangle$ of $G*Y$. Then L is the free product of G and the groups $y_\alpha^{-1}A_\alpha y_\alpha$, as can easily be seen using the Normal Form Theorem for free products. Hence L is isomorphic to the subgroup M of $G*Z$ generated by G and the subgroups $z_\alpha^{-1}A_{-\alpha}z_\alpha$. So we may form the amalgamated free product $K = (G*Y)*_{L=M}(G*Z)$. Let p_α be $y_\alpha z_\alpha^{-1}$, and let H be the subgroup of K generated by G and $\{p_\alpha\}$. An easy sequence of Tietze transformations gives a presentation of K which shows that $K = H*Z$ and that H is the HNN extension $\langle G,p_\alpha;\ p_\alpha^{-1}A_\alpha p_\alpha = A_{-\alpha}\rangle$. It is now easy to obtain Britton's Lemma. By a similar construction we can obtain the Reduced Form Theorem for amalgamated free products from Britton's Lemma and the Normal Form Theorem for free products..

If $h \in H$ is written as in Theorems 31 or 32 we call n the length of h. As before, an element h of length $n > 0$ has infinite order if no conjugate of h has shorter length than h. In particular, h has finite order only if it is conjugate to an element of G.

We have results similar to Propositions 28 and 29.

Proposition 34 *Let H be an HNN extension* $\langle G,p_\lambda;\ p_\lambda^{-1}A_\lambda p_\lambda = A_\lambda\varphi_\lambda\rangle$, *where* φ_λ *is a monomorphism from* A_λ *to G, and* λ *is in some index set* Λ. *Let* M *be a subset of* Λ, *and let B be a subgroup of G such that* $(B \cap A_\mu)\varphi_\mu = B \cap (A_\mu\varphi_\mu)$ *for all* $\mu \in \mathrm{M}$. *Then the subgroup* $\langle B,p_\mu\ (\mu \in \mathrm{M})\rangle$ *is the HNN extension of B with stable letters* p_μ *and associated subgroups* $B \cap A_\mu$ *and* $B \cap A_\mu\varphi_\mu$. *Also,* $\langle B,p_\mu\ (\mu \in \mathrm{M})\rangle \cap G = B$.

Proof Let K be the HNN extension of B with stable letters q_μ and associated subgroups $B \cap A_\mu$ and $B \cap A_\mu \varphi_\mu$. Then the inclusion of B in G and the map sending q_μ to p_μ induce a homomorphism from K to H. This plainly maps onto the subgroup $\langle B, p_\mu \rangle$ of H. The conditions ensure that, in the corresponding map from $B * Q$ to $G * P$, where Q is free on $\{q_\mu\}$, an element with no pinch maps to an element with no pinch. By Britton's Lemma, this tells us both that the homomorphism from K to H has trivial kernel and that an element of K not in B maps to an element of H not in G, which proves the final part.//

Proposition 35 *Let H be the HNN extension $\langle G, p_\lambda ; p_L^{-1} A_\lambda p_\lambda = A_{-\lambda} \rangle$. Let H be finitely generated. Then there are only finitely many λ. If, in addition, each A_λ is finitely generated then G is finitely generated.*

Proof Since the free group on $\{p_\lambda\}$ is an image of H, there can only be finitely many λ.

Write each of the finitely many generators of H in reduced form, and let S be the set of elements of G which occur. Let B be the subgroup $\langle S, A_\lambda, A_{-\lambda}$ (all λ)$\rangle$. Then $H = \langle B, p_\lambda$ (all λ)$\rangle$, so Proposition 34 tells us that $G = B$, giving the result.//

Theorem 36 *Any countable group G can be embedded in a 2-generator group.*

Proof Let G be generated by $\{g_n$, all $n \in \mathbb{N}\}$, where $g_0 = 1$ (with repetitions if G is finite). It is sometimes most convenient to take $\{g_n$, all $n\}$ to be the whole of G, but at other times the use of a generating set is best.

Let F be free with basis $\{a,b\}$. By Corollaries 3 and 1 to Proposition 8, in $G * F$ the subgroups $\langle b^{-n}ab^n \rangle$ and $\langle g_n a^{-n} b a^n \rangle$ are both free with the given elements as basis.

Hence $H = \langle G * F, p;\ p^{-1} b^{-n} a b^n p = g_n a^{-n} b a^n \rangle$ is an HNN extension of $G * F$, and so G embeds in H. But $g_n \in \langle a,b,p \rangle$ and $p^{-1} a p = b$, so that $H = \langle a,p \rangle$.//

We note for use later that in the above construction F embeds in H, so that both a and p have infinite order; also, any element of H with finite order is conjugate to an element of G. Note also that any subgroup of a 2-generator group must be countable.

Proposition 37 *There are exactly* $2^{\aleph_0}$ *non-isomorphic 2-generator groups.*

Proof Since each 2-generator group is a quotient of the countable group $F(a,b)$ which has only $2^{\aleph_0}$ subsets, there cannot be more than $2^{\aleph_0}$ 2-generator groups.

It is enough to show that for any set S of primes there is a 2-generator group H_S such that the prime p is the order of an element of H_S iff $p \in S$, as then there will be $2^{\aleph_0}$ non-isomorphic groups H_S.

Let $G_S = \sum_{p \in S} \mathbb{Z}_p$, the direct sum of cyclic groups. Let H_S be obtained from G_S by the construction of the previous theorem. Since H_S is an HNN extension of $G_S * F$, any element of finite order in H_S is a conjugate of an element of G_S, and the result follows.//

Proposition 38 *Any (countable) group G can be embedded in a (countable) group H such that any two elements of H with the same order are conjugate.*

Proof Let $\{(a_\alpha, b_\alpha)\}$ be the set of all ordered pairs of elements of G with the same order. Then $\langle a_\alpha \rangle$ and $\langle b_\alpha \rangle$ are isomorphic and so the group

$$G^* = \langle G, p_\alpha;\ p_\alpha^{-1} a_\alpha p_\alpha = b_\alpha,\ \text{all } \alpha \rangle$$

is an HNN extensions of G, whence G embeds in G^*.

By construction, any two elements of G with the same order are conjugate in G^*, but two elements of G^* with the same order need not be conjugate (but they will be conjugate when their order is finite). We define groups G_n for all n by $G_0 = G$ and $G_{n+1} = G_n^*$. Let $H = \cup G_n$. Then H is a group which contains G. Any two elements of H will both be in G_n for some n, and if they have the same order they will be conjugate in G_{n+1}, and so will be conjugate in H.//

In the above construction, any element of H with finite order is conjugate to an element of G. This holds because any element of G_n with finite order is conjugate to an element of G_{n-1}.

Theorem 39 *Any countable group G can be embedded in a countable simple group.*

Proof The group $G*\langle x\rangle$, where x has infinite order, can be embedded in a 2-generator group K whose generators have infinite order. Further, K can be embedded in a countable group S in which any two elements of the same order are conjugate. Also any element of S with finite order is conjugate to an element of G.

Let $N \triangleleft S$, with $N \neq 1$. Take any $w \in N$ with $w \neq 1$. If w has finite order, it has a conjugate g in G, and g is in N. Then $x^{-1}g^{-1}xg \in N$, and this element of $G*\langle x\rangle$ has infinite order. Since N has an element of infinite order, any element of infinite order in S is in N, being conjugate to an element of N. In particular, N contains the generators of K, and so $N \geq K$. Then any element of S either has infinite order, and so is conjugate to an element of N, or has finite order and then it is conjugate to an element of G, which is contained in N. As N is normal, we see that $N = S$, so that S is simple.//

Our construction shows that for any set Q of primes there is a countable simple group such that p is the order of an element of the group iff $p \in Q$. We see that there must be $2^{\aleph_0}$ countable simple groups.

Proposition 40 *There is a finitely generated infinite simple group.*

Proof Let G be the group

$$\langle a,b,c,d;\ b^{-1}ab = a^2,\ c^{-1}bc = b^2,\ d^{-1}cd = c^2,\ a^{-1}da = d^2\rangle.$$

Let $H = \langle a,b;\ b^{-1}ab = a^2\rangle$, and let $K = \langle a,b,c;\ b^{-1}ab = a^2,\ c^{-1}bc = b^2\rangle$.

Then H is an HNN extension of the infinite cyclic group $\langle a\rangle$ with stable letter b; hence b has infinite order. It then follows that K is an HNN extension of H with stable letter c. By Britton's Lemma, $\langle a\rangle \cap \langle b\rangle = \{1\}$ in H. By Britton's Lemma again (and Proposition 34) the subgroup $\langle a,c\rangle$ of K is free with basis $\{a,c\}$. Now G can be regarded as the amalgamated free product of two copies of K, so that G is infinite. Precisely, G is the amalgamated free product of K and $\langle x,d,y;\ d^{-1}xd = x^2,\ y^{-1}dy = d^2\rangle$, the amalgamated subgroups being $\langle a,c\rangle$ and $\langle y,x\rangle$.

G is very far from being simple. In fact it is possible to show that G is SQ-universal; that is, any countable group is a subgroup of some quotient of G.

But there is (by Zorn's Lemma) a normal subgroup N maximal with respect to $N \cap \langle a\rangle = \{1\}$. Then G/N is finitely generated, and it is infinite since

$\langle a \rangle$ is infinite and intersects N trivially. Because of our choice of N, in any proper quotient of G/N the image of a will have finite order. We will show that such a quotient is trivial, so that G/N is simple.

Let $\bar{G}$ be a quotient of G in which the image α of a has finite order m. Let β, γ, and δ be the images of b, c, and d. Then $\alpha^{-m}\delta\alpha^{m} = \delta^{2^m}$, so that δ has finite order s which divides $2^m - 1$. Similarly γ and β have finite orders r and n where r, n, and m divide $2^s - 1$, $2^r - 1$, and $2^n - 1$, respectively. If m, n, r, and s are all 1, then $\bar{G}$ is trivial. Otherwise let p be the smallest prime factor of $mnrs$. We may assume that p divides m. Then p divides $2^n - 1$. Also, by Fermat's Theorem, p divides $2^{p-1} - 1$. It follows easily that p divides $2^k - 1$, where k is the greatest common divisor of n and $p - 1$. In particular, $k \neq 1$, so that k has a prime factor smaller than p, contradicting the definition of p. This contradiction proves the result.//

The group of this proposition is finitely generated but not finitely presented. Examples of finitely presented infinite simple groups are known, but their construction is difficult.

Exercise 21 Fill in the details in the proof of the Normal Form Theorem for HNN extensions.

Exercise 22 Show in detail that an element of finite order in an HNN extension must be conjugate to an element of the base group.

Exercise 23 In the construction following Proposition 33, give a sequence of Tietze transformations which shows that the amalgamted free product constructed is the free product of an HNN extension and a free group. Hence obtain Britton's Lemma from the Reduced Form Theorem for amalgamated free products.

Exercise 24 Let A and B be groups, and let C and D be isomorphic subgroups of A and B respectively. Let K be the subgroup $\langle p^{-1}Ap, B\rangle$ of the HNN extension $\langle A * B, p;\ p^{-1}Cp = D\rangle$. Show, without using any reduced form theorems, that K is isomorphic to the amalgamated free product $A *_{C=D} B$. (Use Proposition 30 to show K satisfies the definition of the amalgamated free product.) Hence obtain the Reduced Form Theorem for amalgamated free products from Britton's Lemma and the Normal Form Theorem for free products.

Exercise 25 Let A and B be groups, and let C and D be isomorphic subgroups of A and B respectively. Let G be any group containing (copies of) A

and B and such that $A \cap B = \{1\}$. Show, using Proposition 27, that the subgroup $\langle p^{-1}Ap, B\rangle$ of the HNN extension $\langle G,p; p^{-1}Cp = D\rangle$ is isomorphic to the amalgamated free product $A *_{C=D} B$. (Note that Proposition 27 uses the Reduced Form Theorem, so that the previous exercise cannot be deduced from this one.)

Exercise 26 Let C and D be isomorphic finite subgroups of the residually finite group G. Show that the HNN extension $\langle G,p;\ p^{-1}Cp = D\rangle$ is residually finite.

Exercise 27 Let C and D be isomorphic finite subgroups of the residually finite groups A and B. Show that the amalgamated free product $A *_{C=D} B$ is residually finite. (Use the previous two exercises with $G = A \times B$. We could also take G to be $A * B$, but that would require the corollary to Proposition 22, whereas the suggested approach allows us to deduce that result by taking C to be $\{1\}$.)

Exercise 28 Show that an HNN extension of a residually finite group need not be residually finite when the associated subgroups are infinite. (One possible approach is to argue that if all such HNN extensions were residually finite then the group of Proposition 40 would be residually finite.)

Exercise 29 In this exercise we shall obtain a proof of Britton's Lemma not using the Axiom of Choice.

Let C and D be isomorphic subgroups of the group G. Let M be the set of all finite sequences $(g_0,p^{\varepsilon_0},g_1,\dots,g_{n-1},p^{\varepsilon_{n-1}},g_n)$. Show that M is a monoid whose identity is (1) under the multiplication
$(g_0,\dots,p^{\varepsilon_{n-1}},g_n)(h_0,p^{\eta_0},\dots,h_m) = (g_0,\dots,p^{\varepsilon_{n-1}},g_nh_0,p^{\eta_0},\dots,h_m)$.

Define a relation $\sim$ on M by
$(g_0,p^{\varepsilon_0},\dots,p^{\varepsilon_{n-1}},g_n) \sim (h_0,p^{\varepsilon_0},\dots,p^{\varepsilon_{n-1}},h_n)$ iff there are $a_0,\dots,a_{n-1}$, $b_0,\dots,b_{n-1}$ in G such that $h_0 = g_0a_0^{-1}$, $h_1 = b_0g_1a_1^{-1},\dots,$ $h_n = b_{n-1}g_n$, where $a_i \in C$ if $\varepsilon_i = 1$ and $a_i \in D$ if $\varepsilon_i = -1$, and b_i is the element in D or C respectively which corresponds to a_i. Show that $\sim$ is an equivalence relation. Let $u = (g_0,p^{\varepsilon_0},\dots,g_n)$. Show that if $u \sim v$ where v is $(\dots,p^{\eta},1,p^{-\eta},\dots)$ iff for some i either $\varepsilon_i = 1 = -\varepsilon_{i-1}$ and $g_i \in C$ or $\varepsilon_i = -1 = -\varepsilon_{i-1}$ and $g_i \in D$.

Let M_1 be the set of equivalence classes. Define a relation $\to$ on M_1 by $\alpha \to \beta$ if there is $u = (g_0,p^{\varepsilon_0},\dots,g_n)$ in α with $\varepsilon_i = -\varepsilon_{i-1}$ and $g_i = 1$ and β contains the element $(g_0,\dots,p^{\varepsilon_{i-2}},g_{i-1}g_{i+1},p^{\varepsilon_{i+2}},\dots,g_n)$. Let $\alpha \equiv \beta$ hold iff there are $\alpha_1,\dots,\alpha_k$ with $\alpha_1 = \alpha$, $\alpha_k = \beta$, and, for all i, either $\alpha_{i-1} \to \alpha_i$ or $\alpha_i \to \alpha_{i-1}$. Show that $\equiv$ is an equivalence relation. Show that the set of equivalence classes is a group, and, using the comment following Proposition 30, that this group is the HNN extension $\langle G,p;\ p^{-1}Cp = D\rangle$.

Show that if $\alpha \to \beta_1$ and $\alpha \to \beta_2$ then there is γ such that $\beta_1 \to \gamma$ and $\beta_2 \to \gamma$. Deduce, as in Exercise 4, that if α is equivalent to the identity then there is some β such that $\alpha \to \beta$. Hence obtain Britton's Lemma.

2 SPACES AND THEIR PATHS

2.1 *SOME POINT-SET TOPOLOGY*

We begin with a lemma that will be frequently used, often without explicit mention.

Lemma 1 (Glueing Lemma) (*i*) *Let X and Y be sets, let X_α be subsets of X such that $X = \cup X_\alpha$, and let $f_\alpha : X_\alpha \to Y$ be functions such that $f_\alpha | X_\alpha \cap X_\beta = f_\beta | X_\alpha \cap X_\beta$ for all α and β. Then there is a unique function $f: X \to Y$ such that $f | X_\alpha = f_\alpha$ for all α.*

(*ii*) *Let the conditions of* (*i*) *hold, and let X and Y be topological spaces. Suppose that each f_α is continuous (when X_α is given the subspace topology). Suppose either that there are only finitely many sets X_α each of which is a closed subspace of X or that each X_α is an open subspace of X. Then f is continuous.*

Remark When f is a function from a set X to a set Y and A is a subset of X the notation $f|A$ means the restriction of f to A; that is, the function from A to Y whose value on $a \in A$ is fa.

Proof (i) Let $S \subseteq X \times Y$ be $\{(x,y)$; there is α such that $x \in X_\alpha$ and $y = xf_\alpha\}$. Since $X = \cup X_\alpha$, for every x there is at least one y with $(x,y) \in S$. Suppose that $(x,y) \in S$ and $(x,z) \in S$. Then there are α and β with $x \in X_\alpha$, $y = xf_\alpha$, and $x \in X_\beta$, $z = xf_\beta$. Since $f_\alpha = f_\beta$ on $X_\alpha \cap X_\beta$, by hypothesis, it follows that $y = z$.

This means that we can define a function $f: X \to Y$ by $y = xf$ iff $(x,y) \in S$. By construction, $f|X_\alpha = f_\alpha$, as required. Since $X = \cup X_\alpha$ there can be only one such function.

(ii) Suppose that the first condition holds. Let C be a closed subset of Y. Then $Cf^{-1} = \cup Cf_\alpha^{-1}$. Since each f_α is continuous, Cf_α^{-1} is a closed subset of X_α. Since X_α is closed in X, this implies that Cf_α^{-1} is a closed subset of X.

As there are only finitely many sets X_α, it follows that Cf^{-1} is a closed subset of X. Hence f is continuous.

The case where each X_α is open is similar.//

We shall often be faced with the situation of a set X which is the union of subsets X_α each of which has a topology $\mathfrak{T}_\alpha$, and we will want to define a topology on X. Let $\mathfrak{T}$ be $\{A;\ A\cap X_\alpha \in \mathfrak{T}_\alpha$ for all $\alpha\}$. Then $\mathfrak{T}$ is a topology on X, called the *weak topology induced by* $\{\mathfrak{T}_\alpha\}$.

In particular, if $X_\alpha \cap X_\beta = \emptyset$ for $\alpha \neq \beta$ we call X the *disjoint union* of the spaces X_α, written $\dot{\cup} X_\alpha$. Also, if there is a point p such that $X_\alpha \cap X_\beta = \{p\}$ for all $\alpha \neq \beta$ then we call X the *join* of the spaces X_α (at p), which we write as $\vee X_\alpha$.

It is immediate from the definition that a set C is closed for the topology $\mathfrak{T}$ iff $C\cap X_\alpha$ is closed for $\mathfrak{T}_\alpha$ for all α. In particular, if each $\mathfrak{T}_\alpha$ is a T_1 topology (that is, each point is a closed set) then so is $\mathfrak{T}$.

Let Y be any space, and let $f{:}X \to Y$ be a function. Then f is continuous (using topology $\mathfrak{T}$ on X) if $f|X_\alpha$ is continuous for $\mathfrak{T}_\alpha$ for all α (the converse is obvious). We need only take an open subset V of Y and observe that $Vf^{-1} \cap X_\alpha = V(f|X_\alpha)^{-1}$ is open for $\mathfrak{T}_\alpha$ for all α.

Let $\mathfrak{T}$ be the weak topology induced by the topologies $\mathfrak{T}_\alpha$. Then X_α has two topologies, namely the given topology $\mathfrak{T}_\alpha$ and the topology as a subspace of the topological space X. We would like to ensure that these two topologies are the same. This plainly requires that $\mathfrak{T}_\alpha$ and $\mathfrak{T}_\beta$ induce the same topology on $X_\alpha \cap X_\beta$ for all α and β (since the corresponding property holds for the subspace topologies). In general this necessary condition is not sufficient.

Suppose that this condition holds and that $X_\alpha \cap X_\beta$ is a closed subset of X_α (with the topology $\mathfrak{T}_\alpha$) for all α and β (or that $X_\alpha \cap X_\beta$ is open in X_α for all α and β). Then the two topologies are the same.

For let $C \subseteq X_\alpha$ be closed for $\mathfrak{T}_\alpha$. Then $C \cap X_\beta$ is closed for the topology on $X_\alpha \cap X_\beta$ induced by $\mathfrak{T}_\alpha$. By hypothesis this is also the topology induced by $\mathfrak{T}_\beta$. Since $X_\alpha \cap X_\beta$ is, by hypothesis, a closed subset of X_β with the topology $\mathfrak{T}_\beta$, it follows that $C\cap X_\beta$ is a closed subset of X_β with the topology $\mathfrak{T}_\beta$. Then C is closed for $\mathfrak{T}$, since this holds for all β. The other case is similar.

Proposition 2 *Let X have the weak topology $\mathfrak{T}$ induced by $\{\mathfrak{T}_\alpha\}$, where each $\mathfrak{T}_\alpha$ is a T_1 topology. Suppose that X has a subset A such that $X_\alpha \cap X_\beta = A$ for $\alpha \neq \beta$. Then any compact subset C of X is contained in the union of finitely many X_α.*

Proof For each α such that C meets $X_\alpha - A$ take some $p_\alpha \in C \cap (X_\alpha - A)$. Let P be the set of all these p_α, and let Q be any subset of P. Since $p_\beta \notin X_\alpha$ for $\beta \neq \alpha$, we see that, for each α, $Q \cap X_\alpha$ consists of at most one point, and so is closed for $\mathfrak{T}_\alpha$. Hence Q is closed for $\mathfrak{T}$.

As this holds for every subset Q of P, we see that P is discrete. Also P is compact, being a closed subset of the compact set C. It follows that P is finite, being compact and discrete. This gives the result.//

Let p be a function from a set X onto a set Y. Let $\mathfrak{T}$ be a topology on X. Then we can define a topology $\mathfrak{T}_Y$ on Y by requiring a subset B of Y to be open iff Bp^{-1} is open. This is called the *identification* or *quotient* topology on Y (*induced by p*). We are frequently given an equivalence relation $\equiv$ on X, with Y being the set of equivalence classes and p being the natural map. In this case we write $X/\equiv$ for Y with the identification topology. We may simply refer to this space as being obtained from X by identifying the points of each equivalence class. More generally, given any relation R on X, there is a smallest equivalence relation containing R, and we refer to the corresponding identification space as being obtained by identifying pairs of elements in R.

Proposition 3 *Let $\equiv$ be an equivalence relation on X, let Z be any set and let $f: X \to Z$ be any function. Let Y denote $X/\equiv$ and let $p: X \to Y$ be the projection.*

(i) There is a function $g: Y \to Z$ such that $f = pg$ iff $x_1 f = x_2 f$ whenever $x_1 \equiv x_2$.

(ii) Let X and Z be topological spaces, and let Y have the identification topology. Let g be as in (i). Then g is continuous iff f is continuous.

Proof (i) is obvious.

From the definition of the identification topology, p is continuous. Hence f is continuous if g is.

Conversely, let f be continuous. Let W be an open subset of Z. Then Wf^{-1} is open in X. Since $Wf^{-1} = (Wg^{-1})p^{-1}$, the definition of the identification topology shows that Wg^{-1} is open in Y, as required.//

Let X be a compact space and Y a Hausdorff space and let f be a continuous map from X onto Y. Then Y has the identification topology induced by f. For we may define an equivalence relation on X by $x_1 \equiv x_2$ iff

$x_1f = x_2f$. By the proposition, f induces a continuous bijection from $X/\equiv$ to Y. Since $X/\equiv$ is compact and Y is Hausdorff, this must be a homeomorphism.

Example Let X be any space, and let I be the unit interval $[0,1]$. On $X \times I$, define $(x_1,t_1) \equiv (x_2,t_2)$ iff either $t_1 = 0 = t_2$ (with x_1 and x_2 arbitrary) or both $x_1 = x_2$ and $t_1 = t_2$; more simply expressed, CX is obtained from $X \times I$ by identifying $(x_1,0)$ and $(x_2,0)$ for all x_1 and x_2. The identification space is called the *cone on X*, written CX, and the equivalence class consisting of all $(x,0)$ is called the *vertex of CX*. It is easy to check that the subspace $X \times \{1\}$ of CX is homeomorphic to X, so we usually regard X as a subspace of CX in this way.

Let D^n be $\{(x_1,\ldots,x_n) \in \mathbb{R}^n;\ \Sigma x_i^2 \leq 1\}$, and let S^{n-1} be $\{(x_1,\ldots,x_n) \in \mathbb{R}^n;\ \Sigma x_i^2 = 1\}$; these are called the *n-disk* and $(n-1)$-*sphere*.. Then CS^{n-1} is homeomorphic to D^n. For the continuous map from $S^{n-1} \times I$ to D^n sending $(x_1,\ldots,x_n,t)$ to $(tx_1,\ldots,tx_n)$ obviously defines a bijection from CS^{n-1} to D^n, and this is a homeomorphism since $S^{n-1} \times I$ is compact.

Similarly, if X is any compact subset of $\mathbb{R}^n$ then CX is homeomorphic to the subset of $\mathbb{R}^{n+1}$ which consists of all $(tx_1,\ldots,tx_n,1-t)$ with $(x_1,\ldots,x_n) \in X$. This explains the terminology, as this set is what would be called the cone in geometric situations.

The functor C is not as well behaved as one might expect. For instance, if X is the closed interval $[0,1]$ and A is the open interval $(0,1)$ then the topology of CA as the cone on A is not the same as the topology of this set regarded as a subspace of CX.

Real projective n-space, $\mathbb{R}P^n$, is the space obtained from S^n by identifying $(x_1,\ldots,x_{n+1})$ with $(-x_1,\ldots,-x_{n+1})$. We may map the disk D^n into S^n by sending $(x_1,\ldots,x_n)$ to $(x_1,\ldots,x_n,\sqrt{(1-x_1^2-\ldots-x_n^2)}\ )$. This provides a map from D^n onto $\mathbb{R}P^n$. It follows that $\mathbb{R}P^n$ can also be obtained from D^n by identifying a point $(x_1,\ldots,x_n)$ such that $\Sigma x_i^2 = 1$ with the point $(-x_1,\ldots,-x_n)$. It is easy to see from this that $\mathbb{R}P^1$ is homeomorphic to S^1, although, of course, the identification from S^1 to $\mathbb{R}P^1$ is not even one-one.

Let X and Y be spaces, and let $f: A \to Y$ be a continuous map from a subspace A of X. Then we may form the disjoint union $X \dot{\cup} Y$ and then take the space obtained by identifying a and af for all $a \in A$. The resulting space is written $Y \cup_f X$, and it is called the *space obtained by adjoining* (or *attaching*) *X to Y by f*. In particular, X may be the disjoint union of spaces X_λ, in which case f consists of a family of continuous maps $f_\lambda: A_\lambda \to Y$, and we refer to the space obtained by adjoining the spaces X_λ to Y by the maps f_λ. Note that the join of the spaces X_λ may be regarded as the space obtained from one of them, X_α say, by adjoining all the other spaces X_λ by the maps sending the point p_λ of X_λ to the point p_α of X_α.

In particular, $\mathbb{R}P^n$ is obtained from $\mathbb{R}P^{n-1}$ by adjoining D^n by the identification map from S^{n-1} to $\mathbb{R}P^{n-1}$.

Proposition 4 *Let $f:X \to Y$ be an identification. Let Z be a locally compact space. Then the function $F:X \times Z \to Y \times Z$ given by $(x,z)F = (xf,z)$ is also an identification.*

Proof Let W be a subset of $Y \times Z$ such that WF^{-1} is open We need to show that W is open. So take $(y_0,z_0) \in W$. It is enough to find open subsets U and V of Y and Z with $(y_0,z_0) \in U \times V \subseteq W$.

Choose some x_0 with $x_0 f = y_0$. Since Z is locally compact, there is an open neighbourhood V of z_0 such that $\bar{V}$ is compact and $\bar{V} \subseteq \{z; (x_0,z) \in WF^{-1}\}$, as this latter set is an open subset of Z. Let $A = \{x; \{x\} \times \bar{V} \subseteq WF^{-1}\}$.

Then $Af \times \bar{V} \subseteq W$, and so $Aff^{-1} \times \bar{V} \subseteq WF^{-1}$. Hence $Aff^{-1} = A$. Also, as WF^{-1} is open and $\bar{V}$ is compact, it is well-known (and easy to check) that A is open. Hence Af is the required set U.//

Proposition 5 *Let X be a compact metric space with metric d, and let $\mathfrak{U}$ be an open covering of X. Then $\exists\ \delta > 0$ such that any set of diameter less than δ is contained in some member of $\mathfrak{U}$.*

Remark We call such a δ a *Lebesgue number* of $\mathfrak{U}$.

Proof Since X is compact, we may assume that $\mathfrak{U}$ is finite, say $\mathfrak{U} = \{U_1,\ldots,U_n\}$.

Define $f_i:X \to \mathbb{R}$ by $xf_i = \inf\{d(x,y);\ y \notin U_i\}$. Then f_i is continuous, and $xf_i \neq 0$ for $x \in U_i$, since U_i is open. Hence the function $\max_i f_i$ is continuous and is always positive, since $\mathfrak{U}$ is a covering of X. As X is compact, this function has a positive minimum δ.

Let S be a non-empty set of diameter less than δ. Take some $x \in S$. There is some i with $xf_i \geq \delta$. Then $y \in U_i$ if $d(x,y) < \delta$, so that $S \subseteq U_i$, as required.//

Exercise 1 Let X have the weak topology by the family $(X_\alpha,\mathfrak{T}_\alpha)$ where $\mathfrak{T}_\alpha$ and $\mathfrak{T}_\beta$ induce the same topology on $X_\alpha \cap X_\beta$. If there are only two sets X_α show that $\mathfrak{T}_\alpha$ is the subspace topology on X_α. Show that this need not hold if there are three or more subsets. (There is an example with X having three elements; a slight variation of this will give an example where each X_α is infinite.)

Exercise 2 Let X and $(X_\alpha,\mathfrak{T}_\alpha)$ be as in exercise 1. Suppose that there is a topology $\mathfrak{T}$ on X such that $\mathfrak{T}_\alpha$ is the topology induced on X_α by $\mathfrak{T}$ for all α. Show that $\mathfrak{T}_\alpha$ is also the topology induced by the weak topology.

Exercise 3 Show that the cone on $(0,1)$ is not a subspace of the cone on $[0,1]$.

Exercise 4 Let X be a space, and let $f{:}X\to Y$ and $g{:}Y\to Z$ be onto maps. Let Y be given the identification topology induced by f, and let Z be then given the identification topology induced by g. Show that the topology on Z is the identification topology induced by fg.

Exercise 5 Show that the space Y is a subspace of the space $Y\cup_f X$. Show that $X-A$ is a subspace of $Y\cup_f X$ if A is a closed subset of X.

2.2 *PATHS AND HOMOTOPIES*

We denote the non-negative real numbers by $\mathbb{R}^+$.

Definition A *path* in a space X is a continuous map $f{:}[0,r]\to X$ for some $r\in\mathbb{R}^+$. If $r=1$ we call the path a *standard* path, while if $r=0$ we say the path is *trivial.* If $0f=a$ and $rf=b$, we say the path is a path *from a to b* (or a path *starting at a* and *ending at b*, or other similar phrases). If $a=b$ we say that the path is a *loop based at a*. Finally, if $f{:}[0,r]\to X$ is a path then the *standardisation* of f is the path $f_\sigma{:}I\to X$ defined by $tf_\sigma=(tr)f$.

If the space is one of which we can make an intuitive picture then we can picture a path as a piece of string or elastic. It is then quite reasonable to talk of deforming a path, by stretching, pulling, pushing, and so on, but keeping the end-points fixed. The reader may find it easy to believe that any two paths in the plane $\mathbb{R}^2$ between the same end-points can be deformed into each other in such a way. On the other hand, if we take the upper and lower halves of the unit circle in $\mathbb{R}^2$-(0,0) then these two paths cannot be deformed into each other (we could think of a spike coming out of the plane at (0,0), and any attempt to deform one semicircle into the other would get caught up on the spike). This chapter and Chapter 4 are devoted to making these ideas precise.

The intuitive idea of deformation could be formalised by referring to a continuous family of paths f_u. But we would need to define a topology on the set of paths to make this precise. This approach is useful in some advanced work, but an easier approach is possible.

Observe that a family of functions f_u can be regarded as a function F of two variables by defining $(t,u)F$ to be tf_u; conversely, any F gives rise to a family of functions f_u. We shall replace the idea of a continuous family of functions by continuity of the corresponding F using the definition below. For formal proofs we need to use F and the definition below, but to get an intuitive idea of what is happening it is often useful to think of the family of paths f_u.

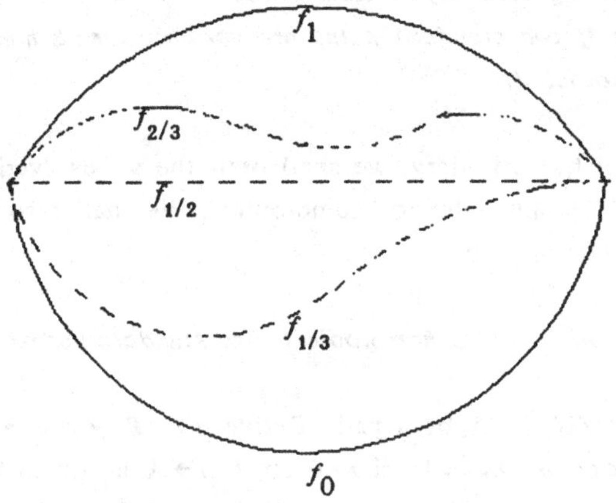

Figure 1, showing some of the paths f_u

Definition Let $f:[0,r] \to X$ and $g:[0,s] \to X$ be paths from a to b. Suppose that there is a continuous function $\alpha: I \to \mathbb{R}^+$ with $0\alpha = r$ and $1\alpha = s$ and a continuous function $F: A \to X$, where $A \subseteq \mathbb{R}^2$ is $\{(t,u);\ 0 \le t \le u\alpha,\ 0 \le u \le 1\}$ such that $(t,0)F = tf$ for $t \le r$, $(t,1)F = tg$ for $t \le s$, $(0,u)F = a$ and $(u\alpha,u)F = b$ for all u. Then we say that f is *variable-length homotopic* to g, and the map F is called a *variable-length homotopy* between f and g. If f and g are standard paths and $u\alpha = 1$ for all u we refer to a *standard homotopy.*

Lemma 6 *Let f be variable-length homotopic to g. Then there is a variable-length homotopy F whose corresponding function α is linear; that is, $u\alpha = us + (1-u)r$ for all u.*

Proof Let G be a variable-length homotopy defined on the set B with corresponding function β. If $r = s = 0$, we define α by $u\alpha = 0$ for all u and we define F by $(0,u)F = a$ for all u, giving the required homotopy.

If $r \ne 0 \ne s$, let A be $\{(t,u);\ 0 \le t \le u\alpha,\ 0 \le u \le 1\}$, where α is given by

$u\alpha = us + (1 - u)r$. Then the function $j{:}A \to B$ given by $(t,u)j = ((u\beta)t/u\alpha, u)$ is obviously continuous. The required homotopy F is just jG.

Suppose that $r = 0 \neq s$ (the remaining case is similar). Define α and A as in the previous paragraph. The previous definition of j makes sense except when $u = 0$ (that is, except at (0,0)) and is continuous except at (0,0). If we define $(0,0)j$ to be (0,0) it is easy to check that j is also continuous at (0,0) because β is continuous. As before, we can now define F to be jG.//

The following corollary is immediate.

Corollary *If two standard paths are variable-length homotopic then they are standard homotopic.//*

Because of this corollary, we shall omit the words "variable-length" and "standard" and will simply refer to "homotopies". We shall also write $f \simeq g$ to mean that f is homotopic to g.

Lemma 7 *Any path is homotopic to its standardisation.*

Proof Let $f{:}[0,r] \to X$ be a path. Define $\alpha{:}I \to \mathbb{R}^+$ by $u\alpha = u + (1 - u)r$, and let A be $\{(t,u);\ 0 \le t \le u\alpha,\ 0 \le u \le 1\}$. If $r = 0$ let $F{:}A \to X$ be given by $(t,u)F = 0f$ for all t and u. If $r \neq 0$ let F be given by $(t,u)F = (tr/u\alpha)f$. Then F is the required homotopy.//

Proposition 8 *Homotopy is an equivalence relation.*

Proof $f \simeq f$ by the homotopy F given by $(t,u)F = tf$, the corresponding α being constant.

Suppose that $f \simeq g$ by the homotopy F and corresponding α. Then $g \simeq f$ by the homotopy G and corresponding β given by $(t,u)G = (t,1 - u)F$ and $u\beta = (1 - u)\alpha$.

Suppose that $f \simeq g$ by the homotopy F and corresponding α, and that $g \simeq h$ by the homotopy G and corresponding β. Then $f \simeq h$ by the homotopy H and corresponding γ given by

$(t,)H = (t,2u)F$ and $u\gamma = (2u)\alpha$ for $u \le 1/2$,
$(t,u)H = (t,2u - 1)G$ and $u\gamma = (2u - 1)\beta$ for $u \ge 1/2$.

The Glueing Lemma assures us that H and γ are continuous, and the other properties needed are obvious.//

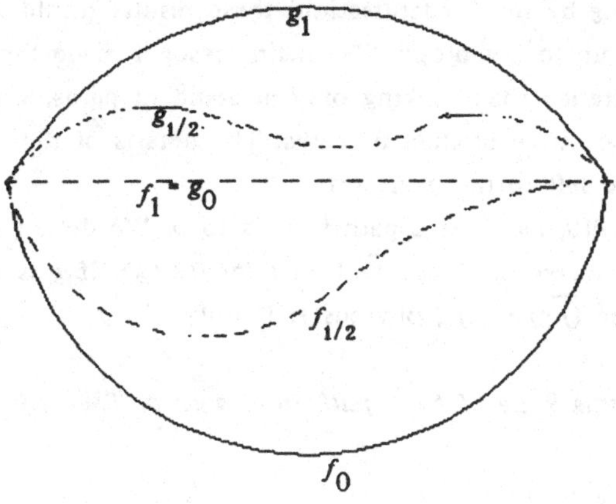

Figure 2, showing some of the maps f_u and g_u

This result is most easy to see pictorially. If we have a continuous family of paths f_u joining f and g, and another continuous family g_u joining g and h, then we get a continuous family joining f and h by following the family f_u by the family g_u. In order to do this in total time 1 and not in total time 2 we need to change u to $2u$. Formalising this gives the definition of H above.

Lemma 7 and the corollary to Lemma 6 tell us that every variable-length homotopy class of paths contains exactly one standard homotopy class of standard paths. It follows that, when considering homotopy classes, we can restrict ourselves to standard homotopy of standard paths whenever this is convenient.

Let $f:[0,r] \to X$ and $g:[0,s] \to X$ be paths such that $0g = rf$. Then we may define a path $f.g$ from $[0,r+s]$ to X by

$$t(f.g) = tf \text{ for } t \leq r,$$
$$t(f.g) = (t-r)g \text{ for } t \geq r.$$

We call $f.g$ the *product* of f and g. We write $f.g$ rather than fg to avoid confusion with the composition of functions.

It is clear that the product of paths is associative; that is, if one of $(f.g).h$ and $f.(g.h)$ is defined then so is the other and they are equal; also these products are defined if both $f.g$ and $g.h$ are defined. Also, if f is a path from a to b and i_a and i_b are the trivial paths at a and b then we have $i_a.f = f = f.i_b$.

If we restricted attention to standard paths (when we would also have to replace $f.g$ by its standardisation) these results would not hold, although they would hold up to homotopy. The main reason I made the definition of paths used here (rather than looking only at standard paths, which is more usual) is to have an associative product of paths. The details of the situation for standard paths are left to the exercises.

Let $f:[0,r] \to X$ be a path from a to b. We define a path $\overline{f}$ from b to a, called the *inverse* of f, by $t\overline{f} = (r-t)f$ for $0 \le t \le r$. If g is another path and $f.g$ is defined then $\overline{(f.g)} = \overline{g}.\overline{f}$, obviously. Plainly, $\overline{\overline{f}} = f$.

Lemma 9 *Let f be a path from a to b. Then $f.\overline{f} \simeq i_a$ and $\overline{f}.f \simeq i_b$.*

Proof Define F by

$(t,u)F = tf$ for $0 \le t \le ru$,
$(t,u)F = (2ru - t)f$ for $ru \le t \le 2ru$.

Then F is a homotopy from i_a to $f.\overline{f}$. The second part follows from the first, replacing f by $\overline{f}$. //

If we think in terms of a family of paths g_u then g_u follows f for a proportion u of the total time taken by f and then follows the same path in reverse.

Lemma 10 *Let $f \simeq g$ and $f' \simeq g'$, and let f' (and g') start where f (and g) end. Then $f.f' \simeq g.g'$ and $\overline{f} \simeq \overline{g}$.*

Proof Let F and F' be the homotopies, with corresponding functions α and α'. Then the homotopy between $f.f'$ and $g.g'$ is $F.F'$ given by

$(t,u)\, F.F' = (t,u)F$ for $t \le u\alpha$,
$(t,u)\, F.F' = (t - u\alpha, u)F'$ for $u\alpha \le t \le u\alpha + u\alpha'$.

Also the homotopy between $\overline{f}$ and $\overline{g}$ is $\overline{F}$ given by $(t,u)\overline{F} = (u\alpha - t, u)F$. //

In terms of a family of paths, given families f_u and f'_u we obtain families $f_u.f'_u$ and $\overline{f}_u$.

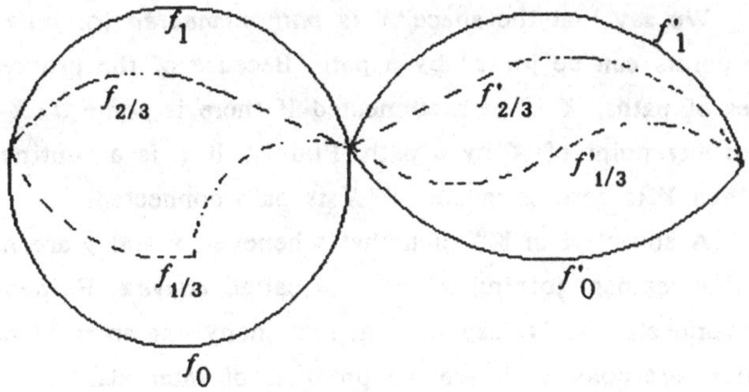

Figure 3, showing some of the maps f_u and f'_u

Lemma 11 *Let f and g be paths from a to b. Then $f \simeq g$ iff $f.\bar{g} \simeq i_a$.*

Proof If $f \simeq g$ then $f.\bar{g} \simeq g.\bar{g}$. As $g.\bar{g} \simeq i_a$, we find that $f.\bar{g} \simeq i_a$. Conversely, if $f.\bar{g} \simeq i_a$ we have $f = f.i_b \simeq f.\bar{g}.g \simeq i_a.g = g$.//

It is now easy to check that the set of homotopy classes of loops based at the point a is a group. This group will be our main link between topology and group theory. The set of all homotopy classes of all paths is an unfamiliar algebraic object which is closely related to a group, but differs from a group in that the product of two elements is not always defined. We shall study these algebraic objects in Chapter 3, and return to their connections with topology in Chapter 4.

Let f be a standard path from a to b, and let g be a standard path from b to c. Let $f \circ g$ denote the standardisation of $f.g$.

Exercise 6 Write down the formula for $f \circ g$.

Exercise 7 Let h be a standard path from c to d. Write down the formulas for $(f \circ g) \circ h$ and for $f \circ (g \circ h)$. Show that these paths are different but are homotopic. Obtain a formula for a homotopy between these paths.

Exercise 8 Let j_a be the standard path given by $tj_a = a$ for $0 \leq t \leq 1$. Show that $j_a \circ f \simeq f$, and obtain a formula for a homotopy between these two paths.

2.3 *PATH-COMPONENTS*

We say that the space X is *path-connected* (or *pathwise connected*) if any two points can be joined by a path. Because of the properties of products and inverses of paths, X is path-connected if there is some $a \in X$ such that a can be joined to any point of X by a path. Plainly, if φ is a continuous map from X onto Y then Y is path-connected if X is path-connected.

A subset A of $\mathbb{R}^n$ such that whenever x and y are in A then so is the whole line-segment joining x and y is called *convex*. Evidently, any convex set is path-connected. It is easy to construct many examples of convex sets. For instance, disks are convex, as are the products of intervals.

More generally, for any space X we may say that a is equivalent to b if a and b can be joined by a path; the properties of products and inverses show that this is an equivalence relation. The equivalence classes are called the *path-components* of X; X is path-connected iff it has only one path-component.

Some authors use the phrase "arcwise connected" rather than "pathwise connected". Strictly speaking, this should mean that any two points can be joined by an arc; that is, by a homeomorphic image of an interval, rather than just by a continuous image. It is a deep theorem of point-set topology that these two notions are in fact the same; fortunately, we shall not need this.

Since the continuous image of an interval is connected, any path-connected space is connected. The converse is false, as is shown by the example below.

Example Let A be $\{(x,y) \in \mathbb{R}^2;\ y = \sin(1/x),\ x > 0\}$. Then A is connected, being the continuous image of an interval, and hence so is its closure B. It is easy to check that $B = A \cup \{(0,y);\ |y| \leq 1\}$. Also A is path-connected.

Now let f be a path in B with $0f = (0,0)$, and let tf be (tf_1, tf_2). By continuity, $0f_1^{-1}$ is closed, and it is plainly non-empty. We shall show that it is also open. This will show that $tf_1 = 0$ for all t, and, in particular, there is no path in B from $(0,0)$ to $(1/\pi, 0)$.

So take c with $cf_1 = 0$, and suppose that $cf_2 \neq 1$. Then there is $\delta > 0$ such that $tf_2 \neq 1$ for $|t - c| < \delta$. It follows that, for $|t - c| < \delta$, $tf_1 \neq 1/(2n + 1/2)\pi$ for any n. Now take t with $|t - c| < \delta$, and assume, if possible, that $tf_1 \neq 0$. Then there will be some n with $1/(2n + 1/2)\pi < tf_1$, and then, by the Intermediate Value Theorem, some u between c and t such that $uf_1 = 1/(2n + 1/2)\pi$. Since this cannot happen, we must have $tf_1 = 0$ for all t with $|t - c| < \delta$. If $cf_2 = 1$ we take $\delta > 0$ such that $cf_2 \neq -1$ for $|t - c| < \delta$ and argue similarly.

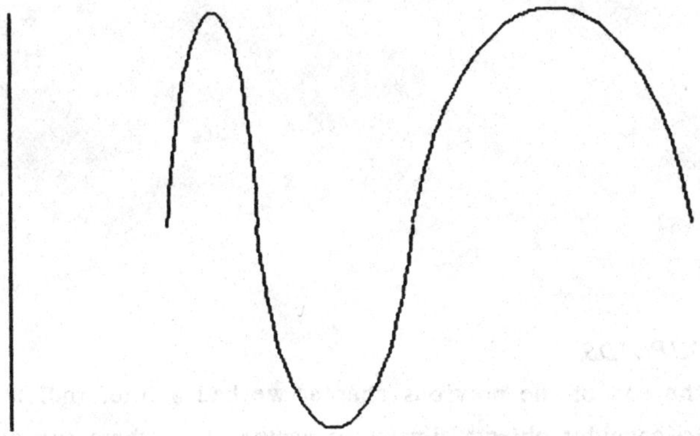

Figure 4, showing part of B

3 GROUPOIDS

3.1 *GROUPOIDS*

At the end of the previous chapter we had a brief indication that it might be useful to consider objects similar to groups, but where the product of two elements is not always defined. In this section we consider these objects in detail.

A *partial multiplication* on a set G is a function from some subset X of $G\times G$ to G. If $(x,y)\in X$ we denote the value of the function on (x,y) by xy or by $x.y$. We say that *xy is defined* to mean that $(x,y)\in X$.

The element e is an *identity* for a partial multiplication if $ex=x$ whenever ex is defined and also $ye=y$ whenever ye is defined. There may be many identities, but it is clear from the definition that if e and f are identities with ef defined then $e=f$.

Definition A *groupoid* is a set G with a partial multiplication such that:

(1) (associative law) if one of $(ab)c$ and $a(bc)$ is defined then so is the other and they are equal; also, if both ab and bc are defined then $(ab)c$ is defined,

(2) (existence of identities) for any a, there are identities e and f with ea and af defined,

(3) (existence of inverses) for any a, and e and f as in (2), there is an element a^{-1} such that $aa^{-1}=e$ and $a^{-1}a=f$.

We nearly always want our groupoids to be non-empty. I leave it to the reader to decide which properties stated should have the empty groupoid given as an exception. It is nonetheless convenient to permit the empty groupoid; for instance, the intersection of subgroupoids of a groupoid (unlike the intersection of subgroups of a group) can be empty.

Examples Plainly any group is a groupoid.

Let Λ be a set, and, for each $\lambda \in \Lambda$ let X_λ be a set. Let $G_{\lambda\mu}$ be the set of all bijections from X_λ to X_μ (this may be empty), and let G be $\cup G_{\lambda\mu}$. Then G is a groupoid under composition. This generalises the group of permutations of a set, which corresponds to the case when Λ has only one member.

Let C be a small category (that is, a category in which the objects form a set, in which case the morphisms also form a set). Let G be the set of all isomorphisms of C. Then G is a groupoid. The previous example is a special case of this one.

Let H be a subgroup of the group G, and let $\{H_\alpha\}$ be the set of right cosets of H. For $g \in G$ let (g,α) denote the bijection from H_α to $H_\alpha g$ obtained by multiplication by g. Then $\{(g,\alpha)$; all g and $\alpha\}$ is a groupoid.

The simplest properties of groupoids are very similar to those of groups, with modifications needed because multiplication is not always defined. We note first that for each a there can only be one identity e such that ea is defined. For suppose that e_1 is also an identity and e_1a is also defined. Then $e_1a = a$, since e_1 is an identity. Hence $e(e_1a)$ is defined. By the associative law, ee_1 is also defined, and we then know that $e = e_1$.

The equation $ax = b$ has a solution only if $a^{-1}b$ is defined, and then its only solution is $a^{-1}b$. First suppose that $a^{-1}b$ is defined. We know that aa^{-1} is defined and equals the identity e. Hence $a(a^{-1}b)$ is defined and equals $(aa^{-1})b = eb = b$, since e is an identity. Conversely, suppose that there is some x such that $ax = b$. Since $a^{-1}a$ is defined and equals the identity f, we know that $a^{-1}(ax)$ is defined and equals $(a^{-1}a)x = fx = x$; that is, $a^{-1}b$ is defined and equals x. Similarly, the equation $ya = b$ has a solution only if ba^{-1} is defined, and then the unique solution is ba^{-1}.

As a particular case of this, the inverse of an element a is unique; in fact there is only one element x such that $ax = e$, where e is the unique identity with ea defined.

Also, any element e such that $e^2 = e$ is an identity. For there is an identity f such that $ef = e$, and then $ee = ef$ shows us that $e = f$.

Notice that ab will be defined iff there is an identity f such that both af and fb are defined. For if this happens then $(af)b$ will be defined, and this is just ab. Conversely, if ab is defined let f be the identity such that af is defined. Since $af = a$, we see that $(af)b$ is defined, and hence fb is defined. In particular, if G has only one identity then ab is always defined, and so G is a group.

We shall frequently refer to the identities of a groupoid G as the *vertices* of G. If $a \in G$ has identities e and f with ea and af defined, we then say that a is an *arrow* or *edge* from e to f, and we say that e is the *start* (or *initial vertex*) of a, while f is the *end* (or *final vertex* or *terminal vertex*) of a. We denote the set of vertices of G by $V(G)$.

It is often more convenient to require the set of vertices of G to be a set bijective with the set of identities of G. When we are given a groupoid G we require for definiteness that $V(G)$ is the set of identities of G. However, when we are constructing a groupoid there is often a natural choice for the vertices. For instance, in the example where H is a subgroup of the group G, it is natural to take the cosets of H as vertices, while in the example of bijections between sets we would naturally take the sets X_λ (or the elements of Λ) to be the vertices.

We have seen that a small category in which all the morphisms are isomorphisms is a groupoid. Conversely, any groupoid is a small category in which all the morphisms are isomorphisms; the objects of this category are the vertices of the groupoid, and its morphisms are the arrows.

Examples Let V be any set. Let $D(V)$ be the groupoid whose vertex set is V and such that, for all $v \in V$ and all $w \in V$ with $w \neq v$, there is exactly one arrow from v to v and there are no arrows from v to w. Let $T(V)$ be the groupoid with vertex set V and with exactly one arrow from v to w for any v and w in V. In both cases the definition of the multiplication is obvious. We call $D(V)$ the *discrete groupoid on* V, and we call $T(V)$ the *tree groupoid on* V.

A *subgroupoid* of a groupoid G is a subset H of G such that if a and b are in H then ab is in H (if ab is defined) and a^{-1} is in H. It then follows that if $a \in H$ and e and f are the identities such that ea and af are defined then e and f are in H. This holds because $e = aa^{-1}$ and $f = a^{-1}a$.

Examples Let G be a group. Then a non-empty subgroupoid of G is the same as a subgroup of G.

The intersection of any collection of subgroupoids of a groupoid is a (possibly empty) subgroupoid.

For any set V, the discrete groupoid on V is a subgroupoid of the tree groupoid on V.

Let G be a groupoid, and let S be a subset of $V(G)$. Define G_S to consist of all arrows which begin and end in S. Then G_S is a subgroupoid of G which we call the *full subgroupoid of G with vertex set S*. If $S = \{e\}$, we write G_e rather than $G_{\{e\}}$. Since G_e has e as its only identity, G_e is a group, which we call the *vertex group of G at e*.

Let e and f be vertices of a groupoid G. Writing $e \equiv f$ iff there is an arrow from e to f, it is easy to check that $\equiv$ is an equivalence relation on $V(G)$. The equivalence classes are called *components* of G, and G is called *connected* if there is only one component. For instance, $T(V)$ is connected and $D(V)$ is totally disconnected (that is, each component consists of only one vertex).

Let a be an arrow from e to f. Then there is a function $a*$ from G_e to G_f defined by $ga* = a^{-1}ga$. It is easily checked that $a*$ is a homomorphism, and that if b is an arrow starting at f then $(ab)* = a*b*$; also $e*$ is the identity. It follows that $a*$ is an isomorphism, since it has $(a^{-1})*$ as its inverse. Notice that if c is an arrow in G_e then $c*$ is just conjugation by c. It follows that if a and b are both arrows from e to f then the isomorphisms $a*$ and $b*$ differ by a conjugation. Because $a*$ is an isomorphism. we obtain the following proposition.

Proposition 1 *If e and f are vertices in the same component then G_e and G_f are isomorphic; in particular, the vertex groups of a connected groupoid are isomorphic.//*

Let φ be a function from the groupoid G to the groupoid H. We call φ a *homomorphism* if whenever ab is defined then $(a\varphi)(b\varphi)$ is defined and $(a\varphi)(b\varphi) = (ab)\varphi$.

Let φ be a homomorphism, and let e be an identity of G. Then $ee = e$, and so $(e\varphi)(e\varphi) = e\varphi$. Hence $e\varphi$ is an identity of H. In particular, φ induces a function, which we still denote by φ, from $V(G)$ to $V(H)$. Also, if a is an arrow from e to f then $a\varphi$ will be an arrow from $e\varphi$ to $f\varphi$ (because $e\varphi$ is an identity such that $(e\varphi)(a\varphi)$ is defined), and then $a^{-1}\varphi = (a\varphi)^{-1}$. If we regard groupoids as categories then a homomorphism is the same as a covariant functor.

If V and W are sets then any function from V to W induces obvious homomorphisms from $D(V)$ to $D(W)$ and from $T(V)$ to $T(W)$. These are functors from the category of sets to the category of groupoids.

Let G be a groupoid and let W be a set. Then any map $\varphi:V(G) \to W$ extends uniquely to a homomorphism $\Phi:G \to T(W)$. For if a is an arrow from e

to f we must define $a\Phi$ to be the unique arrow in $T(W)$ from $e\varphi$ to $f\varphi$ if Φ is to be a homomorphism. And if Φ is defined in this way it is a homomorphism. (In category language, the functors V:Groupoids $\to$ Sets and T:Sets $\to$ Groupoids are adjoint.)

Also, any map $\psi: W \to V(G)$ extends uniquely to a homomorphism $\Psi: D(W) \to G$, but this property is much less important than the previous one.

Let G and H be groupoids, and let S be any subset of $V(G)$. Then any homomorphism $\varphi: G_S \to H$ extends (non-uniquely) to a homomorphism $\psi: G \to H$. First we choose for each $v \in V(G)$ whose component meets S an arrow x_v which ends at v and starts in S; if $v \in S$ we take x_v to be an identity. We also choose some identity i in H. Let g be an arrow of G from v to w. If the component of v (which is also the component of w) meets S then both x_v and x_w are defined. Hence $x_v g x_w^{-1}$ is defined and is in G_S. In this case we define $g\psi$ to be $(x_v g x_w^{-1})\varphi$. If the component of v does not meet S we define $g\psi$ to be i. It is clear that ψ extends φ, and it is easy to check that ψ is a homomorphism.

It is possible to define normal subgroupoids of a groupoid, and to prove isomorphism theorems. We can then develop a theory of generators and relators. These topics are important in the general theory of groupoids, but we will not need them for the applications that concern us. Consequently we shall not develop them further. For the same reason, though we shall need to know something of the behaviour of pushouts of groupoids, we shall not need to show that pushouts always exist.

We may define a partial multiplication on the set $G \times H$ by the rule $(g, h)(g_1, h_1) = (gg_1, hh_1)$. This plainly makes $G \times H$ into a groupoid, which we call the *direct product* of G and H. The element (e, f) is an identity iff e is an identity of G and f is an identity of H. When this happens, the vertex group $(G \times H)_{(e,f)}$ is the direct product of the vertex groups G_e and H_f.

The definition of a pushout of groupoids (or of sets) is similar to that for a pushout of groups. Thus we take groupoids G_0, G_1, and G_2, with homomorphisms $i_1: G_0 \to G_1$ and $i_2: G_0 \to G_2$. Suppose that we also have a groupoid G and homomorphisms $j_1: G_1 \to G$ and $j_2: G_2 \to G$. Then (G, j_1, j_2) is the *pushout* of (i_1, i_2) if $i_1 j_1 = i_2 j_2$ and for any groupoid H and homomorphisms $\varphi_1: G_1 \to H$ and $\varphi_2: G_2 \to H$ such that $i_1\varphi_1 = i_2\varphi_2$ there is a unique homomorphism $\varphi: G \to H$ such that $\varphi_1 = j_1\varphi$ and $\varphi_2 = j_2\varphi$. Replacing groupoids and homomorphisms by sets and maps, we get the corresponding definition of a pushout of sets.

When G_0 is empty, the pushout is just the disjoint union of G_1 and G_2.

If we have a pushout of groupoids in which each groupoid is a group then it is immediate that we have a pushout of groups. Conversely, any pushout of groups is also a pushout of groupoids. To see this, we must observe that a groupoid homomorphism φ from a group G to a groupoid H is just a group homomorphism from G to H_a, where a is 1φ.

All examples of pushouts that are of importance to us have the following extra property; $V(G_0) = V(G_1) \cap V(G_2)$ and $V(G) = V(G_1) \cup V(G_2)$, with the relevant maps being inclusions on the sets of vertices. Pushouts with this property will be called *pointed pushouts.*

Example 1 Let G be a connected groupoid with $V(G) = \{a,b\}$. Then there is a groupoid H such that

$$\begin{array}{ccc} D(a,b) & \rightarrow & T(a,b) \\ \downarrow & & \downarrow \\ G & \rightarrow & H \end{array}$$

is a pushout square.

Further, H is connected, $V(H) = \{a,b\}$, the map $G \rightarrow H$ is a monomorphism, which we regard as inclusion, and the vertex group $H_a = G_a * \langle t \rangle$, where t has infinite order. We show this by constructing a groupoid H with these properties, and proving that it is the pushout. Note that these properties define H up to isomorphism.

Let $x \in T(a,b)$ be the arrow from a to b, and let u be an arrow in G from a to b. We define a groupoid H as follows: it has $\{a,b\}$ as vertex set, the arrows in H from a to a form the group $G_a * \langle t \rangle$, where t has infinite order, and we denote this group by H_a; the arrows in H from a to b form a set bijective with H_a, and we denote the element corresponding to h by hu; similarly the arrows in H from b to a form a set bijective with H_a, the element corresponding to h being written as $u^{-1}h$, while the arrows from b to b are also bijective with H_a, the element corresponding to h being written $u^{-1}hu$. The multiplication in H is given by the obvious definition, and there is an obvious inclusion of G into H. The required homomorphism from $T(a,b)$ to H is given by requiring it to send x to tu. We have to show that H, constructed in this way, is the pushout.

Take any groupoid K, and homomorphisms $\varphi: G \rightarrow K$ and $\psi: T(a,b) \rightarrow K$ such that $\varphi = \psi$ on $D(a,b)$. Hence $a\varphi = a\psi$ and $b\varphi = b\psi$. Let k be $x\psi$, which is an arrow from $a\varphi$ to $b\varphi$. Define $\vartheta: H \rightarrow K$ as follows. First define ϑ on H_a to be the homomorphism from the group H_a to the group $K_{a\varphi}$ which is φ on G_a and which sends t to $k(u\varphi)^{-1}$. Now define ϑ on all of H by requiring that, for all $h \in H_a$ we have $(hu)\vartheta = (h\vartheta)(u\varphi)$, $(u^{-1}h)\vartheta = (u\varphi)^{-1}(h\vartheta)$, and $(u^{-1}hu)\vartheta = (u\varphi)^{-1}(h\vartheta)(u\varphi)$. Then ϑ is easily checked to be a homomorphism with the required properties, and to be the only such homomorphism.

In particular, if G is also $T(a,b)$ then H_a is infinite cyclic, generated by xu^{-1}.

Example 2 We can generalise the previous example, and this generalisation will be useful in Chapter 4. Let G and H be groupoids with $V(G) = \{a,b\} = V(H)$, and let H be connected. Let K be the disconnected groupoid with $V(K) = \{a,b\}$ and with $K_a = H_a$ and $K_b = H_b$. Given any homomorphism from K to G there is a groupoid L such that we have a pushout

$$\begin{array}{ccc} K & \rightarrow & G \\ \downarrow & & \downarrow \\ H & \rightarrow & L \end{array}.$$

The homomorphism from K to G can be regarded as two homomorphisms $\alpha: H_a \rightarrow G_a$ and $\beta: H_b \rightarrow G_b$. Since H is connected, we can choose an arrow $u \in H$ from a to b. We define $\beta': H_a \rightarrow G_b$ by $h\beta' = (u^{-1}hu)\beta$. If G is connected we also choose an arrow $v \in G$ from a to b, and we define $\beta'': H_a \rightarrow G_a$ by $h\beta'' = v(h\beta')v^{-1}$.

We shall construct a groupoid L and prove that it is the pushout. We require L to be connected with $\{a,b\}$ as its set of vertices. Hence L is determined by its vertex group L_a. We specify an arrow x from a to b in L.

If G is disconnected we define L_a to be the pushout of the homomorphisms α and β'. If G is connected we define L_a to be the pseudo-HNN extension corresponding to α and β''.

Suppose that we have homomorphisms from G and H to a groupoid M such that the homomorphisms $K \rightarrow G \rightarrow M$ and $K \rightarrow H \rightarrow M$ are the same. First suppose that G is disconnected. Then the homomorphism from G to M consists of two homomorphissm $\varphi: G_a \rightarrow M_c$ and $\psi: G_b \rightarrow M_d$. The homomorphism $\vartheta: H \rightarrow M$ coincides with $\alpha\varphi$ on H_a and with $\beta\psi$ on H_b. Let w be $u\vartheta$. For $h \in H_a$ we have $h\beta'\psi = (u^{-1}hu)\beta\psi = (u^{-1}hu)\vartheta = w^{-1}(h\alpha\varphi)w$. It follows that $\alpha\varphi = \beta'\psi(w^{-1})\#$. Since L_a is the pushout of α and β', it follows that there is a homomorphism $\mu: L_a \rightarrow M_c$ such that $\lambda\mu = \varphi$ and $\lambda'\mu = \psi(w^{-1})\#$, where λ and λ' are the homomorphisms from G_a and G_b to L_a. We have a homomorphism from G to L which is λ on G_a and which is $\lambda' x\#$ on G_b. Similarly we have a homomorphism from H to L which is $\alpha\lambda$ on H_a and which sends u to x. Then L with these two homomorphisms is the required pushout, because we can extend μ to a homomorphism from L to M by requiring it to send x to w, and this is the only possible homomorphism satisfying the relevant conditions.

Now let G be connected, and choose an arrow $v \in G$ from a to b. A homomorphism Φ from G to M is given by a homomorphism $\varphi: G_a \to M_c$ and an element $y = v\Phi$ of M. Let $\vartheta: H \to M$ be a homomorphism such that $K \to G \to M$ is the same as $K \to H \to M$. Then $\vartheta = \alpha\varphi$ on H_a. Also, for $h \in H_a$, we have $(u^{-1}hu)\vartheta = (u^{-1}hu)\beta\Phi = [v^{-1}(h\beta'')v]\Phi = h\beta''\varphi y\#$. We see that $h\alpha\varphi = h\beta''\varphi(yw^{-1})\#$. The remainder of the proof that the stated groupoid L is the required pushout will be left to the reader.

Example 3 Suppose that we have a pointed pushout

$$\begin{array}{ccc} G_0 & \xrightarrow{i_1} & G_1 \\ {\scriptstyle i_2}\downarrow & & \downarrow{\scriptstyle j_1} \\ G_2 & \xrightarrow[j_2]{} & G_3 \end{array}$$

and let S_3 be a subset of $V(G_3)$. For $r = 0,1,2$, let S_r be $S_3 \cap V(G_r)$, and suppose that S_r meets every component of G_r. Let H_r be $(G_r)_{S_r}$ for $r = 0,1,2,3$. Then we have a pushout square

$$\begin{array}{ccc} H_0 & \to & H \\ \downarrow & & \downarrow \\ H_2 & \to & H_3 \end{array} .$$

In particular, suppose that G_r is connected for $r = 0,1,2$ (which implies that G_3 is also connected), and let $a \in V(G_0)$. Then we have a pushout square of groups

$$\begin{array}{ccc} (G_0)_a & \to & (G_1)_a \\ \downarrow & & \downarrow \\ (G_2)_a & \to & (G_3)_a \end{array} .$$

To see this, take a groupoid H and homomorphisms $\varphi_r: H_r \to H$ for $r = 0,1,2$ such that on H_0 we have $i_1\varphi_1 = \varphi_0 = i_2\varphi_2$. We have already shown how to extend φ_r to a homomorphism $\psi_r: G_r \to H$. We would like to do this so that $i_1\psi_1 = \psi_0 = i_2\psi_2$.

In the construction we needed to take for each vertex $v \in G_r$ an arrow $x_{rv} \in G_r$ from a vertex of S_r to v, and the only condition in the general construction is that x_{rv} must be an identity if $v \in S_r$. Having made this choice, ψ_r is uniquely defined because S_r meets every component of G_r. We begin by choosing suitable x_{0v}, and then we make the additional requirements that for any $v \in V(G_0)$ the arrow x_{rv} must be $x_{0v}i_r$ for $r = 1,2$. This ensures that $i_1\psi_1 = \psi_0 = i_2\psi_2$.

It follows that there is a homomorphism $\psi_3:G_3 \to H$ such that $j_1\psi_3 = \psi_1$ and $j_2\psi_3 = \psi_2$. Let φ_3 be the restriction of ψ_3 to H_3. Then, for $r = 1,2$, on H_r we have $j_r\varphi_3 = \varphi_r$, as needed to show that H_3 is the pushout wanted.

We still have to show that φ_3 is unique. It is enough to show that any φ_3 can be extended to a corresponding ψ_3 with $j_r\psi_3 = \psi_r$ on G_r. For ψ_3 will then be unique, since G_3 is a pushout, and so φ_3 will also be unique. Because our pushout is pointed, we can find uniquely for each $v \in V(G)$ an arrow x_{3v} such that $x_{3v} = x_{rv}j_r$ for $v \in V(G_r)$ for $r = 1,2$. Using these arrows to define ψ_3 from φ_3, we see that ψ_3 has the required properties.

Exercise 1 Let S_1 and S_2 be sets. Show that we have a pushout square of sets

$$\begin{array}{ccc} S_1 \cap S_2 & \to & S_1 \\ \downarrow & & \downarrow \\ S_2 & \to & S_1 \cup S_2 . \end{array}$$

Exercise 2 Let the groupoid G_3 be the pushout of $G_0 \to G_1$ and $G_0 \to G_2$. Show that the set $V(G_3)$ is the pushout of $V(G_0) \to V(G_1)$ and $V(G_0) \to V(G_2)$.

Exercise 3 Fill in the missing details in Example 2.

3.2 *DIRECT LIMITS*

We are used to limiting processes in analysis, but they are less familiar in algebra. We shall define a procedure which will enable us to obtain properties of infinite objects (for instance, a group which is not finitely presented) from related finite objects.

Definition A *directed set* is a set Λ together with a relation $\leq$ on Λ such that:

(1) if $\lambda \leq \mu$ and $\mu \leq \nu$ then $\lambda \leq \nu$,

(2) $\lambda \leq \lambda$ for all λ,

(3) if $\lambda \leq \mu$ and $\mu \leq \lambda$ then $\lambda = \mu$,

(4) for all λ and μ there is some ν wih $\lambda \leq \nu$ and $\mu \leq \nu$.

Examples The natural numbers $\mathbb{N}$ form a directed set with the usual meaning of $\leq$. The positive integers also form a directed set when $\lambda \leq \mu$ means that λ divides μ.

Let X be any set, and let Λ be the set of all subsets of X. Then Λ is a directed set, with $\leq$ being inclusion.

Similarly, the set of all subgroups of a group G (or of all finitely generated subgroups of G) is a directed set under inclusion.

Definition Let Λ be a directed set. For all $\lambda \in \Lambda$ let G_λ be a group. For all $\lambda$and μ in Λ with $\lambda \leq \mu$ let $\varphi_{\lambda\mu}:G_\lambda \to G_\mu$ be a homomorphism such that $\varphi_{\lambda\lambda}$ is the identity map for all λ and such that $\varphi_{\lambda\mu}\varphi_{\mu\nu} = \varphi_{\lambda\nu}$ whenever $\lambda \leq \mu \leq \nu$. Then the collection of groups G_λ and homomorphisms $\varphi_{\lambda\mu}$ is called a *direct system* of groups (and homomorphisms) indexed by Λ.

Replacing groups by groupoids gives the notion of a direct system of groupoids. Replacing groups and homomorphisms by sets and maps, we also have the notion of a direct system of sets.

Examples Let Λ be the set of all finitely generated subgroups of a group G. Then we have a direct system of groups over Λ, obtained by assigning to each member of Λ (which is, by definition, a subgroup regarded as an index) the subgroup itself, and to each $\lambda \leq \mu$ the corresponding inclusion. It is notationally confusing to use the same symbol for a subgroup regarded as a set and for the same subgroup regarded as an index. Consequently, in this and other examples, we shall refer to an index set Λ with subgroup G_λ corresponding to λ.

Definition Suppose that we have a direct system of groups over Λ, and let G be a group with homomorphisms $\psi_\lambda:G_\lambda \to G$ for all λ. Then $(G,\{\psi_\lambda\})$ is called the *direct limit* of the direct system if $\psi_\lambda = \varphi_{\lambda\mu}\psi_\mu$ for all $\lambda \leq \mu$ and if for any group H and homomorphisms $f_\lambda:G_\lambda \to H$ such that $f_\lambda = \varphi_{\lambda\mu}f_\mu$ for all $\lambda \leq \mu$ thre is a unique homomorphism $f:G\to H$ such that $f_\lambda = \psi_\lambda f$ for all λ. We write $G = \varinjlim G_\lambda$, omitting mention of the ψ_λ. Similarly we can define the direct limit of a direct system of groupoids or of sets.

As usual the direct limit is unique up to isomorphism.

Proposition 2 *Any direct system of groups has a direct limit.*

Proof Let K be $*G_\lambda$, and let N be the normal subgroup of K generated by $\{(g_\lambda\varphi_{\lambda\mu})g_\lambda^{-1};$ all $\lambda\le\mu$ and all $g_\lambda\in G_\lambda\}$. The inclusion of G_λ in K induces a homomorphism $\psi_\lambda:G_\lambda\to K/N$. Suppose that we have homomorphisms $f_\lambda:G_\lambda\to H$ such that $f_\lambda=\varphi_{\lambda\mu}f_\mu$ for all $\lambda\le\mu$. Then we have a homomorphism $\vartheta:K\to H$, which is f_λ on G_λ. Now $((g_\lambda\varphi_{\lambda\mu})g_\lambda^{-1})\vartheta=(g_\lambda\varphi_{\lambda\mu}f_\mu)(g_\lambda f_\lambda)^{-1}=1$, and so ϑ induces a homomorphism $f:K/N\to H$. It is now clear that K/N has the required properties.//

Proposition 3 *Let* $G=\lim\limits_{\to}G_\lambda$. *Then* $G=\cup G_\lambda\psi_\lambda$. *Also* g_λ *satisifes* $g_\lambda\psi_\lambda=1$ *iff there is some* $\mu\ge\lambda$ *such that* $g_\lambda\varphi_{\lambda\mu}=1$.

Corollary 1 *If each* $\varphi_{\lambda\mu}$ *is a monomorphism then each* ψ_λ *is a monomorphism.*

Corollary 2 *Let H be a group, and let* $\vartheta_\lambda:G_\lambda\to H$ *be homomorphisms. Suppose that* $\vartheta_\lambda=\varphi_{\lambda\mu}\vartheta_\mu$ *for* $\lambda\le\mu$, *that* $H=\cup G_\lambda\vartheta_\lambda$ *and that* $g_\lambda\vartheta_\lambda=1$ *iff there is some* $\mu\ge\lambda$ *such that* $g_\lambda\varphi_{\lambda\mu}=1$. *Then* $H=\lim\limits_{\to}G_\lambda$.

Proof Corollary 1 is obvious from the proposition.

Under the hypotheses of Corollary 2, there is a homomorphism $\vartheta:G\to H$ which is seen by the proposition to be an isomorphism.

We may take G to be $(*G_\lambda)/N$, where N is the normal subgroup generated by all $(g_\lambda\varphi_{\lambda\mu})g_\lambda^{-1}$. Then any $g\in G$ is a product $(g_{\lambda_1}\psi_{\lambda_1})\dots(g_{\lambda_r}\psi_{\lambda_r})$ where $g_{\lambda_i}\in G_{\lambda_i}$. Since Λ is directed, $\exists$ μ with $\mu\ge\lambda_i$ for all i. Then $g_{\lambda_i}\psi_{\lambda_i}=g_{\lambda_i}\varphi_{\lambda_i\mu}\psi_\mu$, by the definition of N, and so $g\in G_\mu\psi_\mu$, as required for the first part of the proposition.

Now take $g_\lambda\in G_\lambda$ such that $g_\lambda\psi_\lambda=1$. Then, in the free product $*G_\alpha$, we have $g_\lambda=\Pi(u_i^{-1}v_iu_i)$, where u_i is arbitrary and $v_i=(g_{\lambda_i}\varphi_{\lambda_i\mu_i})g_{\lambda_i}^{-1}$ for some λ_i and μ_i with $\lambda_i\le\mu_i$ and $g_{\lambda_i}\in G_{\lambda_i}$. So $\exists\nu$ with $\nu\ge\lambda$ and $\nu\ge\mu_i$ for all i. There is a homomorphism from $*G_\alpha$ to G_ν which is $\varphi_{\alpha\nu}$ for all $\alpha\le\nu$ and trivial for all other α. Under this homomorphism g_λ maps to $g_\lambda\varphi_{\lambda\nu}$ and v_i maps to 1. Hence $g_\lambda\varphi_{\lambda\nu}=1$, as required.//

Examples Let G be any group. We have seen that the finitely generated subgroups of G form a direct system of groups. By Corollary 2, G is the direct limit of this direct system. More generally, if $\mathfrak{m}$ is any infinite cardinal, the collection of subgroups of G with cardinality less than $\mathfrak{m}$ is a direct system with direct limit G.

Let $G=\langle X;R\rangle$. Consider pairs (Y,S) where Y is a finite subset of X and S is a finite subset of $R\cap F(Y)$. We obtain a directed set by defining

$(Y,S) \leq (Y_1,S_1)$ if $Y \subseteq Y_1$ and $S \subseteq S_1$. As usual, it is more convenient to denote this set by Λ, and to denote the corresponding pair by (Y_λ,S_λ). Let G_λ be the group $\langle Y_\lambda;S_\lambda\rangle$. There are obvious homomorphisms from G_λ to G_μ if $\lambda \leq \mu$, and from G_λ to G. With these maps, G is the direct limit of the direct system $\{G_\lambda\}$.

This can be proved using Corollary 2 above. However, it is simpler to prove the result directly. By Proposition 2 the direct limit has a presentation whose set of generators is the disjoint union of the sets Y_λ and whose set of relators is the disjoint union of the sets S_λ together with extra relators which equate an element of Y_λ with the corresponding element of Y_μ for $\lambda \leq \mu$. An easy use of Tietze transformations shows that this group also has the presentation $\langle \cup Y_\lambda; \cup S_\lambda\rangle$; that is, it is G.

There are many variations on this example. For instance, let G be $\langle X;R\rangle$. Take $X_0 \subseteq X$ and $R_0 \subseteq R \cap F(X_0)$. Then, by essentially the same argument, G is the direct limit of the direct system of groups $\langle Y_\lambda;S_\lambda\rangle$, where each Y_λ is the union of X_0 and a finite subset of X, while $R_0 \subseteq S_\lambda \subseteq R \cap F(Y_\lambda)$ (and $S_\lambda - R_0$ finite, if we wish) for all λ. We gave, after the proof of Britton's Lemma, an alternative proof in which it was necessary to reduce from the general case to the case of finitely many stable letters. That reduction could be regarded as applying Proposition 3 to this example.

Exercise 4 Let $\{S_\lambda\}$ be a direct system of sets. If each $\varphi_{\lambda\mu}$ is an inclusion show that $\cup S_\lambda$ is the direct limit.

Exercise 5 Let the groupoid G be the direct limit of the direct system of groupoids $\{G_\lambda\}$. Show that the set $V(G)$ is the direct limit of the sets $\{V(G_\lambda)\}$.

Exercise 6 Fill in the details of the proof of Proposition 2.

4 THE FUNDAMENTAL GROUPOID AND THE FUNDAMENTAL GROUP

4.1 *THE FUNDAMENTAL GROUPOID AND THE FUNDAMENTAL GROUP*

We have seen that the set of paths in a space X has a partial multiplication, and that this multiplication is associative and has identities. It does not have inverses, but to each path f there is a path $\bar{f}$ such that both $f.\bar{f}$ and $\bar{f}.f$ are homotopic to identities. Further, homotopy is preserved by taking products. It follows that there is an induced partial multiplication on the set of homotopy classes $\gamma(X)$, and that $\gamma(X)$ is a groupoid under this multiplication. We call $\gamma(X)$ the *fundamental groupoid of* X.

The vertex set of $\gamma(X)$ is just X itself (more precisely, it is a set bijective with X in a natural way; however, it is more convenient to regard it as being X). The vertex group of $\gamma(X)$ at the vertex a is called the *fundamental group of* X *with base-point* a; it is denoted by $\pi_1(x,a)$. It consists, by definition, of the homotopy classes of loops based at a. For both $\gamma(X)$ and $\pi_1(X,a)$ we may restrict attention to standard paths (or loops) and standard homotopies, and this will usually be done without further comment.

The components of $\gamma(X)$ are the path-components of X. In particular, X is path-connected iff $\gamma(X)$ is connected. In this case $\pi_1(X,a)$ is isomorphic to $\pi_1(X,b)$ for any a and b in X, and we usually denote the group by $\pi_1(X)$ unless there is a need to look at the base-point. The isomorphism is λ_*, where λ is the homotopy class of some path from a to b; if λ is the class of f we may denote the isomorphism by f_*. If we take a different class μ of paths from a to b it is easy to see that μ_* is λ_* followed by an inner automorphism; in particular, if $\pi_1(X)$ is abelian we obtain a unique isomorphism from $\pi_1(X,a)$ to $\pi_1(X,b)$.

Let $\varphi: X \to Y$ be a continuous map (we will usually omit the word "continuous", and just refer to a "map"). If f and g are homotopic paths in X

then $f\varphi$ and $g\varphi$ are homotopic paths in Y (the homotopy between $f\varphi$ and $g\varphi$ is $F\varphi$, where F is the homotopy between f and g). Hence φ gives rise to a homomorphism $\varphi^*:\gamma(X)\to\gamma(Y)$. It is easy to see that if φ is the identity function from X to X then φ^* is the identity homomorphism from $\gamma(X)$ to $\gamma(X)$. Also, if $\psi:Y\to Z$ is another continuous map then $(\varphi\psi)^* = \varphi^*\psi^*$ (so that we have a covariant functor). Plainly, if $a\varphi = b$ then we also have a homomorphism φ^* from $\pi_1(X,a)$ to $\pi_1(Y,b)$.

Readers are warned that when i is the inclusion of a subspace A into X the corresponding i^* need not be one-one. We will see examples of this later. For the moment we just observe that if i^* were one-one then any loop in A which is homotopic to the trivial loop in X would have to be homotopic to the trivial loop in A, and there is no reason to expect this.

If X is homeomorphic to Y then $\gamma(X)$ is isomorphic to $\gamma(Y)$. For we have maps $\varphi:X\to Y$ and $\psi:Y\to X$ such that $\varphi\psi$ is the identity on X and $\psi\varphi$ is the identity on Y. Hence $\varphi^*\psi^*$ is the identity on $\gamma(X)$ and $\psi^*\varphi^*$ is the identity on $\gamma(Y)$. We will give later on a weaker relation between X and Y which makes $\gamma(X)$ isomorphic to $\gamma(Y)$.

Proposition 1 *$\gamma(X\times Y)$ is isomorphic to $\gamma(X)\times\gamma(Y)$ and $\pi_1(X\times Y,(x,y))$ is isomorphic to $\pi_1(X,x)\times\pi_1(Y,y)$.*

Proof Let p and q be the projections of $X\times Y$ onto X and Y. Then $p^*\times q^*$ is a homomorphism from $\gamma(X\times Y)$ to $\gamma(X)\times\gamma(Y)$. Let f and g be standard paths in X and Y, respectively. Define h by $th = (tf,tg)$. Then h is a standard path in $X\times Y$ such that $hp = f$ and $hq = g$. It follows that $p^*\times q^*$ is onto.

To show that this map is one-one, it is enough to show that if h and h' are standard loops in $X\times Y$ such that hp and hq are standard homotopic to $h'p$ and $h'q$ respectively then h is standard homotopic to h'. Let F and G be the standard homotopies from hp to $h'p$ and hq to $h'q$ respectively. Then the standard homotopy from h to h' is evidently H, where $(t,u)H = ((t,u)F,(t,u)G)$.

The result for the fundamental group follows at once.//

The map from I to the unit circle S^1 sending t to $(\cos 2\pi t, \sin 2\pi t)$ is continuous, is one-one on $[0,1)$, and sends 0 and 1 to $p_0 = (1,0)$. It is an identification, being a map from a compact space to a Hausdorff space. Consequently, a standard loop in X based at x can be regarded as the same thing as a map from S^1 to X sending p_0 to x.

The map from $S^1 \times I$ to the unit disk D^2 which sends (p,t) to $tp_0+(1-t)p$ is continuous, is one-one on $(S^1 - \{p_0\}) \times [0,1)$, sends $S^1 \times 1$ to p_0 and $(p,0)$ to p. For the same reason as before it is an identification. The map from I to S^1 induces a map from $I \times I$ to $S^1 \times I$, which will also be an identification. The composite of these two is an identification from $I \times I$ to D^2. It then follows that a standard loop based at x is standard homotopic to the constant standard loop at x iff the corresponding map from S^1 to X can be extended to a map from D^2 to X. Also, two maps $f,g:S^1 \to X$ will give rise to homotopic standard loops iff there is some map $F:S^1 \times I \to X$ such that $(p_0,t) = x$ for all t, and $(p,0)F = pf$, $(p,1)F = pg$ for all p.

Let f and g be standard loops in X, based at a and b respectively. We say that f is *freely homotopic* to g if there is $F:I \times I \to X$ such that $(t,0)F = tf$ and $(t,1)F = tg$ for $0 \le t \le 1$, and $(0,u)F = (1,u)F$ for $0 \le u \le 1$. In terms of a family of paths, this amounts to saying that there is a continuous family f_u of standard loops (whose base-point depends on u) such that f_0 is f and f_1 is g.

We have another identification map from $S^1 \times I$ to D^2, this time sending (p,t) to tp; again, this leads to an identification map from $I \times I$ to D^2. Using this identification, it is easy to check that a map from S^1 to X gives rise to a standard loop which is freely homotopic to a constant standard loop iff it extends to a map from D^2 to X. It follows that a standard loop based at x which is freely homotopic to a constant standard loop is actually homotopic to the constant standard loop based at x. The corollary to the next lemma deals with the case of arbitrary standard loops which are freely homotopic.

Lemma 2 *Let F be a map from $I \times I$ to X. Let f_0, f_1, g_0, and g_1 be the standard paths given by $tf_0 = (t,0)F$, $tf_1 = (t,1)F$, $tg_0 = (0,t)F$, and $tg_1 = (1,t)F$. Then $f_0 . g_1 \simeq g_0 . f_1$.*

Proof It is enough to show that $i_a \simeq f_0 . g_1 . \bar{f}_1 . \bar{g}_0$, where $a = (0,0)F$. In terms of a family of paths, at time u we apply F to the boundary of a square of side u. Formally, the homotopy G is defined by

$(t,u)G = (t,0)F$ for $0 \le t \le u$,
$(t,u)G = (u,t-u)F$ for $u < t \le 2u$,
$(t,u)G = (3u-t,u)F$ for $2u \le t \le 3u$,
$(t,u)G = (0,4u-t)F$ for $3u \le t \le 4u$. //

Figure 1 gives an indication of what is happening. The map sending t to $(t,u)G$ is, for $u=1$, 2/3, and 1/3, the image by F of the paths in $I \times I$ which are the boundaries (starting at A) of the squares $ABCD$, $APQR$, and $ALMN$ respectively.

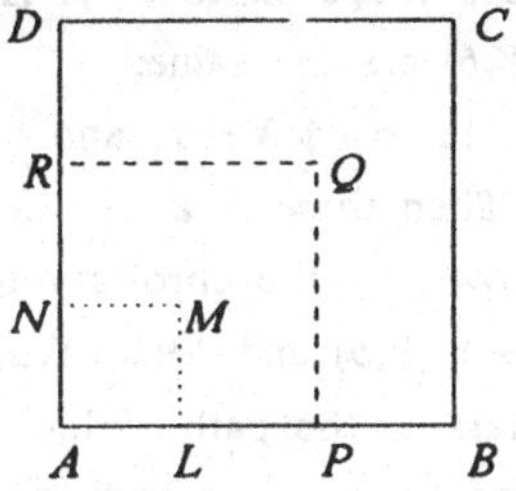

Figure 1

Corollary *Let F be a free homotopy between the standard loops f and g. Let the classes of f and g be α in $\pi_1(X,a)$ and β in $\pi_1(X,b)$. Let λ be the class of the path h from a to b which sends t to $(0,t)F$. Then $\beta = \alpha\lambda*$.*

Proof By the lemma, $f.h \simeq h.g$, so that $g \simeq \bar{h}.f.h$. This gives the result, by the definition of $\lambda*$.//

Let φ and ψ be maps from X to Y. We say that φ is *homotopic* to ψ if there is a map $\Phi: X \times I \to Y$ such that $(x,0)\Phi = x\varphi$ and $(x,1)\Phi = x\psi$ for all $x \in X$.

Homotopy of maps is an equivalence relation, by the same argument as in Proposition 2.8. We will again write $\varphi \simeq \psi$ when φ is homotopic to ψ.

Let $\varphi_0 \simeq \varphi_1: X \to Y$ and let Φ be the homotopy, and let $\psi: Y \to Z$ and $\vartheta: W \to X$ be maps. Then $\vartheta\varphi_0\psi \simeq \vartheta\varphi_1\psi$, with the homotopy sending (w,t) to $(w\vartheta,t)\Phi\psi$. Further, if $\psi_0 \simeq \psi_1: Y \to Z$ then $\varphi_0\psi_0 \simeq \varphi_1\psi_1$, since $\varphi_0\psi_0 \simeq \varphi_0\psi_1 \simeq \varphi_1\psi_1$.

Let C be a subset of X. If there is a homotopy Φ from φ to ψ such that $(c,u)\Phi = c\varphi$ for all $c \in C$ (in particular, $c\psi = c\varphi$ for $c \in C$) we say that φ and ψ are homotopic *relative to* C, and we write $\varphi \simeq \psi$ rel C. In this notation, a standard homotopy of standard paths is a homotopy rel $\{0,1\}$ of maps from I to X.

Let $\varphi: X \to Y$ be any map. We write $\varphi:(X,A_i) \to (Y,B_i)$, where i is in some index set and $A_i \subseteq X$ and $B_i \subseteq Y$, to mean that $A_i\varphi \subseteq B_i$ for all i (and we use similar simpler notations where relevant). We say that φ and ψ are homotopic as maps from (X,A_i) to (Y,B_i), and write $\varphi \simeq \psi:(X,A_i) \to (Y,B_i)$ to mean that there is a homotopy Φ such that $(A_i \times I)\Phi \subseteq B_i$ for all i. Note that we may have maps

$\varphi,\psi:(X,A) \to (Y,B)$ such that $\varphi \simeq \psi: X \to Y$ but for which we do not have $\varphi \simeq \psi:(X,A) \to (Y,B)$, since the latter imposes an extra condition on the homotopy.

Let $\varphi \simeq \psi:(X,a) \to (Y,b)$. Let f be a standard loop based at a. Then $f\varphi$ and $f\psi$ are homotopic standard loops based at b. Hence the homomorphisms φ^* and ψ^* from $\pi_1(X.a)$ to $\pi_1(Y,b)$ are the same.

More generally, let $\varphi \simeq \psi: X \to Y$, and let $a\varphi = b$ and $a\psi = c$. Let Φ be the homotopy from φ to ψ. Then there is a free homotopy from $f\varphi$ to $f\psi$ which sends (t,u) to $(tf,u)\Phi$. It follows, by the corollary to Lemma 2, that the homomorphisms $\varphi^*:\pi_1(X,a) \to \pi_1(Y,b)$ and $\psi^*:\pi_1(X.a) \to \pi_1(Y,c)$ are related by $\psi^* = \varphi^*\lambda\#$, where λ is the class of the path which sends u to $(a,u)\Phi$. We shall not investigate the connection between φ^* and ψ^* as homomorphisms from $\gamma(X)$ to $\gamma(Y)$, as this is a little messy.

Let 1_X and 1_Y be the identity maps on X and Y. A map $\varphi: X \to Y$ is called a *homotopy equivalence* if there is a map $\psi: Y \to X$ such that $\varphi\psi \simeq 1_X$ and $\psi\varphi \simeq 1_Y$. We then say that X is *homotopy equivalent to* Y, and that ψ is a *homotopy inverse* to φ. It is easy to see that homotopy equivalence is an equivalence relation. Plainly homeomorphic spaces are homotopy equivalent. We shall give later examples of many different spaces all of which are homotopy equivalent to a point (and hence to each other).

Suppose that X is homotopy equivalent to Y and that X is path-connected. Then Y is also path-connected. For the homotopy between $\psi\varphi$ and 1_Y shows that any $y \in Y$ is joined by a path to $y\psi\varphi$, and $X\varphi$ is path-connected because X is.

Proposition 3 *Let* $\varphi: X \to Y$ *be a homotopy equivalence, and let* $a\varphi = b$. *Then* $\varphi^*:\pi_1(X,a) \to \pi_1(Y,b)$ *is an isomorphism.*

Proof We have to be careful because φ need not be a homotopy equivalence from (X,a) to (Y,b).

So let $b\psi = c$, and let $c\varphi = d$. We have $\varphi^*:\pi_1(X,a) \to \pi_1(Y,b)$ and $\psi^*:\pi_1(Y,b) \to \pi_1(X,c)$; we also have a homomorphism from $\pi_1(X,c)$ to $\pi_1(Y,d)$ induced by φ, and this will be denoted by φ_1^*. By the previous remarks, there is a homotopy class λ of paths in X from a to c and a homotopy class μ of paths in Y from b to d such that $\varphi^*\psi^* = \lambda\#$ and $\psi^*\varphi_1^* = \mu\#$. Since $\lambda\#$ and $\mu\#$ are isomorphisms, it follows first that ψ^* is an isomorphism, and then that φ^* is an isomorphism.//

A space X is called *simply connected* if it is path-connected and $\pi_1(X)$ is trivial. Plainly a space consisting of a single point is simply connected; we will give some other examples of simply connected spaces shortly. Examples of spaces which are path-connected but not simply connected will be given in the next section.

Let X be simply connected. By Lemma 2.11, since any loop is homotopic to the trivial loop, we see that any two paths from a point a to a point b are homotopic. Hence there is exactly one homotopy class of paths from a to b, for all a and b. In other words, $\gamma(X)$ is the tree groupoid on the vertex set X.

We say that the space X is *contractible* to the point a if there is a map $F: X \times I \to X$ such that $(x,0)F = x$ and $(x,1)F = a$ for all $x \in X$. Equivalently, X is contractible to a if the inclusion $\{a\} \to X$ and the trivial map $X \to \{a\}$ are homotopy inverses, and so X is homotopy equivalent to $\{a\}$.. It follows that a contractible space X is simply connected, using Proposition 3 and the remark before it. It also follows that X is contractible to each of its points. We refer to the map F as a *contraction.*

We say that X is *contractible relative to a* (or *rel a*) if there is a map $F: X \times I \to X$ such that $(x,0)F = x$ and $(x,1)F = a$ for all $x \in X$ and $(a,t)F = a$ for all t. We shall see that a space may be contractible relative to one of its points but not relative to another.

Example 1 A subset X of $\mathbb{R}^n$ is called *star-shaped from* $a \in X$ if for every $x \in X$ the line-segment joining x to a lies in X. Examples of star-shaped sets include $\mathbb{R}^n$ itself, the closed disk D^n, the open disk which is the interior of D^n, the n-dimensional cube I^n, and many others. It is easy to see that a set is convex iff it is star-shaped from each of its points. Figure 2 shows an example of a star-shaped space (it is star-shaped from V) which explains the name. This space is not convex, since the line-segment joining A to B does not lie entirely in the set. In a star-shaped art gallery there is a place from which all the pictures can be seen!

The line-segment joining x to a is $\{ta + (1-t)x;\ 0 \le t \le 1\}$. It follows that if X is star-shaped from a then X is contractible, the contraction being given by $(x,t)F = ta + (1-t)x$.

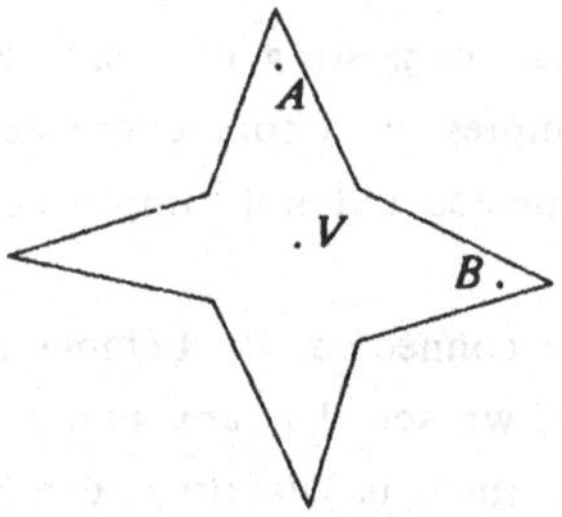

Figure 2

Example 2 Let X be any space, and CX the cone on X. Then CX is contractible relative to its vertex v. We have a map $F:(X\times I)\times I\to X\times I$ given by $(x,t,u)F=(x,tu)$. This induces a function G from $CX\times I$ to CX. Now G is plainly continuous at all points other than the points (v,u), and G will be the required contraction if it is continuous at these points also.

By definition of CX as an identification space, a subset of CX containing v is open iff it is the image of an open set U of $X\times I$ with $X\times 0\subseteq U$. Given such a set U, consider the set $V=\{(x,t);\ x\times[0,t]\subseteq U\}$ which contains $X\times 0$. If $(x,t)\in V$ then the compactness of $[0,t]$ ensures that there is an open set W of X contining x such that $W\times[0,t]\subseteq U$. It is now easy to see that V is open. Further, the image of $V\times I$ in $CX\times I$ plainly maps by G into the image of U. Hence G is continuous at the points (v,u), as required.

Example 3 It is easy to check that $S^n-\{(0,\dots,0,1)\}$ is homeomorphic to $\mathbb{R}^n$, and that, for any $p\in S^n$, $S^n-\{p\}$ is homeomorphic to $S^n-\{0,\dots,0,1)\}$. It follows that $S^n-\{p\}$ is simply connected for any p.

Then S^n is path-connected if $n\neq 0$. If f is a standard loop such that $If\neq S^n$ it follows from the previous paragraph that f is homotopic to the trivial loop. Unfortunately, even when $n>1$, it is possible to find loops f with $If=S^n$ (these are the space-filling curves due to Peano and others). However, the argument at the start of the proof of van Kampen's Theorem in the next section shows that any loop f may be written as a product $f_1\cdot\ldots\cdot f_k$ where each f_i is either a path in $S^n-\{(0,\dots,0,1)\}$ or in $S^n-\{(0,\dots,0,-1)\}$. Since each of these spaces is simply connected, each f_i is homotopic to a path g_i whose underlying set is a circular arc. Then $f\simeq g$, where Ig is the union of a finite number of circular arcs. Hence $Ig\neq S^n$ when $n>1$. Then g is homotopic to a trivial loop, and hence so is f. We deduce that S^n is simply connected for $n>1$.

A little geometric intuition should persuade the reader that S^n cannot be contractible. A proof of this would require consideration of the higher homotopy groups $\pi_n(S^n)$ or the homology groups $H_n(S^n)$, neither of which will be discussed in this book.

Example 4 The *Fox comb* is the subspace X of $\mathbb{R}^2$ given by $X = (0 \times I) \cup (I \times 1) \cup \bigcup\{(1/n,t);\ 0 \le t \le 1$, all positive integers $n\}$.

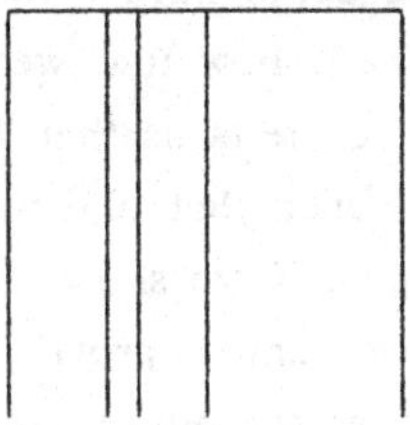

Figure 3, showing (part of) the Fox comb.

It is easy to check that X is contractible relative to (1,0), and even that it is contractible relative to any point not in $0 \times [0,1)$.

However X is not contractible relative to (0,0), and (by a similar argument) is not contractible relative to any point in $0 \times [0,1)$. For let $F{:}X \times I \to X$ be a map such that $(p,0)F = p$ for all p and $(0,0,t)F = (0,0)$ for all t.

Let U be $\{(x,y) \in \mathbb{R}^2;\ |y| < 1/2\}$. Since $((0,0) \times I)F \subseteq U$, there is some $\delta > 0$ such that $(p \times I)F \subseteq U$ if $\|p\| < \delta$ (for instance, using uniform continuity of the second coordinate of F and the fact that this coordinate has an attained maximum on $(0,0) \times I$ which is less than 1/2). In particular, if n is large enough, $((1/n,0) \times I)F$ must lie in that path-component of $X \cap U$ which contains $(1/n,0)$. This path-component is easily seen to be $\{(1/n,t);\ 0 \le t < 1/2\}$. Since this does not contain (0,0), we cannot have $(1/n,0,1)F = (0,0)$, and so F is not a contraction.

Let $-X$ be $\{(x,y);\ (-x,-y) \in X\}$. Then $-X$ is also a Fox comb. Let $Y = X \cup -X$, which is the union of two Fox combs with (0,0) as their only common point; that is, Y is the join of two copies of the Fox comb.. Then Y is not contractible.

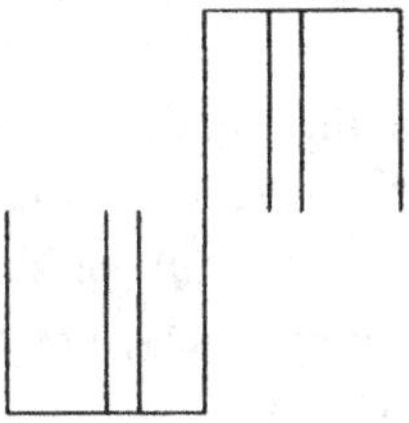

Figure 4, showing (part of) the join of two Fox combs

First note that the function $r:Y \to X$ given by $pr = p$, $(-p)r = (0,0)$ for all $p \in X$ is continuous, by the Glueing Lemma. It follows that if we had a contraction F of Y relative to $(0,0)$ then Fr, restricted to $X \times I$, would give a contraction of X relative to $(0,0)$, which we know is impossible.

Suppose that we could have a function $F:Y \times I \to Y$ such that $(p,0)F = p$ for all $p \in Y$. We shall show that we must have $(0,0,t)F = (0,0)$ for all t, and the previous paragraph then tells us that F is not a contraction.

The set of all c such that $(0,0,t)F = (0.0)$ for all $t \leq c$ is plainly a closed subset of I containing 0. If we show that it is also open it must be the whole of I, which is what we want to prove.

Suppose that c has the above property. Then we can find $d > c$ such that $(0,0,t) \in U$ for $t \leq d$. Then, as before, we can find $\delta > 0$ such that $(p,t) \in U$ if $\|p\| < \delta$ and $t \leq d$. As before, we find that, for n large enough, $((1/n,0) \times [0,d])F$ is contained in $\{(1/n,t);\ 0 \leq t < 1/2\}$. In particular, for any $t \in [0,d]$, continuity of F tells us that $(0,0,t)F = (0,y)$ for some $y \geq 0$. Similarly, looking at $(-1/n,0)$, we see by continuity that, for any $t \in [0,d]$, $(0,0,t)F = (0,z)$ for some $z \leq 0$. Thus $(0,0,t)F$ must be $(0,0)$ for all $t \leq d$, which gives the result.

Exercise 1 With X and U as in example 4, show that the path-component of $(1/n,0)$ is $\{(1/n,t);0 \leq t \leq 1/2\}$.

Exercise 2 Find a homeomorphism from $\mathbb{R}^n$ to $S^n - \{(0,\ldots,0,1)\}$.

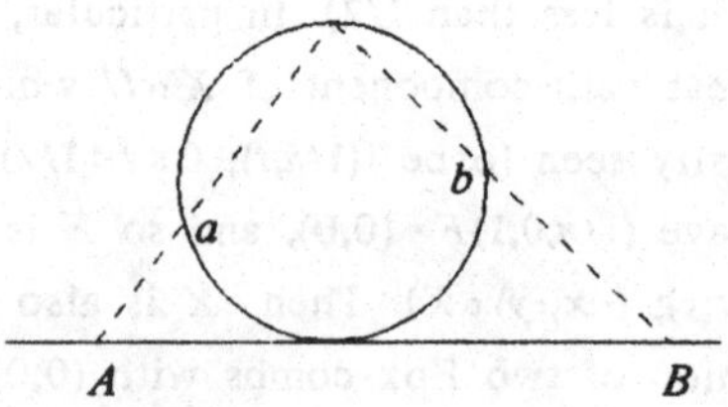

Figure 5, showing a homeomorphism from $\mathbb{R}^1$ to $S^1 - \{(0,1)\}$. *A* and *B* are mapped to *a* and *b*, respectively.

Exercise 3 Show that there is a homeomorphism from $S^n - \{p\}$ to $S^n - \{(0,\ldots,0,1)\}$ for any $p \in S^n$. (You might want to show first that there is a homeomorphism from $S^n - \{p\}$ to $S^n - \{q\}$ for some suitable q which has one more coordinate zero than p has. Alternatively, use the fact that any unit vector is part of an orthonormal basis.)

Exercise 4 Show that homotopy equivalence between spaces is an equivalence relation.

Exercise 5 Can you think of any reason why we might expect $\pi_1(S^1)$ to be non-trivial? Why we might expect it to be cyclic? To be infinite cyclic? Don't try to find a convincing argument; just see if you can see anything which suggests what kind of answer we should look for.

4.2 *VAN KAMPEN'S THEOREM*

So far we have not been able to calculate any non-trivial fundamental groups. In this section we shall find a way of doing this, and we will show that any group occurs as the fundamental group of some space.

Theorem 4 (van Kampen's Theorem) *Let X_1 and X_2 be subspaces of a space X, and let X be the union of the interiors of X_1 and X_2. Then the inclusions induce a pushout diagram of groupoids*

$$\begin{array}{ccc} \gamma(X_1 \cap X_2) & \xrightarrow{i_1*} & \gamma(X_1) \\ i_2* \downarrow & & \downarrow j_1* \\ \gamma(X_2) & \xrightarrow[j_2*]{} & \gamma(X) \end{array} .$$

Proof Let U_n be the interior of X_n for $n=1,2$. Let $f:[0,r]\to X$ be a path. Let δ be the Lebesgue number (see Proposition 2.5) of the covering $\{U_1f^{-1},U_2f^{-1}\}$ of $[0,r]$. Choose a partition $\mathfrak{P}$ of $[0,r]$ of mesh less than δ; that is, choose points $t_0=0<t_1<t_2<\ldots<t_k=r$ for some k such that $t_i-t_{i-1}<\delta$ for all i. Then f is the product of paths $f_1,\ldots,f_k$ such that each f_i is either a path in X_1 or a path in X_2, where f_i is defined on $[0,t_i-t_{i-1}]$ by $tf_i=(t+t_{i-1})f$. This property was used in Example 3 of the previous section.

Let $\varphi_1:\gamma(X_1)\to G$ and $\varphi_2:\gamma(X_2)\to G$ be homomorphisms into a groupoid G such that $i_1*\varphi_1=i_2*\varphi_2$. We can lift each φ_n $(n=1,2)$ to a multiplicative map ψ_n from the paths of X_n to G, and these two will agree on the paths of $X_1\cap X_2$. It follows that to each partition $\mathfrak{P}$ as in the previous paragraph there corresponds an element of G, namely the product of the elements $f_i\psi_{n_i}$ (where f_i is a path in X_{n_i}). We need to be sure that this product is defined. This follows because the trivial path at the point t_if_i is a right identity for f_i and a left identity for f_{i+1}, and so its image in G will be a right and a left identity for the elements of G corresponding to f_i and f_{i+1}, and the product is therefore defined.

Because each ψ_n is multiplicative, it is immediate that adding one extra division point to the partition $\mathfrak{P}$ does not change the resulting element of

G. Inductively, adding any number of new division points does not change the element. Now let $\mathfrak{P}$ and $\mathfrak{Q}$ be any partitions, and let $\mathfrak{R}$ be the partition which has the division points of $\mathfrak{P}$ together with those of $\mathfrak{Q}$. Then the element of G corresponding to $\mathfrak{P}$ is the same as that corresponding to $\mathfrak{R}$, which is in turn the same as that corresponding to $\mathfrak{Q}$. It follows that the element of G depends only on the path f and not on the partition $\mathfrak{P}$ chosen (provided that $\mathfrak{P}$ satisfies the required condition). We denote this element of G by $f\psi$. In particular, if f is a path in X_n then $f\psi = f\psi_n$. To make the notation easier to read, we shall write ψ instead of ψ_n in future. Also, ψ is plainly multiplicative.

If we can show that ψ is constant on each homotopy class it will induce a homomorphism φ from $\dot{\gamma}(X)$ to G. By construction, $\varphi_n = j_n{}^* \varphi$, and φ is plainly the only homomorphism with this property. This will show that we have a pushout, as required.

We first observe that if f is the product of the paths $f_1,\ldots,f_k$ as above then its standardisation f_σ is the product of paths $g_1,\ldots,g_k$. Here each g_i is obtained from the corresponding f_i by a change of scale, and so $g_i \simeq f_i$. Since ψ (on each subspace X_n) is induced from a function defined on homotopy classes, we see that $g_i\psi = f_i\psi$ Hence $f_\sigma\psi = f\psi$. It follows that we need only show that ψ is constant on standard homotopy classes.

Let f and f' be standard paths, and let F be a standard homotopy between them. Let δ be the Lebesgue number of the covering $\{U_1F^{-1}, U_2F^{-1}\}$ of $I \times I$. Take an integer k with $\sqrt{2}/k < \delta$. Then F maps any square of side $1/k$ into one of X_1 and X_2.

Let D_{ij} be the square of side $1/k$ whose lower left vertex is $(i/k, j/k)$. Let f_{ij} and g_{ij} be the paths defined on $[0,1/k]$ given by $tf_{ij} = (t+i/k, j/k)F$ and $tg_{ij} = (i/k, t+j/k)F$ respectively. Let e_{ij} and e'_{ij} be the horizontal and vertical edges of D_{ij} beginning at $(i/k, j/k)$, which we may regard as paths of length $1/k$. Then f_{ij} and g_{ij} are the composites of these paths with F. In Figure 6, the square $D_{2,1}$ and the edges $e_{2,1}$, $e_{2,2}$, $e'_{2,1}$, and $e'_{3,1}$ are indicated.

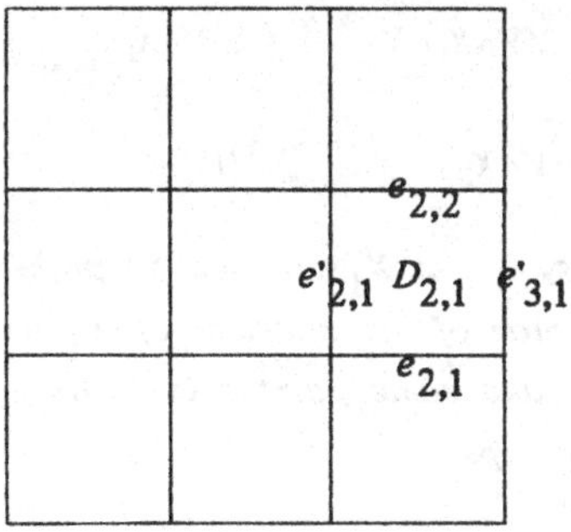

Figure 6

By Lemma 2 (with a change of scale) applied to D_{ij}, we know that $f_{ij}g_{i+1,j} \simeq g_{ij}f_{i,j+1}$, the homotopy taking place entirely in X_1 or entirely in X_2. Since ψ is constant on homotopy classes of paths in X_1 or X_2, we see that $(f_{ij}\psi)(g_{i+1,j}\psi) = (g_{ij}\psi)(f_{i,j+1}\psi)$ for $0 \le i, j < k$.

An easy induction shows that, for $r < k$, we have

$$(\Pi_{i=0}^{i=r} f_{ij}\psi)(g_{r+1,j}\psi) = (g_{0j}\psi)(\Pi_{i=0}^{i=r} f_{i,j+1}\psi).$$

We use this with $r = k-1$, and a further easy induction shows that, for $s < k$, we have

$$(\Pi_{r=0}^{r=k-1} f_{i0}\psi)(\Pi_{j=0}^{j=s} g_{kj}\psi) = (\Pi_{j=0}^{j=s} g_{0j}\psi)(\Pi_{i=0}^{i=k-1} f_{i,s+1}\psi).$$

Now f is the product of the f_{i0}, while f' is the product of the f_{ik}. Also each g_{0j} is the constant path at the point a, while each g_{kj} is the constant path at the point b, and so their homotopy classes are identities of either the groupoid $\gamma(X_1)$ or the groupoid $\gamma(X_2)$. Hence each $g_{0j}\psi$ and $g_{kj}\psi$ is an identity in the groupoid G. Applying the formula above with $s = k-1$ now shows that $f\psi = f'\psi$. This completes the proof of the theorem.//

The following corollaries are immediate, using Example 3 on pushouts of groupoids.

Corollary 1 *Let X_1 and X_2 be subspaces of X such that X is the union of the interiors of X_1 and X_2. Let V be a subset of X such that $V \cap X_1$, $V \cap X_2$, and $V \cap X_1 \cap X_2$ meet every path-component of X_1, X_2, and $X_1 \cap X_2$ respectively. Then the inclusions induce a push-out square*

$$\begin{array}{ccc} \gamma(X_1 \cap X_2)_{V \cap X_1 \cap X_2} & \to & \gamma(X_1)_{V \cap X_1} \\ \downarrow & & \downarrow \\ \gamma(X_2)_{V \cap X_2} & \to & \gamma(X)_V \end{array} \quad . \; //$$

Corollary 2 *Let X_1 and X_2 be path-connected subspaces of X such that X is the union of the interiors of X_1 and X_2. Let $X_1 \cap X_2$ be path-connected, and take some point a in $X_1 \cap X_2$. Then the inclusions induce a push-out square of groups*

$$\begin{array}{ccc} \pi_1(X_1 \cap X_2, a) & \to & \pi_1(X_1, a) \\ \downarrow & & \downarrow \\ \pi_1(X_2, a) & \to & \pi_1(X, a) \end{array} \quad . \; //$$

Most authors do not develop the theory of groupoids, and consequently they discuss the fundamental group but not the fundamental groupoid. As a result, Corollary 2 is usually referred to as van Kampen's Theorem. Our version of the theorem is slightly stronger than this one. For instance, we will be able to find $\pi_1(X)$ even when $X_1 \cap X_2$ is not path-connected. Our version, though complicated to prove, is nonetheless easier to prove than the more usual version. If we do not wish to use groupoids at all, it is neccesary to have various auxiliary paths in order to replace our paths f_{ij} and g_{ij} by loops. Our proof does not get rid of these auxiliary paths entirely, but it separates out the topological and algebraic aspects of the proof; we used such auxiliary paths when we discussed Example 3 on pushouts of groupoids.

It was primarily for their use in van Kampen's Theorem that we introduced groupoids. Most of our results in future will be expressed in terms of the fundamental group. When X is path-connected, the fundamental groupoid of X is determined by the fundamental group of X together with X itself (as a set).

Example Let X be S^1, let a be (1,0), and let b be (-1,0). Let X_1 be $S^1 - \{(0,-1)\}$, and let X_2 be $S^1 - \{(0,1)\}$. We cannot apply corollary 2, since $X_1 \cap X_2$ is not path-connected. However, X_1 and X_2 are contractible, while $X_1 \cap X_2$ consists of two contractible components, one containing a and the other containing b. It follows that $\gamma(X_1)_{\{a,b\}}$ and $\gamma(X_2)_{\{a,b\}}$ are both tree groupoids on $\{a,b\}$, while $\gamma(X_1 \cap X_2)_{\{a,b\}}$ is the discrete groupoid on $\{a,b\}$. Corollary 1 can now be used to give a push-out square whose structure has already been analysed (see the special case of Example 1 on pushouts of groupoids).

We find from this that $\pi_1(S^1,a)$ is infinite cyclic, and its generator is the class of $f.g$, where f is a path in X_1 from a to b and g is a path in X_2 from b to a. In particular, the generator is the class of the standard path which sends t to $(\cos 2\pi t, \sin 2\pi t)$. We will obtain another proof of this result, and related stronger properties, after we have developed the theory of covering spaces.

Notice that the inclusion $i:S^1 \to D^2$ does not have $i*$ one-one, since $\pi_1(D^2)$ is trivial but $\pi_1(S^1)$ is not trivial.

The subspace A of the space X is called a *retract* of X if there is a map $r:X \to A$ such that $ar = a$ for all $a \in A$; the map r is then called a *retraction*.

Proposition 5 *Let A be a retract of X, and let $i:A \to X$ be the inclusion. Then $i*:\pi_1(A) \to \pi_1(X)$ is one-one.*

Proof Since ir is the identity on A, we see that $i*r*$ is the identity on $\pi_1(A)$. //

Theorem 6 (No Retraction Theorem) *S^1 is not a retract of D^2.* //

Theorem 7 (Brouwer's Fixed-point Theorem) *Let $f:D^2 \to D^2$ be a continuous function. Then there is some $x \in D^2$ such that $xf = x$.*

Proof Suppose not. Define $g:D^2 \to S^1$ by requiring xg to be the point where the line segment from xf to x, when extended, meets S^1. Because

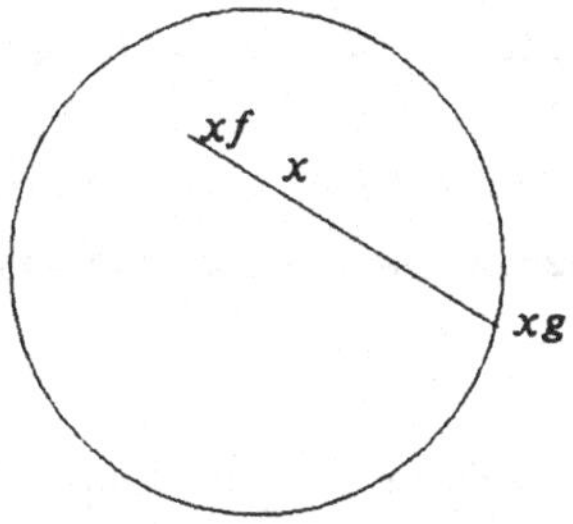

Figure 7

we are assuming that xf is always distinct from x, the function g is defined for all $x \in D^2$. Plainly $xg = x$ if $x \in S^1$. It is easy to check that g is continuous (for instance, by finding a formula for g). This contradicts the previous theorem. //

More generally, S^{n-1} is not a retract of D^n, and any continuous function from D^n to itself has a fixed point. These are proved by almost identical methods, but using the higher homotopy or homology groups.

When $X = X_1 \cup X_2$ but X is not the union of the interiors of X_1 and X_2, the conclusion of van Kampen's Theorem need not apply. For instance, we shall obtain in the last section two contractible spaces whose join is not even simply connected.

We frequently need to use van Kampen's Theorem when $X = X_1 \cup X_2$ with X_1, X_2, and $X_1 \cap X_2$ all path-connected but X is not the union of the interiors of X_1 and X_2. This can often be done by enlarging the spaces X_1 and X_2. More precisely, suppose that we have open subspaces Y_1 and Y_2 of X such that $X_n \subseteq Y_n$ for $n = 1, 2$, with Y_1, Y_2, and $Y_1 \cap Y_2$ all path-connected, and such that the inclusion of X_1 into Y_1, of X_2 into Y_2, and of $X_1 \cap X_2$ into $Y_1 \cap Y_2$ all induce isomorphisms of the fundamental groups. Then we can apply Corollary 2 to Y_1 and Y_2, and then use these isomorphisms to get the corresponding result for X_1 and X_2. In particular, this holds if we can find Y_1 and Y_2 such that X_n is a deformation retract of Y_n (defined below) for $n = 1, 2$, and $X_1 \cap X_2$ a deformation retract of $Y_1 \cap Y_2$.

Definition The subspace A of the space X is a *deformation retract of* X if there is a retraction $r{:}X \to A$ such that $ri \simeq 1_X$, where i is the inclusion and 1_X is the identity on X. If the homotopy is relative to A we say that A is a *strong deformation retract of* X.

Examples X is contractible to p iff p is a deformation retract of X, and X is contractible rel p iff p is a strong deformation retract of X.

For any space X, $X \times 1$ is a strong deformation retract of $X \times I$ and of $X \times (0,1]$.

Lemma 8 *Let A be a strong deformation retract of X_2, and let $f{:}A \to X_1$ be a map. Then X_1 is a strong deformation retract of the space $X_1 \cup_f X_2$.*

Proof Let $\varphi{:}X_2 \times I \to X_2$ be a map such that $(x,0)\varphi = x$, $(x,1)\varphi \in A$ for all $x \in X_2$, and $(a,t)\varphi = a$ for all $a \in A$ and $t \in I$.

Since $X_1 \cup_f X_2$ is obtained from the disjoint union of X_1 and X_2 by identifying a and af for all $a \in A$, it follows from Proposition 2.4 that $(X_1 \cup_f X_2) \times I$ is obtained from the space obtained from the disjoint union of $X_1 \times I$ and $X_2 \times I$ by identifying (a,t) and (af,t) for all $a \in A$ and $t \in I$.

Then, by Proposition 2.3, we obtain a continuous function on $(X_1 \cup_f X_2) \times I$ by defining continuous functions on $X_1 \times I$ and $X_2 \times I$ which are compatible with this identification. Thus we get a homotopy proving the result by taking the functions which send (x_1, t) to x_1 and (x_2, t) to $(x_2, t)\varphi$ respectively.//

Example Let X be the join of the two path-connected spaces X_1 and X_2 with the common point p. It follows from the definition of the join that each of X_1 and X_2 is a retract of X. In particular, the inclusions induce monomorphisms, which we treat as inclusions, of $\pi_1(X_1, p)$ and $\pi_1(X_2, p)$ into $\pi_1(X.p)$.

If X_2 is contractible rel p we find that X_1 is a strong deformation retract of X. Further, if both X_1 and X_2 are contractible rel p, it follows from this that X is contractible rel p.

Now suppose that X_1 and X_2 have open sets U_1 and U_2 containing p which are contractible rel p. Then $X_1 \cup U_2$ and $X_2 \cup U_1$ are open sets of X whose union is X, and they have X_1 and X_2 respectively as strong deformation retracts. Also their intersection is $U_1 \cup U_2$, which is contractible. Applying van Kampen's Theorem to these sets, and using the relevant isomorphisms, we find that $\pi_1(X, p)$ is the free product of $\pi_1(X_1, p)$ and $\pi_1(X_2, p)$. We will give another condition later for this to hold, and we will show that it does not hold in general.

This example can be generalised. Let X be the join of n path-connected spaces X_i, the common point being p, and suppose that for all i there is an open set of X_i which is contractible rel p. It is easy to check that X is the join of $X_1 \vee \ldots \vee X_{n-1}$ and X_n, so that we can prove results by induction. In particular, we find that there is, for all r, an open subset of $X_1 \vee \ldots \vee X_r$ which is contractible rel p. We also see that $\pi_1(X)$ is the free product of the $\pi_1(X_i)$. We shall look at infinite joins later in this section.

As a special case of this, the join of n circles is free of rank n.

As a further example, let Y be the join of a path-connected space X and the interval I, where $1 \in I$ is identified with some point of X. Then X is a strong deformation retract of Y, and so $\pi_1(Y)$ is isomorphic to $\pi_1(X)$. The reason for looking at Y rather than X is that the point 0 in Y has (arbitrarily small) open neighbourhoods which are contractible rel 0, while there may be no points with a similar property in X.

Definition Let S be a subset of a group G. Then the group $G/\langle S\rangle^G$ is said to be obtained from G by *cancelling* S. (I do not want to say that it is obtained by factoring out S, as this phrase is best used only when S itself is a normal subgroup of G. The process is often referred to as "killing S ", but there is enough violence in the world already without introducing it into mathematics).

Proposition 9 *Let A and X be path-connected spaces, and let $f:(A,a_0)\to(X,x_0)$ be a map. Let Y be $X\cup_f CA$, where A is regarded as the subset $A\times\{1\}$ of the cone CA. Then $\pi_1(Y,x_0)$ is $\pi_1(X,x_0)$ with $\pi_1(A,a_0)f*$ cancelled.*

Proof Let Y_1 be the open subset $X\cup_f(A\times(0,1])$, and let Y_2 be the open subset consisting of all (a,t) with $t<1$. Let $g:A\to Y_1\cap Y_2$ be the map given by $ag=(a,1/2)$, and let b_0 be a_0g.

Then g is a homotopy equivalence (in fact $Y_1\cap Y_2$ is homeomorphic to $A\times(0,1)$), and Y_2 is contractible. Hence van Kampen's Theorem tells us that $\pi_1(Y,b_0)$ is obtained from $\pi_1(Y_1,b_0)$ by cancelling $\pi_1(A,a_0)g*$.

Now let λ be the class of the path which sends t to $(a_0,t+1/2)$. Then $\lambda\#$ is an isomorphism from $\pi_1(Y_1,b_0)$ to $\pi_1(Y_1,a_0)$. Also, since $(a,1)$ is identified with af, we see that $g\simeq f$, where the homotopy sends (a,u) to $(a,(u+1)/2)$. By the corollary to Lemma 2 it follows that $g*\lambda\#=f*$. Hence $\pi_1(Y,x_0)$ is obtained from $\pi_1(Y_1,x_0)$ by cancelling $\pi_1(A,a_0)f*$.

Finally, Lemma 8 tells us that X is a strong deformation retract of Y_1, and so the inclusion induces an isomorphism from $\pi_1(X,x_0)$ to $\pi_1(Y_1,x_0)$, giving the result.//

Remark Let a and b be points of the path-connected space X. Now $S^0=\{-1.1\}$, so we may regard CS^0 as the interval $[-1,1]$. Let $f:S^0\to X$ be given by $(-1)f=a$ and $1f=b$. A similar proof at the groupoid level shows that $\pi_1(X\cup_f CS^0,a)=\pi_1(X,a)*\langle\alpha\rangle$, where α has infinite order and is the class of the path $g.h$, where g is the obvious path from a to b in CS^0, while h is some path from b to a in X.

Proposition 10 *Let X be a space. Let Λ be a directed set, and for each $\lambda\in\Lambda$ let X_λ be a subspace of X such that $X_\lambda\subseteq X_\mu$ if $\lambda<\mu$. Suppose that for every compact subset of X there is some λ such that X_λ contains this subset. Then $\gamma(X)$ is the direct limit of the groupoids $\gamma(X_\lambda)$, the* *homomorphisms being induced by inclusions. Further, if there is a point* x_0 *such that* $x_0 \in X_\lambda$ *for all* λ *then* $\pi_1(X, x_0)$ *is the direct limit of the groups* $\pi_1(X_\lambda, x_0)$.

Proof The result for the fundamental groups is identical to that for fundamental groupoids, replacing arbitrary paths by loops based at x_0.

The proof is similar to the proof of van Kampen's Theorem. Let P be the set of paths of X, and let P_λ be the set of paths of X_λ. If $f \in P$ then $[0,1]f$ is compact, and so, by asssumption, it lies in some X_λ. Hence $P = \bigcup P_\lambda$, and $P_\lambda \subseteq P_\mu$ if $\lambda < \mu$. For the same reason, any homotopy between two standard paths takes place entirely in some X_λ.

Let G be a groupoid and let $\varphi_\lambda : \gamma(X_\lambda) \to G$ be homomorphisms such that $\varphi_\lambda = i_{\lambda\mu *} \varphi_\mu$ if $\lambda < \mu$, where $i_{\lambda\mu}$ is the inclusion of X_λ in X_μ.

Then φ_λ lifts to a multiplicative map $\psi_\lambda : P_\lambda \to G$, and, for $\lambda < \mu$, ψ_λ is the restriction of ψ_μ to P_λ. For any λ and μ there is some ν with $\nu \geq \lambda$ and $\nu \geq \mu$. On $P_\lambda \cap P_\mu$ we then see that $\psi_\lambda = \psi_\mu$, since both equal ψ_ν. It follows that we may define uniquely a map $\psi : P \to G$ by requiring ψ to be ψ_λ on P_λ.

Let f and g be paths in X such that $f.g$ is defined. We take λ and μ with f and g in P_λ and P_μ, respectively. We can find ν with $\nu \geq \lambda$ and $\nu \geq \mu$, and we can then regard f and g as elements of P_ν. Since ψ_ν is multiplicative, it follows that $(f.g)\psi = (f\psi)(g\psi)$.

If ψ induces a function $\varphi : \gamma(X) \to G$ it will be the required homomorphism (and is easily seen to be the only such homomorphism). So we need only show that ψ is constant on homotopy classes. Let f and g be homotopic paths in X. Then there is some λ such that f and g are homotopic in X_λ. Then $f\psi = f\psi_\lambda$ and $g\psi = g\psi_\lambda$, and $f\psi_\lambda = g\psi_\lambda$, by definition of ψ_λ, because f and g are homotopic in X_λ.//

Remark Let X be the union of subspaces X_α for α in some index set A. Let Λ be the set of all finite subsets of A. When λ is $\{\alpha_1, \dots, \alpha_n\}$ let X_λ be $X_{\alpha_1} \cup \dots \cup X_{\alpha_n}$. Then Λ is directed and $X_\lambda \subseteq X_\mu$ if $\lambda < \mu$. Proposition 2.2 can often be used to show that every compact subset of X is contained in some X_λ. In particular, let X be obtained by adjoining a collection of spaces A_α to a space B, and let X_α be obtained by adjoining A_α to B. Then X has the weak topology defined by the subspaces X_α, and Proposition 2.2 applies.

Example Let X be the join of the spaces X_α, the common point being p. Suppose that each X_α is a T_1-space and that each X_α has an open subset contractible to p. With the above notation, when λ is $\{\alpha_1, \dots, \alpha_n\}$, we have already seen that the group $\pi_1(X_\lambda, p)$ is the free product of the groups $\pi_1(X_{\alpha_i}, p)$ for $i = 1,\ldots,n$. Also $\pi_1(X, p)$ is the direct limit of the groups $\pi_1(X_\lambda, p)$. It follows that $\pi_1(X,p)$ is the free product of the groups $\pi_1(X_\alpha, p)$.

In particular, suppose that each X_α is a circle. In this case we call X a *bunch of circles*. The discussion above shows that $\pi_1(X)$ is free of rank $|A|$, where A is the index set.

Warning The subspace of $\mathbb{R}^2$ consisting of those (x,y) such that $x^2+y^2 = 2x/n$ for some positive integer n is the union of countably many circles; namely, the circles with centres $(1/n,0)$ and radius $1/n$. However its topology is not that of the join of countably many circles, and its fundamental group is complicated; it will be discussed further in the last section of this chapter.

Proposition 11 *Let X be a path-connected space, and let the spaces A_i (for i in some index set) be path-connected. Let $f_i:(A_i, a_i) \to (X, x_0)$ be maps. Let Y be obtained from X by adjoining the spaces CA_i by the maps f_i. Suppose either that the index set is finite or that each A_i is a T_1-space. Then $\pi_1(Y, x_0)$ is obtained from $\pi_1(X, x_0)$ by cancelling all the subgroups $\pi_1(A_i, a_i)f_i{*}$.*

Proof If the index set has only one element the result holds by Proposition 9. The result for a finite index set follows by induction. The general case holds using Proposition 10 and the remark following it.//

The result extends to the case where we do not require that $a_i f_i = x_0$. In that case we must use the isomorphism induced by a path to map $\pi_1(A_i, a_i)f_i{*}$ into $\pi_1(X, x_0)$. A different path will change this homomorphism by an inner automorphism, and this will make no difference when we cancel.

As a particular case, we can take each A_i to be S^1, so that CA_i is D^2. The map f_i can be regarded as a loop whose homotopy class will be denoted by α_i. Since $\pi_1(S^1)$ is infinite cyclic, we see that $\pi_1(A_i)f_i{*}$ is the subgroup generated by α_i. So the following corollary is immediate.

Corollary *Let X be a path-connected space, and let $f_i:(S^1, a) \to (X, x_0)$, where a is $(1,0)$, be maps representing the elements α_i of $\pi_1(X, x_0)$. Adjoin copies of the disk D^2 to X using the maps f_i, and let Y be the resulting space. Then $\pi_1(Y, x_0)$ is obtained from $\pi_1(X, x_0)$ by cancelling all the elements α_i.//*

When we wish to calculate homotopy groups, Proposition 11 and its corollary are often used in preference to using van Kampen's Theorem.

For instance, we may regard $I \times I$ as sa square with vertices A, B, C, and D, and we may obtain spaces by identifying edges of the square. If we identify the edges AB and DC, and the edges AD and BC, we obtain the *torus*, while if we identify AB and CD, and AD and BC, we obtain the *Klein bottle*. These are shown in the figure below. The torus is easily seen to be homeomorphic to $S^1 \times S^1$, so we know, by Proposition 1, that its fundamental group is $\mathbf{Z} \times \mathbf{Z}$. But we may also use the discussion above. The space obtained from the edges of $I \times I$ after the identifications is just the join of two circles, one coming from AB and the other from AD. The fundamental group of this is free of rank 2, with generators x and y, say. The torus is obtained from this by adjoining (a homeomorph of) D^2, and the element which is to be cancelled is plainly $xyx^{-1}y^{-1}$. Thus the fundamental group of the torus has presentation, as we would expect, $\langle x,y;\ xyx^{-1}y^{-1}\rangle$. The Klein bottle is left as an exercise.

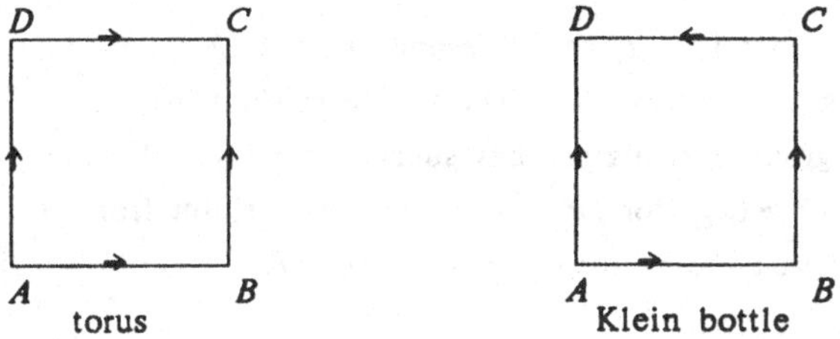

torus Klein bottle

Figure 8

We can now see that groups and their homomorphisms can be represented by spaces and their maps.

Theorem 12 *Any group is the fundamental group of some T_1-space.*

Proof Let G have presentation $\langle A;R\rangle$. Let X_1 be a bunch of circles, one for each element of A. Then $\pi_1(X_1)$ is (isomorphic to) the free group $F(A)$.

Now obtain X by adjoining to X_1 copies of D^2 using maps from S^1 to X_1 representing the elements of R. It is easy to check that X is a T_1-space. The corollary above shows that $\pi_1(X)$ has presentation $\langle A;R\rangle$.//

Theorem 13 *Let Y be a path-connected space, and let G be a group. Let $\varphi:\pi_1(Y,y_0)\to G$ be a homomorphism. Then there is a space X containing Y, which is a T_1-space if Y is a T_1-space, such that $\pi_1(X,y_0)$ is G and the inclusion of Y in X induces φ.*

Proof Take a T_1-space X_0 with $\pi_1(X_0)$ being G. Let Y_1 be the join of Y and I with y_0 being identified with 1, and let X_1 be the join of X_0 and I. Let Z be the join of X_1 and Y_1 at the point 0 of I. We have seen that Y_1 and X_1 have the same fundamental groups as Y and X_0, and that $\pi_1(Z)$ is the free product of $\pi_1(Y)$ and G (since G is $\pi_1(X_0)$). Let B be a set of generators of $\pi_1(Y, y_0)$. By adding discs to Z we obtain a space X (which is T_1 if Y is T_1) such that $\pi_1(X, y_0))$ is obtained from $\pi_1(Y, y_0) * G$ by cancelling the elements $b(b\varphi)^{-1}$ for all $b \in B$. Thus $\pi_1(X, y_0)$ is just G, and the homomorphism induced by the inclusion of Z in X sends b to $b\varphi$. Hence the inclusion of Y in X induces φ.//

Corollary 1 *Let G and H_i (for i in some index set) be groups, and let $\varphi_i : H_i \to G$ be homomorphisms. Then there is a T_1-space X and subspaces Y_i with $Y_i \cap Y_j = \{x_0\}$ for $i \neq j$ and such that $\pi_1(X, x_0)$ is G, $\pi_1(Y_i, x_0)$ is H_i, and the inclusion of Y_i in X induces φ_i.*

Proof Let H_i have presentation $\langle A_i ; R_i \rangle$, where $A_i \cap A_j = \emptyset$ for $i \neq j$. Then $*H_i$ has presentation $\langle \cup A_i ; \cup R_i \rangle$. The space Y we constructed with fundamental group $*H_i$ plainly has subspaces Y_i with fundamental group H_i and such that $Y_i \cap Y_j = \{x_0\}$ for $i \neq j$, because A_i is disjoint from A_j. Now embed Y in a space X so that the inclusion induces $*\varphi_i$.//

Corollary 2 *Let G be a group. Let Y_i be spaces such that there is $y_i \in Y_i$ with an open set which is contractible rel y_i. If the index set is infinite suppose that each Y_i is a T_1-space. Let $\varphi_i : \pi_1(Y_i, y_i) \to G$ be homomorphisms. Then there is a space X (which is T_1 if the Y_i are T_1) containing subspaces (homeomorphic to) Y_i such that the union of the Y_i is their join and such that $\pi_1(X)$ is G and the inclusion of Y_i in X induces φ_i.*

Proof Let Y be the join of the Y_i, the common point being y_i in Y_i. The conditions ensure that $\pi_1(Y)$ is the free product of the $\pi_1(Y_i)$. We then have the homomorphism $*\varphi_i$ from $\pi_1(Y)$ to G. The result now follows from the theorem.//

Corollary 3 *Let G be a group. Let Y_i be spaces with base-points y_i, and let $\varphi_i : \pi_1(Y_i, y_i) \to G$ be homomorphisms. Then there is a space X (which is T_1 if the Y_i are T_1) with base-point x_0 containing disjoint subspaces (homeomorphic to) Y_i such that $\pi_1(X, x_0)$ is G and such that φ_i is induced by* *inclusion followed by* $f_i{}^*$ *for some suitably chosen path* f_i *from* y_i *to* x_0.

Proof Take the spaces Z_i which are the joins of Y_i and the interval I, and apply Corollary 2 to these.//

Let A_0 and A_1 be path-connected subspaces of the space X such that there is a homeomorphism $\varphi: A_0 \to A_1$. Let Y be obtained from X by adding a *handle* from A_0 to A_1; that is, $Y = X \cup_f (A_0 \times I)$ where $f: A_0 \times \{0,1\} \to X$ is given by $(a,0)f = a$ and $(a,1)f = a\varphi$. Let Z be obtained from X by identifying a with $a\varphi$ for all $a \in A_0$. There is an obvious map from Y to Z which is known to be a homotopy equivalence subject to some fairly mild restrictions on the position of A_0 and A_1 in X.

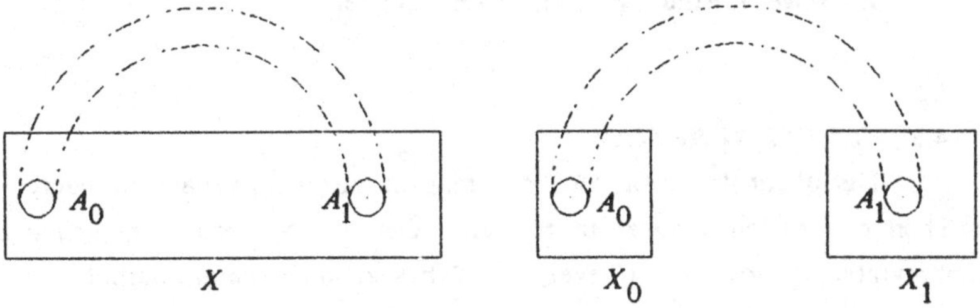

Figure 9, showing two cases of attaching a handle

We can apply van Kampen's Theorem (in the groupoid form, and using the corollary with a pair of vertices $(a_0,1/3)$ and $(a_0,2/3)$) to the subspaces $A_0 \times (0,1)$ and $Y - A_0 \times \{1/2\}$. This situation was considered in Example 2 of pushouts of groupoids. Since X is a deformation retract of $Y - A_0 \times \{1/2\}$ we obtain the following results.

If X is path-connected then $\pi_1(Y)$ is the pseudo-HNN extension of $\pi_1(X)$ with associated subgroups the images of $\pi_1(A_0)$ and $\pi_1(A_1)$. Note here that the images are only determined up to conjugacy (since we have to use the isomorphism induced by a path to refer all the groups to the same base-point, and this isomorphism is only known up to conjugacy); but we know that this makes no difference to the pseudo-HNN extension.

Suppose that A_0 and A_1 are in different path-components X_0 and X_1 of X. In this case $\pi_1(Y, a_0)$ is the pushout of the homomorphisms $\pi_1(A_0) \to \pi_1(X_0)$ and $\pi_1(A_0) \to \pi_1(A_1) \to \pi_1(X_1)$.

Theorems 12 and 13 ensure that any pseudo-HNN extension or pushout may be obtained in this way from suitable spaces. Thus the notion of adding a handle is the topological analogue of these group-theoretic concepts.

Let W be obtained from X by adjoining $A_0 \times [0,1/2]$ and $A_1 \times [1/2,1]$. Then Y is obtained from W by identifying $A_0 \times \{1/2\}$ and $A_1 \times \{1/2\}$. We could also refer to W as obtained from Y by *cutting along* $A_0 \times \{1/2\}$.

Conversely, suppose that we are given a space Y with a subspace A such that there is an open set U with (U,A) homeomorphic to $(A \times (-1,1),A)$. With slight restrictions on A, we can cut Y along A to obtain a space W. We can also regard Y as being obtained from $Y - A \times (-1/2,1/2)$ by adding a handle from $A \times \{-1/2\}$ to $A \times \{1/2\}$. Such a space Y will then have its fundamental group either a pseudo-HNN extension or a pushout (but the pushout could be the trivial kind in which all the homomorphisms are the identity).

Exercise 6 Prove the remark after Proposition 9.
Exercise 7 Find π_1 of the Klein bottle.

4.3 *COVERING SPACES*

Definition Let $p:\tilde{X} \to X$ be a map of non-empty path-connected spaces. Then p is called a *covering map* and $\tilde{X}$ is called a *covering space of X* (both abbreviated to *covering*) if every $x \in X$ has an open neighbourhood U such that the components of Up^{-1} are open and are mapped homeomorphically onto U by p. Such a neighbourhood U is called an *elementary neighbourhood* of x.

If U is an elementary neighbourhood then $Up^{-1}p$ must be U unless $Up^{-1} = \emptyset$ (the definition permits this, as then Up^{-1} has no components). Hence $U \subseteq \tilde{X}p$ if $U \cap \tilde{X}p \neq \emptyset$. It follows that $\tilde{X}p$ is both open and closed, and so $X = \tilde{X}p$, since X is connected. In particular, we cannot have $Up^{-1} = \emptyset$.

Note that each point of xp^{-1} has an open neighbourhood which maps homeomorphically onto U, and so does not contain any other point of xp^{-1}. Hence xp^{-1} is discrete. In particular (this is a special case of Lemma 14 below), if f is a map from a connected space to $\tilde{X}$ such that fp is constant then f itself is constant.

We shall give later a simple topological condition which ensures that every component of Up^{-1} is open.

Examples Regard S^1 as the set of complex numbers $\{z; |z| = 1\}$. Then $\mathbb{R}$ is a covering of S^1, with the covering map sending t to $e^{2\pi it}$. We only need two elementary neighbourhoods, namely $S^1 - \{1\}$ and $S^1 - \{-1\}$. This example, which is one of the simplest and most important, will be referred to as *unwrapping the circle*.

The map of S^1 to itself which sends z to z^n is also a covering map, with the same elementary neighbourhoods.

Let X be the join of two circles. The subset W of $\mathbb{R}^2$ consisting of $\{(x,y)$; either x or y is an integer$\}$ is a covering of X, the horizontal lines mapping to one circle and the vertical lines to the other. (See Figure 10)

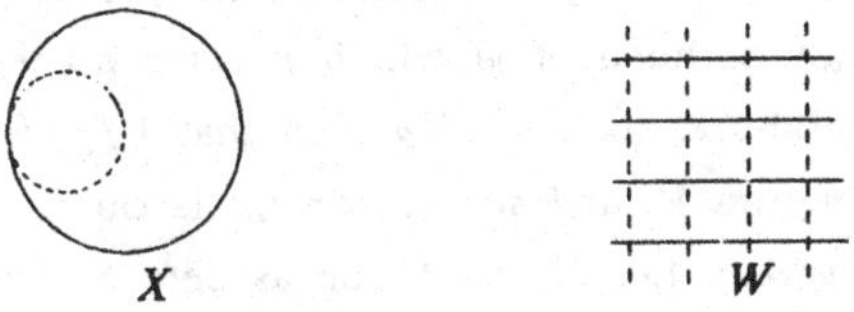

Figure 10

Let $\tilde{X}$ be obtained from X by unwrapping one circle; that is, $\tilde{X}$ consists of $\mathbb{R}$ with a circle attached at each integer point (see Figure 11). We may obtain a covering Y of $\tilde{X}$ by unwrapping one circle, and another covering Z of $\tilde{X}$ by unwrapping every circle. Both Y and Z are also coverings of X. They are illustrated in Figure 11 (the picture of Z is approximate, as it is difficult to draw in such a way that intervals remain disjoint).

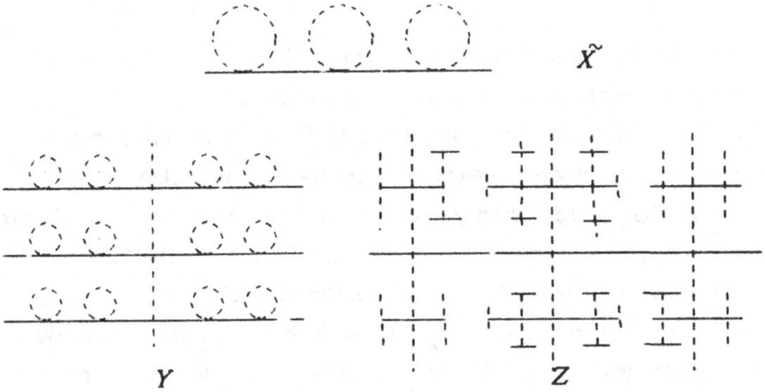

Figure 11

Let $p:\mathbb{R} \to S^1$ be the unwrapping. Let f and g be maps from a connected space Y to $\mathbb{R}$ such that $fp = gp$. Then $f - g$ is a continuous map which is always an integer. Hence $f - g$ is constant, and so $f = g$ if there is some point y_0 with $y_0 f = y_0 g$. This is a general property of coverings, as is shown in

Lemma 14. The proof for the special case is so simple that I feel it was worthwhile giving it separately.

Lemma 14 *Let $p:\tilde{X} \to X$ be a covering. Let f and g be maps from a connected space Y to $\tilde{X}$ such that $fp = gp$ and $y_0f = y_0g$ for some y_0. Then $f = g$.*

Proof Since Y is connected, it is enough to show that $\{y;\ yf = yg\}$ is both open and closed.

Take y with $yf = yg$. Let U be an elementary neighbourhood of yfp ($= ygp$). Let $\tilde{U}$ be an open neighbourhood of yf which p maps homeomorphically to U. Then there is an open neighbourhood V of y such that $Vf \subseteq \tilde{U}$ and $Vg \subseteq \tilde{U}$. Since $fp = gp$, it follows that $f = g$ on V, and so $\{z;\ zf = zg\}$ is open.

Now take y with $yf \neq yg$. Let U and $\tilde{U}$ be as before, and let $\tilde{U}_1$ be an open neighbourhood of yg which maps homeomorpically to U. Then there will be an open neighbourhood V of y such that $Vf \subseteq \tilde{U}$ and $Vg \subseteq \tilde{U}_1$. Now $\tilde{U}$ and $\tilde{U}_1$ are components of Up^{-1} which must be distinct, since they contain distinct points with the same image. They are therefore disjoint, from which we see that $V \subseteq \{z;\ zf \neq zg\}$, and so this set is also open.//

The next result is one of the key properties of coverings.

Proposition 15 *Let $p:\tilde{X} \to X$ be a covering. Let $f:[0,r] \to X$ be a path, and let $\tilde{a}$ be a point of $\tilde{X}$ such that $\tilde{a}p = 0f$. Then there is a unqiue path $\tilde{f}:[0,r] \to \tilde{X}$ such that $0\tilde{f} = \tilde{a}$ and $\tilde{f}p = f$.*

Remark The path $\tilde{f}$ will be called a *lift* of f.

Proof Uniqueness is immediate from Lemma 14.

Let δ be the Lebesgue number of $\{Uf^{-1};\ U$ is an elementary neighbourhood$\}$, which is an open cover of $[0,r]$ by the definition of covering spaces. Take n such that $r < n\delta$. Then f maps each interval $[ir/n, (i+1)r/n]$ into an elementary neighbourhood. Suppose that we have lifted f on the interval $[0,kr/n]$ to a path $\tilde{f}_k$. Let U be an elementary neighbourhood such that $[kr/n, (k+1)r/n]f \subseteq U$, and let $\tilde{b}$ be $(kr/n)\tilde{f}_k$. Then $\tilde{b}$ has an open neighbourhood which p maps homeomorphically onto U. It follows that there is a map $g:[kr/n, (k+1)r/n] \to \tilde{X}$ such that $(kr/n)g = \tilde{b}$ and $gp = f$ on this interval. By the Glueing Lemma, the function which is $\tilde{f}_k$ on $[0,kr/n]$ and g on $[kr/n, (k+1)r/n]$ is continuous. Inductively, we obtain the required lift of f.//

Plainly the standardisation of a lift of f is a lift of the standardisation of f.

Proposition 16 *Let $p:\tilde{X} \to X$ be a covering. Let f and g be homotopic paths in X. Let $\tilde{f}$ and $\tilde{g}$ be lifts of f and g beginning at the same point. Then $\tilde{f}$ and $\tilde{g}$ are homotopic.*

Corollary *With the above notation, $\tilde{f}$ and $\tilde{g}$ end at the same point.*

Proof $\tilde{f}$ and $\tilde{g}$ will be homotopic if their standardisations are homotopic. Since f and g are homotopic, the standardisations of f and g will also be homotopic. Hence it is enough to prove the result for standard maps, and we may take a standard homotopy F from f to g.

We now take δ to be the Lebesgue number of the open cover of $I \times I$ consisting of $\{UF^{-1}; U$ is an elementary neighbourhood$\}$. Take an integer n with $\sqrt{2} < n\delta$, so that any square with side $1/n$ is mapped by F into an elementary neighbourhood. Divide $I \times I$ into n^2 squares of side $1/n$. Let D_{ij} be the square whose bottom left corner is $(i/n, j/n)$, and let C_{ij} be the union of the squares which are either lower than D_{ij} or on the same level as it but to its left.

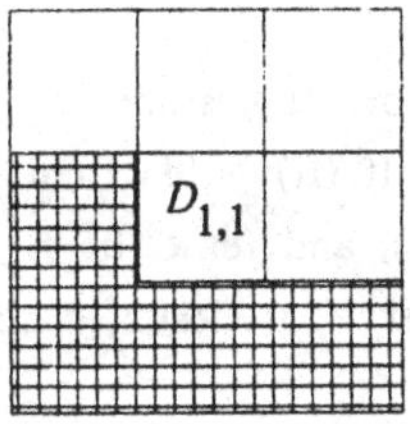

The shaded area is $C_{1,1}$.

Figure 12

Let f and g start at a and end at b, and let $\tilde{f}$ and $\tilde{g}$ start at $\tilde{a}$. Suppose that we have found a map $\tilde{F}_{ij}:C_{ij} \to \tilde{X}$ such that $\tilde{F}_{ij}p = F$ on its domain, and with $(0,0)\tilde{F}_{ij} = \tilde{a}$. We will obtain a map $H:D_{ij} \to \tilde{X}$ such that $Hp = F$ on its domain and such that H agrees with $\tilde{F}_{ij}$ where they are both defined.

Because F maps D_{ij} into an elementary neighbourhood, it is immediate, as in the previous proposition, that there is a map $H:D_{ij} \to \tilde{X}$ with $Hp = F$ on D_{ij} and with $(i/n, j/n)H = (i/n, j/n)\tilde{F}_{ij}$. Suppose that i and j are greater than 0 (the other cases are simpler). Then $C_{ij} \cap D_{ij}$ is the union of the horizontal line-segment $[i/n, (i+1)/n] \times \{j/n\}$ and the vertical line-segment $\{i/n\} \times [j/n, (j+1)/n]$. This set is connected, and $Hp = \tilde{F}_{ij}p$ on it, and there is one point of it where H and $\tilde{F}_{ij}$ are the same. Hence H agrees with $\tilde{F}_{ij}$ where they are both defined, by

Lemma 14. Then the Glueing Lemma gives a continuous function defined on $C_{ij} \cup D_{ij}$. Inductively we will obtain a map $\tilde{F}: I \times I \to \tilde{X}$ such that $(0,0)\tilde{F} = \tilde{a}$ and $\tilde{F}p = F$. Now $(0,u)\tilde{F}p = a$ for all u and $(0,0)\tilde{F} = \tilde{a}$. Hence $(0,u)\tilde{F}$ must be $\tilde{a}$ for all u. Similarly, $(1,u)\tilde{F}$ must also be constant. Since $(t,0)\tilde{F}p = (t,0)F = tf = t\tilde{f}p$ and $(0,0)\tilde{F} = 0\tilde{f}$, we find that $(t,0)\tilde{F} = t\tilde{f}$, by Lemma 14. For the same reasons, since we know that $(0,1)\tilde{F} = \tilde{a}$, we find that $(t,1)\tilde{F} = t\tilde{g}$ for all t. Hence $\tilde{F}$ is a homotopy from $\tilde{f}$ to $\tilde{g}$.

The corollary is now immediate.//

In our definition of a covering space we required the components of Up^{-1} to be open. The lemma below gives a condition under which the components of open sets are open.

Lemma 17 *The following properties of a space X are equivalent:*
(i) the path-components of any open set are open;
(ii) for any open U and $x \in U$ there is a path-connected open V with $x \in V \subseteq U$;
(iii) for any open U and $x \in U$ there is an open V with $x \in V \subseteq U$ such that any two points of V can be joined by a path in U.

Proof If (i) holds so does (ii), since we need only take V to be the path-component of U containing x. If (ii) holds then (iii) obviously holds.

Suppose that (iii) holds, and let C be a path-component of the open set U. Take $x \in C$. Then the open set V of (iii) lies completely in C. Hence C is open.//

A space with the properties of this lemma is called *locally path-connected.* In particular, the path-components of a locally path-connected space are open. Also, if U is an open set in a locally path-connected space then the path-components of U are disjoint open connected sets; they must therefore be the components of U. As a special case of this, a connected and locally path-connected space must be path-connected; thus our example of a space which is connected but not path connected provides an example of a space which is not locally path-connected.

Evidently, if $\tilde{X}$ is a covering of X then $\tilde{X}$ is locally path-connected iff X is. Also, in this case, if U is an elementary neighbourhood and V is any open set then the components of $U \cap V$ are also elementary neighbourhoods. Further, for any open neighbourhood $\tilde{W}$ of a point $\tilde{v} \in \tilde{X}$, that component of $\tilde{W} \cap Up^{-1}$ (where U is an elementary neighbourhood of $\tilde{v}p$) which contains $\tilde{v}$ will be mapped homeomorphically by p onto an elementary neighbourhood in X.

Exercise 8 Show that the identification map from S^2 to $\mathbb{R}P^2$ is a covering.

Exercise 9 Give an example of a space which is locally path-connected but not path-connected.

Exercise 10 Give an example of a space which is path-connected but not locally path-connected. (You could remove a suitable infinite set of line-segments from $\mathbb{R}^2$.)

4.4 *THE CIRCLE AND THE COMPLEX PLANE*

In this section we shall investigate the properties of $\pi_1(S^1)$ further, giving a new proof that it is infinite cyclic. We shall use these properties to obtain results about the complex numbers; in particular, we will prove the Fundamental Theorem of Algebra. Because we are working with the complex numbers $\mathbb{C}$, we will regard S^1 as $\{z; |z| = 1\}$ and D^2 as $\{z; |z| \leq 1\}$.

Let $p:\mathbb{R} \to S^1$ be given by $tp = e^{2\pi it}$. In the previous section we saw that to any path f in S^1 and any α with $\alpha p = 0f$ there is a unique path φ in $\mathbb{R}$ such that $f = \varphi p$ and $0\varphi = \alpha$. We refer to φ as a *lift* of f. Note that two different choices of α differ by an integer, and hence two different lifts of f have their difference a constant integer function. We also saw that if we lift homotopic paths in S^1 to paths in $\mathbb{R}$ with the same start then these lifts are homotopic; in particular, they have the same end-point.

The proofs of these results were very similar to the proof of van Kampen's Theorem. However, van Kampen's Theorem applied to S^1 would enable us to show that every path was homotopic to a path which has a lift, but would not be enough to tell us that every path lifts. Several of the results can be proved using the discussion of $\pi_1(S^1)$ after van Kampen's Theorem if the reader prefers; the proofs would be similar to those given here, but would need a little more care.

Let $f:[0,r] \to S^1$ be a loop. Since $rf = 0f$, $r\varphi$ and 0φ must differ by an integer. This integer does not depened on the choice of φ. We call it the ***degree*** of f.

Theorem 18 *For any $a \in S^1$, the degree is an isomorphism from $\pi_1(S^1,a)$ to $\mathbb{Z}$.*

Proof Let f and g be homotopic loops. Let φ and ψ be lifts of f and g, chosen so that $0\varphi = 0\psi$. Then we know that φ and ψ have the same endpoint. Hence f and g have the same degree, and so the degree can be regarded as a function from $\pi_1(S^1)$ to $\mathbb{Z}$..

Now let $f{:}[0,r] \to S^1$ be a loop with degree n, and let $g{:}[0,s] \to S^1$ be a loop with degree m, both loops being based at a. Let φ be a lift of f, with $0\varphi = \alpha$, and hence with $r\varphi = \alpha + n$. Since $0g = a = 0f$, we may take a lift ψ of g with $0\psi = \alpha + n$. As g has degree m, we find that $s\psi = \alpha + n + m$. Now $\varphi.\psi$ is continuous, by the Glueing Lemma, and it plainly lifts $f.g$. Hence $f.g$ has degree $n+m$, showing that the degree defines a homomorphism.

This homomorphism is onto, since if φ is the standard path sending t to nt then φp has degree n. Also, if f has degree 0 then φ is a loop in $\mathbb{R}$, and so f represents an element of $\pi_1(\mathbb{R}.\alpha)p*$. As $\pi_1(\mathbb{R})$ is trivial, this tells us that a loop of degree 0 represents the identity element of $\pi_1(S^1)$.//

Lemma 19 *Let f be a path in S^1 from a to b, and let g be a path in S^1 from b to a. Then the loops $f.g$ and $g.f$ have the same degree.*

Proof Let φ be a lift of f starting at α and ending at β, and let ψ be a lift of g starting at β. Then ψ ends at $\alpha + n$ for some n. As $\varphi.\psi$ is a lift of $f.g$, the loop $f.g$ has degree n. Let φ' be given by $t\varphi' = t\varphi + n$. Then $g.f$ lifts to $\psi.\varphi'$, which shows that $g.f$ also has degree n.//

Corollary 1 *Let f be a loop in S^1 based at a, and let g be a path in S^1 from a to b. Then $\bar{g}.f.g$ has the same degree as f.*

Proof By the lemma, $\bar{g}.f.g$ has the same degree as $f.g.\bar{g}$, and this is homotopic to f, hence it has the same degree as f. An easy direct proof can also be given.//

Corollary 2 *Let f and g be freely homotopic loops in S^1. Then f and g have the same degree.*

Proof By Lemma 2 and its corollary there is a path h such that f is homotopic to $\bar{h}.g.h$. The result is now immediate.//

Lemma 20 *Let f and g be maps from $[0,r]$ to S^1. Define h by $th = (tf)(tg)$. Then degree h = degree f + degree g.*

Proof Let φ and ψ be lifts of f and g respectively. Then $\varphi+\psi$ is a lift of h, and the result follows.//

We can extend the above results to $\mathbb{C}-\{0\}$. For $\mathbb{C}-\{0\}$ is homeomorphic to $S^1 \times \mathbb{R}^+$, the homeomorphism sending z to $(z/|z|, |z|)$. Since $\mathbb{R}^+$ is contractible, S^1 is a deformation retract of $\mathbb{C}-\{0\}$, the retraction sending z to $z/|z|$. Consequently we can define the degree of a loop in $\mathbb{C}-\{0\}$ as the degree of the corresponding loop in S^1, and obtain similar results immediately.

Thus $\pi_1(\mathbb{C}-\{0\})$ is infinite cyclic, with an isomorphism given by the degree. Freely homotopic loops have the same degree, and two loops with the same basepoint and the same degree are homotopic. Also, if f and g are two loops in $\mathbb{C}-\{0\}$ with the same domain, and h is given by $th=(tf)(tg)$, then degree h = degree f + degree g.

The degree of a loop f in $\mathbb{C}-\{0\}$ is also called the *winding number of f about* 0. More generally, if f is a loop in $\mathbb{C}$ not going through the point a, then the winding number of f about a is the degree of the loop g given by $tg=tf-a$.

Any map f from S^1 to any space X gives rise to the standard loop which is the restriction of pf to I, and any standard loop is of this form. It follows that we can extend our definitions to give the degree of a map f from S^1 to $\mathbb{C}-\{0\}$, and similar results hold. The basepoint of this loop may be taken as $1f$ for definiteness, but Corollary 1 to Lemma 19 tells us that the choice of basepoint does not affect the degree.

As an example, the map of S^1 to itself which sends z to z^n is easily seen to have degree n.

Theorem 21 (Rouché's Theorem) *Let f and g:$[0,r]\to\mathbb{C}$ be loops such that $|tg|<|tf|$ for all t. Then f and $f+g$ have the same degree.*

Remark The conditions ensure that f and $f+g$ both map into $\mathbb{C}-\{0\}$.

Proof f and $f+g$ are freely homotopic in $\mathbb{C}$, the homotopy F being given by $(t,u)F=tf+(tg)u$. It is easy to check that F is never 0, so F is a homotopy in $\mathbb{C}-\{0\}$, giving the result.//

Theorem 22 (Fundamental Theorem of Algebra) *Every non-constant polynomial with complex coefficients has a complex root.*

Proof Let f be the polynomial given by $zf = \Sigma_0^n a_k z^k$, with $a_n \neq 0$. For any real number $r \geq 0$, let $f_r : S^1 \to \mathbb{C}$ be given by $zf_r = (rz)f$. Then, for any r, f_r is freely homotopic in $\mathbb{C}$ to f_0, the homotopy sending (z,u) to $(urz)f$. If f were never zero then this would be a free homotopy in $\mathbb{C} - \{0\}$, and so f_r and f_0 would have the same degree.

Now f_0 has degree 0, since it is constant. Suppose that, for some R, $|\Sigma_0^{n-1} a_k w^k| < |a_n w^n|$ for all w with $|w| = R$. Then, by Rouché's Theorem, f_R will have the same degree as the function on S^1 sending z to $a_n R^n z^n$, which is n. The theorem would then be proved.

Now $|\Sigma_0^{n-1} a_k w^k| \leq \Sigma_0^{n-1} |a_k||w|^k$, which is at most $(\Sigma_0^{n-1} |a_k|)|w|^{n-1}$, provided $|w| \geq 1$. So we have the property needed provided that $R > \max(1, \Sigma_0^{n-1} |a_k|/|a_n|)$.//

The degree of the loop f in $\mathbb{C} - \{0\}$ can be described as $1/2\pi$ times the change in the argument of z as z moves round the image of f. If f is sufficiently well-behaved, this can also be described as $1/2\pi i$ times the integral of $1/z$ round f. The degree (or the winding number, which is essentially the same) plays an important part in complex variable theory.

Lemma 23 *Let f be a standard loop in S^1 such that $(t+1/2)f = -tf$ for $0 \leq t \leq 1/2$. Then f has odd degree.*

Remark In terms of maps from S^1 to itself, the lemma tells us that a map $g: S^1 \to S^1$ has odd degree if $(-z)g = -(zg)$ for all $z \in S^1$.

Proof Let φ be a lift of f on $[0,1/2]$. Since $(1/2)f = -0f$, there is an odd integer m such that $(1/2)\varphi = 0\varphi + m/2$. Since $(t+1/2)f = -tf$ for $0 \leq t \leq 1/2$, f lifts on $[1/2,1]$ to the function ψ such that $(t+1/2)\psi = t\varphi + m/2$. Then $\varphi.\psi$ is a lift of f on $[0,1]$. As $1\psi = (1/2)\varphi + m/2 = 0\varphi + m$, the loop f has odd degree m.//

Theorem 24 (Borsuk-Ulam Theorem, or Ham Sandwich Theorem) *Let $f: S^2 \to \mathbb{R}^2$ be a map such that $(-p)f = -(pf)$ for all p. Then there is some q such that $qf = (0,0)$.*

Proof Suppose not. Then the retraction of $\mathbb{R}^2 - \{(0,0)\}$ onto S^1 gives a map $g: S^2 \to S^1$ such that $(-p)g = -(pg)$ for all p. We can then define a map $h: D^2 \to S^1$ by $(x,y)h = (x, y, \sqrt{(1 - x^2 - y^2)})g$, and we find that $(-p)h = -(ph)$ for all $p \in S^1$. By Lemma 23 and the remark following it, the map h restricted to S^1 will then have odd degree. But this loop is freely homotopic to the constant loop, since h is defined on D^2, and so its degree is 0. This contradiction shows that f must take on the value (0,0) somewhere.//

The corresponding result for maps from S^n to $\mathbb{R}^n$ also holds. For $n=1$ it is just a version of the Intermediate Value Theorem; for general n a significantly more difficult proof (for instance, involving cohomology rings) is needed.

I now explain why the name "Ham Sandwich Theorem" is used. First consider a nice set A in the plane. (The precise definition of a "nice" set would require some knowledge of measure theory; what we need is that some intuitive properties of area and volume are valid for the sets we look at.) Perpendicular to any direction there is a line which divides A into two parts of equal area. To see this, consider the difference between the areas of the portion of A above the line $x\cos\alpha + y\sin\alpha = c$ and the portion below it. This is a continuous function of c, which is negative for large positive c and positive for large negative c. Thus it must be zero somewhere. For nice A there will be only one such line, which we call $l(\alpha)$.

This gives c as a function of α. This function can be shown to be continuous (since, if we take a disc containing A, the lines corresponding to α and β must meet inside this disc).

Now take a second nice set B in the plane. Consider the area of the portion of B above $l(\alpha)$ minus the area of the portion of B below it. This will be a continuous function of α. If we replace α by $\alpha+\pi$ the value of this function will be multiplied by -1, since $l(\alpha+\pi)$ is the same line as $l(\alpha)$, but "above" and "below" get interchanged. By the Intermediate Value Theorem, this function is zero somewhere. This means that there is a line which simultaneously bisects A and B.

We now extend these results to sets in $\mathbb{R}^3$. Let A be a nice set in $\mathbb{R}^3$. We can regard a direction u as a point of S^2. As before, perpendicular to any direction there is a unique plane, which we call $p(u)$, dividing A into two pieces of equal volume, and this plane varies continuously with the direction.

Let B and C be two further nice subsets of $\mathbb{R}^3$. We get a continuous function $f:S^2 \to \mathbb{R}^2$ whose components are the volume of B and C respectively above $p(u)$ minus the volume of the same set below it. Since $p(-u)$ is the same plane as $p(u)$, but with above and below interchanged, $(-u)f = -(uf)$. Hence the theorem tells us that there is some direction perpendicular to which there is a plane dividing A,B, and C simultaneously into two equal halves. Now take the sets to be the two pieces of bread and the filling in a ham sandwich!

Exercise 11 Let f and g be loops in S^1 of the same degree, but with different basepoints. Show that f is freely homotopic to g.

Exercise 12 Let f and g be maps of S^1 to itself. Show that the degree of their composition is the product of their degrees. (You may want to take maps freely homotopic to f and g for which the result is easy.)

Exercise 13 Show that a homeomorphism of S^1 has degree ±1.

Exercise 14 Let U be an open neighbourhood of z_0 in $\mathbb{C}$, and let f be a map from $U - \{z_0\}$ to $\mathbb{C} - \{0\}$. Define f_r on S^1 by $zf_r = (z_0 + rz)f$ for positive real r. Show that the degree of f_r is constant for r small enough. We will call this the *order of f at* z_0.

Exercise 15 Let g be a map from an open neighbourhood of z_0 to $\mathbb{C}$ such that $z_0 g \neq 0$. Let f be given by $zf = (z - z_0)^k (zg)$ for some $k \in \mathbb{Z}$. Show that the order of f at z_0 is k.

Exercise 16 Let $z_1, \dots, z_n$ be points of $D^2 - S^1$. Let r be a small enough positive real number. Let c_k be the loop given by $zc_k = (z_k + rz)$ for $z \in S^1$, and let c_0 be the loop given by $zc_0 = z$. Show that c_0 is homotopic in $D^2 - \{z_1, \dots, z_n\}$ to the product $\Pi g_k.c_k.\bar{g}_k$ for some paths g_k. (You may want to break $D^2 - \{z_1, \dots, z_n\}$ into pieces and use induction with the help of homeomorphisms. Or you could break D^2 minus some small discs into the union of star-shaped sets. The proof may be rather messy.)

Exercise 17 Let $z_1, \dots, z_n$ be points of $D^2 - S^1$, and let f be a map from $D^2 - \{z_1, \dots, z_n\}$ to $\mathbb{C} - \{0\}$. Suppose that f has an order at each z_k. Show that the degree of the restriction of f to S^1 is the sum of the orders of f at the points $z_1, \dots, z_n$. (Use the previous exercise.)

Exercise 18 Let A be a set which has two binary operations $\times$ and $*$, each of which has a two-sided identity. Suppose that $(w \times x) * (y \times z) = (w * y) \times (x * z)$ for all w, x, y, and z. Show that $\times$ and $*$ coincide, and that the operation is commutative and associative. (Begin by proving that the identities for the two operations must be the same.)

Exercise 19 Use the preceding exercise to show, for standard loops f and g in S^1 and their product fg defined by $r(fg) = (rf)(rg)$ for $r \in I$, that $\deg(fg) = \deg f + \deg g$.

4.5 *JOINS AND WEAK JOINS*

In this section we consider some matters of topological interest which are not relevant to the group-theoretical applications. We begin by looking at a space closely related to the join of spaces and discuss aspects of its fundamental group, which is very complicated. From this we are able to obtain two contractible spaces whose join is not even simply connected. We then obtain some new conditions under which the fundamental group of a join is the free product of the fundamental groups of the spaces.

Let X be the union of the spaces X_i, with $X_i \cap X_j = \{p\}$ for $i \neq j$, and let $\{p\}$ be closed in each X_i. The join topology on X can be given the following description. If $x \neq p$ there is exactly one i with $x \in X_i$, and the open neighbourhoods of x in X_i provide a basis for the neighbourhoods of x in X. A basis for the neighbourhoods of p in X is given by the sets $\cup U_i$ where the U_i are open neighbourhoods of p in X_i.

When the index set is infinite, there is another interesting topology on X, called the *weak join* of the X_i. We take the same neighbourhood basis as before for points other than p. But the basis for the neighbourhoods of p is now to consist of those $\cup U_i$ with U_i an open neighbourhood of p in X_i and, in addition, $U_i = X_i$ for all but finitely many i.

The subspace of $\mathbb{R}^2$ which is the union of the circles S_n, where $S_n = \{(x,y); (x-1/n)^2+y^2 = 1/n^2\}$, is often called the *Hawaian earring* (look at the picture!). It is easily seen to be the weak join of countably many circles.

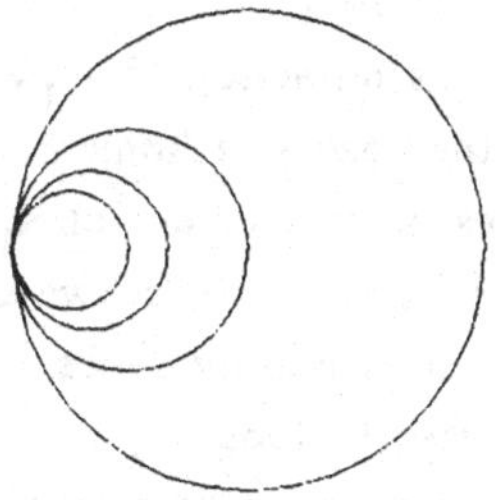

Figure 13, showing (part of) the Hawaian earring

Let X be the weak join of countably many spaces X_i. Then $\pi_1(X)$ is complicated. It is easy to see that it contains standard loops f such that $(1-1/n)f = p$ for all n, with f mapping $[1-1/n, 1-1/(n+1)]$ into X_n. We could regard such an element as amounting to an infinite product $\alpha_1\alpha_2\ldots$ with $\alpha_n \in \pi_1(X_n, p)$ for all n. But $\pi_1(X)$ would also contain products of the form $\ldots\alpha_2\alpha_1$, which are inverses of the previous infinite products, as well as more complicated elements obtained by multiplication from these.

We will now proceed to give some properties of $\pi_1(X)$ when X is the weak join of reasonably nice spaces. We shall assume that we have only countably many spaces X_n, each of which has non-trivial fundamental group. We assume that $\{p\}$ is closed in each X_n, and that the spaces are such that the fundamental group of the join of any finite number of them is the free product of the fundamental groups of the corresponding spaces.

Let X_+ be the weak join of those X_n with n even, and let X_- be the weak join of those X_n with n odd. Then X is the join of X_+ and X_-. Let Y_+ be the cone on X_+ and let Y_- be the cone on X_-. Let Y be the join at p of Y_+ and Y_-. Then Y_+ and Y_- are contractible. Applying Proposition 9 twice, we find that $\pi_1(Y)$ is obtained from $\pi_1(X)$ by cancelling $\pi_1(X_+)$ and $\pi_1(X_-)$. We shall show that this group is uncountable.

Let $G_n = \pi_1(X_n)$, which we are assuming non-trivial. Let H_n, H_{n+}, and H_{n-} be the free products, respectively, of all G_i with $i \leq n$, of those G_i with $i \leq n$ and i even, and of those G_i with $i \leq n$ and i odd.

There is a homomorphism from H_{n+1} onto H_n which is trivial on G_{n+1} and which maps H_n identically. Let H be the group consisting of all infinite sequences $(h_1, h_2, \ldots)$ such that, for all n, $h_n \in H_n$ and h_n is the image of h_{n+1} in the homomorphism of the previous sentence, with the multiplication being componentwise. This group is known as the *inverse limit* of the H_n. The subgroup H_+ consists of those sequences for which every h_n is in H_{n+}, and similarly for H_-. The subset (not subgroup) L consists of those sequences for which $h_n = g_1 g_2 \ldots g_n$ with $g_i \in G_i$ for all i.

Now X has an obvious retraction to $X_1 \vee \ldots \vee X_n$ for all n. This provides, by our assumption on the spaces, a homomorphism from $\pi_1(X)$ to H_n for all n. The sequence of images of an element of $\pi_1(X)$ under these homomorphisms is evidently an element of H, so we obtain a homomorphism from $\pi_1(X)$ to H. This homomorphism plainly sends $\pi_1(X_+)$ into H_+ and $\pi_1(X_-)$ into H_-. An element of L gives rise to loops $\alpha_n \in X_n$ for all n, and the infinite product $\alpha_1 \alpha_2 \ldots$ discussed above is an element of $\pi_1(X)$ which maps to the given element of L.

Hence our description of $\pi_1(Y)$ shows that it is uncountable if we can find uncountably many elements of L, no two of which are in the same coset of the subgroup $H_+ H_- [H,H]$. The proof of this will take some time.

Fix n for the moment. Consider an element u of H_n. Then u is a product $x_1 \ldots x_k$ with each x_i in some G_j, $x_i \neq 1$, and x_i and x_{i+1} in different factors. If $x_i \in G_p$ and $x_{i+1} \in G_q$ we say that (p,q) *occurs at* i. Let $u\alpha_{pq}$ be the number of occurrences of (p,q), and let β_{pq} be $\alpha_{pq}-\alpha_{qp}$. Finally, let $u\lambda$ be $|\{r;\, u\beta_{2r-1,2r}\neq 0\}|$.

Let $h\in H$ have component $h_n\in H_n$. We shall show that $h_n\lambda$ is a bounded function on n if $h\in H_+H_-[H,H]$. We shall find an uncountable subset M of L such that $h_n\lambda$ is not a bounded function of n if $1\neq h\in M^{-1}M$. This will prove the result. We begin by obtaining some properties of α and β.

Plainly $u^{-1}\alpha_{pq}=u\alpha_{qp}$, and so $u^{-1}\beta_{pq}=-u\beta_{pq}$.

Given $u_1,\ldots,u_s$ in H_n let γ_{pq} (which is a function of $u_1,\ldots,u_s$, but we shall omit these arguments) be given by $\gamma_{pq}=(u_1\ldots u_s)\beta_{pq}-\Sigma_i\, u_i\beta_{pq}$. Then we will show that $\Sigma_{p,q}|\gamma_{pq}|\leq 6(s-1)$.

This follows by an easy induction from the case $s=2$, which we now prove. So let $u=x_k\ldots x_1$ (note the order) and let $v=y_1\ldots y_m$. There is some j so that $x_j\ldots x_1=(y_1\ldots y_j)^{-1}$ but $x_{j+1}\neq y_{j+1}{}^{-1}$. Let $x_{j+1}\in G_a$, $x_j\in G_b$ (and so $y_j\in G_b$), and $y_{j+1}\in G_c$. Then $a\neq b\neq c$; we may or may not have $a=c$. Let u' be $x_k\ldots x_{j+1}$, $u''=x_j\ldots x_1$, $v'=y_1\ldots y_j$, and $v''=y_{j+1}\ldots y_m$.

It is easy to check that $u\alpha_{ab}=u'\alpha_{ab}+u''\alpha_{ab}+1$, while $u\alpha_{pq}=u'\alpha_{pq}+u''\alpha_{pq}$ for all other pairs (p,q). Similarly for v, the special pair being (b,c). Also, if $a\neq c$ we have $(uv)\alpha_{ac}=u'\alpha_{ac}+v''\alpha_{ac}+1$, while for all pairs (p,q), except (a,c) when $a\neq c$, we have $(uv)\alpha_{pq}=u'\alpha_{pq}+v''\alpha_{pq}$. As $u''=v'^{-1}$, we have $u''\alpha_{pq}=v'\alpha_{qp}$ for all (p,q). Combining these gives the result.

Now suppose that $h=xy\Pi_1^s\,[z_i,w_i]$, where $x\in H_+$, $y\in H_-$, and z_i and w_i are in H. Then, with h_n, x_n, etc, being the components of h, x. etc, in H_n, we have

$$h_n=x_ny_nz_{1n}{}^{-1}w_{1n}{}^{-1}z_{1n}w_{1n}\ldots z_{sn}{}^{-1}w_{sn}{}^{-1}z_{sn}w_{sn}.$$

Plainly $\alpha_{2r-1,2r}$ and $\alpha_{2r,2r-1}$ are both zero on H_{n+} and on H_{n-}, so the same holds for $\beta_{2r-1,2r}$. The formula for β on an inverse, and the formula relating β on a product to β on the individual terms and to γ, now tells us that $h_n\beta_{2r-1,2r}=\gamma_{2r-1,2r}$, where $\Sigma_{p,q}|\gamma_{pq}|\leq 6(4s+1)$. Hence $h_n\lambda\leq 6(4s+1)$ for all n.

Choose a_n to be a non-identity element of G_n for all n. An element of L is given by a sequence $\{g_n\}$ where $g_n\in G_n$ for all n. Let S be an infinite set of positive integers. Let m_S be the element of L whose entry in G_{2n} is a_{2n} if $n\in S$ and is 1 if $n\notin S$ while its entry in G_{2n-1} is a_{2n-1} if $n\in S$ and is 1 if $n\notin S$. Let $h=m_S{}^{-1}m_T$, where T is another infinite set. If k is the smallest integer which is in exactly one of S and T, it is easy to check that in computing h_n cancellation cannot reach beyond G_{2k-2}. We find easily that $h_n\beta_{2r-1,2r}\neq 0$ whenever $2r\leq n$ and r is in exactly one of S and T. Hence $h_n\lambda$ is an unbounded function of n if the set of elements in exactly one of S and T is infinite.

So we want to find uncountably many subsets of the positive integers, any two of which are either equal or have finite intersection. The easiest way to show this is to prove the same result for the rational numbers, and then use the bijection between the positive integers and the rationals. To each real number r we take an (increasing) sequence of rationals which tends to r, and let S_r be the set of members of this sequence. Then the sets S_r for all r have the required property.

It is possible to show that the homomorphism from $\pi_1(X)$ to H is one-one. Also the image of this homomorphism consists of those h for which, for every r, the number of occurrences of an element of G_r in h_n is a bounded function of n. We shall not prove these results, which are difficult.

We now investigate circumstances under which the fundamental group of a join of spaces is the free product of the fundamental groups of the spaces

The space X is called *almost simply connected at p* if $X - \{p\}$ is the union of disjoint open sets U_i (for i in some index set) such that each $U_i \cup \{p\}$ is simply connected. Such a space is plainly pathwise connected. Also $\{p\}$ must be closed. We call X *locally almost simply connected at p* if $\{p\}$ is closed and each open neighbourhood of p contains an almost simply connected at p open neighbourhood of p. Notice that the join at p of two spaces both (locally) almost simply connected at p is itself (locally) almost simply connected at p. As remarked above, in contrast, the join of two simply connected spaces need not be simply connected.

We say that X is *first countable at p* if there are open neighbourhoods U_n of p (for n a positive integer) such that every open neighbourhood of p contains U_n for some n. The join of two spaces both first countable at p is itself first countable at p.

Let $f:[0,r] \to X$ be a loop based at p, where $\{p\}$ is closed. Then $(X - \{p\})f^{-1}$ is an open subset of $(0,r)$. It will then be the union of countably many disjoint open intervals. This holds because the components of an open set must be disjoint connected open sets; that is, they must be disjoint open intervals. There can only be countably many of them, since each interval contains a rational number, and no rational number is in two of the intervals. Note that the endpoints of these intervals must be in pf^{-1}. If there are only finitely many such intervals we say that f is *geometrically finite*.

Lemma 25 *Let X be a space which is first countable at p and locally almost simply connected at p. Suppose that $\{p\}$ is the intersection of open sets. Then every loop based at p is homotopic to a geometrically finite loop. Also X will be simply connected if it is almost simply connected.*

Proof We may as well assume that f is a standard loop. Let $(X - \{p\})f^{-1}$ be the union of the disjoint open intervals J_i.

We first show that for any open neighbourhood U of p the set Uf^{-1} contains all but finitely many of the J_i. Suppose not. Then we can find, for all n, distinct $a_n < c_n < b_n$ with $c_n f \notin U$ and (a_n, b_n) being one of the intervals J_i, so that $a_n f = p$. Taking subsequences if necessary, we may assume that the sequence a_n converges; let a be its limit. Since the intervals do not overlap, $\sum(b_n - a_n) \leq 1$, so that $b_n - a_n \to 0$. Hence the sequence c_n also converges to a. But this contradicts continuity of f at a, since $a_n f = p$ and $\{p\}$ is closed, while $c_n f$ is in the closed set $X - U$ which does not contain p.

The hypotheses ensure that there are open neighbourhoods U_n of p, such that each U_n is almost simply connected, every neighbourhood of p contains U_n for some n, and such that $U_{n+1} \subseteq U_n$ and $\cap U_n = \{p\}$. Let U_0 be X.

Take an interval J_i. Then $J_i f$ is contained in U_0 but is not contained in $\cap U_n$. There must then be some n_i such that $J_i f$ is contained in U_n iff $n \leq n_i$. In particular, the result we have just shown tells us that $n_i = 0$ for only finitely many i. If $n_i \neq 0$ then $U_{n_i} - \{p\}$ is the union of disjoint open sets each of which has its union with $\{p\}$ simply connected. Now $J_i f$, being connected, lies in one of these sets, U' say, and $\overline{J_i} f \subseteq U' \cup \{p\}$, so that the restriction of f to $\overline{J_i}$ is homotopic in U_{n_i}, relative to the endpoints, to the constant map. If X itself is almost simply connected this holds for $n_i = 0$ also. Let F_i be such a homotopy.

Now define F on $I \times I$ as follows. If $tf = p$ then $(t,u)F = p$ for all u. If $t \in \overline{J_i}$ with $n_i \neq 0$ then $(t,u)F = (t,u)F_i$; when X is almost simply connected this definition is used even when $n_i = 0$. When X is not almost simply connected and $t \in \overline{J_i}$ with $n_i = 0$ then $(t,u)F = tf$ for all u. This does define a function, since when t is an endpoint of J_i we have $tf = p$ and $(t,u)F_i = p$. If we can show that F is continuous it will provide a homotopy between f and the loop sending t to $(t,1)F$. By definition, $(t,1)F = p$ unless $t \in J_i$ with $n_i = 0$ (and everywhere when X is almost simply connected). Hence this loop is geometrically finite, as required.

Now F is plainly continuous at any (t,u) with $t \in J_i$ for some i, since F_i and f are continuous. So let $tf = p$, so that $(t,u)F = p$. Let V be any neighbourhood of p. Take n with $U_n \subseteq V$. Then for all but finitely many i we have $J_i f \subseteq U_n$, and hence $n_i \geq n$. In particular, there is some δ such that $(t - \delta, t + \delta)$

meets only those J_i with $n_i \geq n$ and those (at most two) J_i with t as an end-point. Then, for $t' \in (t-\delta, t+\delta)$ and any u', either $t'f = p$ and so $(t',u')F = p$, or $t' \in J_i$ with $n_i \geq n$, or (for at most two values of i) both t and t' are in $\bar{J}_i$. In the second case, $(t',u')F = (t',u')F_i \in U_{n_i} \subseteq U_n \subseteq V$. In the last case, the continuity of F_i on $\bar{J}_i$ shows that $(t',u')F \in V$ if (t',u') is close enough to (t,u). This shows F is continuous.//

Proposition 26 *Let X be the join at p of the spaces X_i. Let each X_i be locally almost simply connected at p and first countable at p, and let $\{p\}$ be the intersection of open sets in each X_i. If the index set is finite, or if each X_i is a T_1-space then $\pi_1(X,p)$ is the free product of the $\pi_1(X_i,p)$.*

Remark If a space is T_1 then $\{p\}$ is the intersection of all the open sets containing p.

Proof The usual direct limit argument (Proposition 10) gives the result for infinitely many spaces when we have proved it for finitely many. Since the join of two spaces with the stated properties also has these properties, the result holds by induction for any finite number of spaces once we have proved it for two spaces.

So let X be $X_1 \vee X_2$, with the conditions holding for X_1 and X_2, and so holding for X. We already know that the inclusion induces monomorphisms of $\pi_1(X_1,p)$ and $\pi_1(X_2,p)$ into $\pi_1(X,p)$. By Lemma 25, any element of $\pi_1(X,p)$ can be represented by a geometrically finite loop. Such a loop is the product of loops in X_1 and loops in X_2. Hence $\pi_1(X)$ is generated by the images of $\pi_1(X_1)$ and $\pi_1(X_2)$.

In particular, if X_2 is simply connected (which, by Lemma 25, holds if it is almost simply connected) then the inclusion induces an isomorphism from $\pi_1(X_1)$ to $\pi_1(X)$.

In the general case, there are almost simply connected neighbourhoods U_1 and U_2 of p in X_1 and X_2 respectively. So the previous paragraph tells us that the inclusions induce isomorphisms of $\pi_1(X_1)$ in $\pi_1(X_1 \vee U_2)$ and of $\pi_1(X_2)$ in $\pi_1(U_1 \vee X_2)$, while $U_1 \vee U_2$ is simply connected. The result now follows by van Kampen's Theorem.//

5 COMPLEXES

5.1 *GRAPHS*

A *graph* Γ consists of two disjoint sets, the set $V(\Gamma)$ of *vertices* of Γ and the set $E(\Gamma)$ of *edges* of Γ, together with two functions $\sigma: E(\Gamma) \to V(\Gamma)$ and $^{-}: E(\Gamma) \to E(\Gamma)$ such that, for all $e \in E(\Gamma)$, $e \neq \bar{e}$ and $e = \bar{\bar{e}}$. We define another function $\tau: E(\Gamma) \to V(\Gamma)$ by $e\tau = \bar{e}\sigma$. We call $e\sigma$ the *start* (or *beginning*) of e and $e\tau$ the *end* (or *finish*) of e, referring to them both as *end-points* of e. We call $\bar{e}$ the *inverse* of e. We call e a *circle* if $e\tau = e\sigma$ (such an edge is often called a "loop", but this name could lead to confusion later). We simply refer to E and V, rather than $E(\Gamma)$ and $V(\Gamma)$, where no confusion is likely. It is sometimes convenient to allow a graph to be empty, and sometimes inconvenient; I will leave it to the reader to decide which statements require the graphs considered to be non-empty.

An alternative definition is often used, in which edges do not come in mutually inverse pairs. This approach is sometimes easier, but at some stage we would need to introduce some version of the inverse of an edge. It is really a matter of personal preference which definition to use.

Our definition does differ from other definitions of a graph in two further ways. We permit edges to have both end-points the same. We also permit more than one edge with a given starting point and finishing point. We will find it very useful to allow this generality.

A *path of length* $n > 0$ in Γ is a finite sequence $e_1, \ldots, e_n$ of edges such that $e_i\tau = e_{i+1}\sigma$ for all $i < n$. This path *starts* at $e_1\sigma$ and *ends* at $e_n\tau$, and the *vertices of the path* are $e_i\sigma$ for all $i \leq n$ and also $e_n\tau$. A path of length 0 is just a vertex v, and v is the start and end of this path. A *loop* is a path which ends at the point at which it starts. A path is *simple* if its vertices are all distinct. A *simple loop* or *circuit* is a loop of positive length for which the n vertices $e_i\sigma$ are all distinct, and which is not of the form $e, \bar{e}$. A single edge is a circuit iff it is a circle. If $e_1, \ldots, e_n$ is a circuit or a loop then so are all its cyclic permutations $e_i, e_{i+1}, \ldots, e_n, e_1, \ldots, e_{i-1}$. When it is necessary to distinguish these paths and loops from the topological ones we shall call them *edge-paths* and *edge-loops*.

Let f be the path $e_1,\ldots,e_n$ and let g be the path $e'_1,\ldots,e'_m$. When $e'_1\sigma = e_n\tau$ we define the product $f.g$ of f and g to be the path $e_1,\ldots,e_n,e'_1,\ldots,e'_m$. When $e_1\sigma$ is v and $e_n\tau$ is w, we also define $v.f$ and $f.w$ to be f. The *inverse* $\bar{f}$ of f is $\bar{e}_n,\ldots,\bar{e}_1$; if the path is a vertex v, then the inverse of f is also v.

A *subgraph* Γ_1 of Γ consists of subsets V_1 of V and E_1 of E such that, for any $e \in E_1$, we have $\bar{e} \in E_1$ and $e\sigma \in V_1$; evidently $e\tau$ is also in V_1, and Γ_1 is a graph. Evidently the intersection and the union of arbitrary subgraphs of Γ are themselves subgraphs of Γ. When V_1 is a subset of V, the *full subgraph* on V_1 is the subgraph whose vertex set is V_1 and whose edge-set E_1 is given by $E_1 = \{e; e\sigma$ and $e\tau$ are in $V_1\}$.

We say that Γ is *connected* if given any two vertices of Γ there is a path joining them. More generally, for any Γ we can define a relation on V by requiring v to be related to w iff there is a path starting at v and ending at w. This is easily seen to be an equivalence relation. The full subgraphs whose vertex sets are the equivalence classes are called the *components* of Γ.

Note that the union of connected graphs Γ_i will be connected provided that $\cap\Gamma_i \neq \emptyset$. We need only take some $v_0 \in \cap\Gamma_i$ and observe that every vertex of $\cup\Gamma_i$ can be joined to v_0 by a path in some Γ_i, so that the component of v_0 in $\cup\Gamma_i$ contains every vertex.

Lemma 1 *Let Γ be connected. Let S be a non-empty subset of V such that, for any edge e, if $e\sigma$ is in S then $e\tau$ is also in S. Then $S = V$.*

Proof Take $v \in S$. Let w be any vertex. Since Γ is connected, there is a path $e_1,\ldots,e_n$ from v to w. Since $e_1\sigma = v$ is in S, an easy induction shows that the end-points of all e_i are in S. In particular, $w = e_n\tau$ is in S. As w was an arbitrary vertex, we see that $S = V$.//

A path is called *reducible* if there is some i such that $e_{i+1} = \bar{e}_i$; otherwise it is called *irreducible*. Evidently a reducible path is not simple. If $n > 2$ and the path is reducible then $e_1,\ldots,e_{i-1},e_{i+2},\ldots,e_n$ is also a path. In particular, if two vertices can be joined by a path (equivalently, if they are in the same component) then they can be joined by an irreducible path.

A *forest* is a graph all of whose loops of positive length are reducible. A *tree* is a connected forest. Plainly any subgraph of a forest is a forest. It is also clear that a graph is a forest iff all its components are trees.

Lemma 2 *A graph is a forest iff it contains no circuits.*

Proof Plainly a forest contains no circuits. Suppose that Γ is not a forest. Then it contains an irreducible loop of positive length $e_1,\ldots,e_n$. Take r and s with $s-r$ as small as possible subject to $1\le r\le s\le n$ and $e_r\sigma=e_s\tau$. (Such r and s exist, since $e_1\sigma=e_n\tau$.) Then $e_r,e_{r+1},\ldots,e_s$ is a circuit, since it is a loop of positive length by definition of r and s, it is irreducible because the original loop is irreducible, and the beginnings of its edges are distinct by the choice of r and s.//

Lemma 3 *A graph is a forest iff, for all vertices v and w, there is at most one irreducible path from v to w.*

Proof If the graph is not a forest it contains a circuit $e_1,\ldots,e_n$. Then $e_1,\ldots,e_{n-1}$ and $\bar{e}_n$ are both irreducible paths from $e_1\sigma$ to $e_n\sigma$ (when $n=1$ the first of these paths is a single vertex).

Now suppose that the graph contains two distinct irreducible paths joining some pair of vertices. Take irreducible paths f and g which start at the same point and end at the same point and with lengths n and m such that $m+n$ is as small as possible. Let f be $e_1,\ldots,e_n$ and let g be $e'_1,\ldots,e'_m$. If $m=0$ and $n=1$ then f is a circle, and hence f is a circuit. If $m=0$ and $n>1$, then f is itself an irreducible loop.

If we had $e_n=e'_m$ then the paths $e_1,\ldots,e_{n-1}$ and $e'_1,\ldots,e'_{m-1}$ would be distinct irreducible paths from $e_1\sigma$ to $e_n\sigma$. This would contradict the minimality of $m+n$, and so $e_n\ne e'_m$. The path $f.\bar{g}$ is therefore an irreducible loop.//

Lemma 4 *If* Γ *is a tree then any two vertices of* Γ *can be joined by exactly one irreducible path. Conversely, if* Γ *has a vertex* v_0 *such that* v_0 *can be joined to every vertex by exactly one irreducible path then* Γ *is a tree.*

Proof Let Γ be a tree. As Γ is connected, any two vertices may be joined by an irreducible path. As Γ is a forest, there cannot be more than one irreducible path joining two given vertices.

Conversely, suppose that the vertex v_0 can be joined to every vertex by exactly one irreducible path. Then Γ is connected, since every vertex is in the component of v_0. Suppose that Γ is not a tree, so that it contains a circuit $e_1,\ldots,e_n$. Let f be the shortest path from v_0 to a vertex of this circuit. Since cyclic permutations of circuits are circuits, we may assume f ends at $e_1\sigma$.

Also, the last edge of f cannot be e_n, else we could omit this edge from f and get a path from v_0 to $e_n\sigma$ which is shorter than f. For the same reason, the last edge of f is not $\bar{e}_1$. Let g be the path $e_1,\ldots,e_{n-1}$. Then $f.g$ and $f.\bar{e}_n$ are two distinct irreducible paths from v_0 to $e_n\sigma$.//

Lemma 5 *Let T be a tree. Let T_i be subtrees of T (for i in some index set) such that $\cap T_i \neq \emptyset$. Then both $\cup T_i$ and $\cap T_i$ are trees.*

Proof Since each T_i is connected and $\cap T_i \neq \emptyset$, we know that $\cup T_i$ is connected. As a subgraph of a tree, $\cup T_i$ must be a forest. Hence it is a tree.

As a subgraph of a tree, $\cap T_i$ must be a forest. We must show that it is connected. Take two vertices v and w of $\cap T_i$. For each i there is an irreducible path f_i from v to w in T_i. Since T is a tree, there is only one irreducible path in T from v to w. Hence the f_i for differing i are all the same, and they are therefore a path in $\cap T_i$ from v to w.//

Lemma 6 *Let T_i be trees such that there is a vertex v with $T_i \cap T_j = \{v\}$ for $i \neq j$. Then $\cup T_i$ is a tree.*

Proof As $\cap T_i \neq \emptyset$, $\cup T_i$ is connected. Suppose that it contains a circuit. Taking a cyclic permutation if necessary, we may assume that the circuit is $e_1,\ldots,e_n$ and that $e_r\sigma \neq v$ except possibly for $r = 1$. For $r < n$, if $e_r \in T_i$ and $e_{r+1} \in T_j$ we must have $i = j$. For otherwise $e_{r+1}\sigma = e_r\tau$ would be in $T_i \cap T_j$, which is $\{v\}$. It then follows that the circuit must lie completely in one T_i. This is impossible, as T_i is a tree.//

A *maximal forest* in a graph Γ is a subgraph F of Γ such that F is a forest and no subgraph of Γ which properly contains F is a forest.

Proposition 7 *Let the forest F be a subgraph of* Γ. *Then there is a maximal forest of* Γ *which contains F. In particular (take F empty) every graph contains a maximal forest.*

Proof Suppose that we have a collection of forests F_i such that for all i and j either $F_i \subseteq F_j$ or $F_j \subseteq F_i$. Then $\cup F_i$ is a forest. Otherwise the union contains a circuit. This circuit would be in the union of finitely many of the F_i, and the assumed property ensures that any finite union equals one of them.

Thus some F_i would contain a circuit, which is impossible.

The result now follows by Zorn's Lemma.//

Proposition 8 *Let F be a maximal forest in Γ, and let Γ_1 be a component of Γ. Then F contains every vertex of Γ. Also $F \cap \Gamma_1$ is a maximal forest of Γ_1 and is a tree.*

Proof If there were a vertex v of Γ not in F then $F \cup \{v\}$ would be a larger forest.

$F \cap \Gamma_1$ is a forest in Γ_1, being a subgraph of F. If it were not a maximal forest in Γ_1 it would be properly contained in a maximal forest F_1 of Γ_1. Then $F \cup F_1$, being the disjoint union of F_1 and $F - \Gamma_1$, would be a forest of Γ properly containing F.

We have to show that $F \cap \Gamma_1$ is connected. By Lemma 1, it is enough to show that any edge of Γ_1 has both its vertices in the same component of F. Suppose not. Take an edge e such that $e\sigma$ and $e\tau$ are in different components.

We shall show that $F \cup \{e, \bar{e}\}$ is a forest. It is certainly a graph, since F contains all the vertices of Γ. If it contained a circuit, we could assume (taking cyclic permutations, if necessary) that the circuit is $e_1, \dots, e_n$ where e_r is not e or $\bar{e}$ if $r > 1$. If e_1 is not e or $\bar{e}$ we would have a circuit in F. If e_1 is e or $\bar{e}$ then the path $e_2, \dots, e_n$ would show that $e\sigma$ and $e\tau$ are in the same component of F, contrary to assumption.

Since F is a maximal forest, we have a contradiction, and so $F \cap \Gamma_1$ must be connected.//

In particular, a maximal forest in a connected graph is a tree (which we refer to as a *maximal tree*). It is remarkable that a maximal forest contains only one tree! The following corollary is immediate.

Corollary *Any forest in a connected graph is contained in a maximal tree.*

Lemma 9 *Let Γ be a graph, and let T be a tree in Γ which contains every vertex. Then T is a maximal tree.*

Proof Let Δ be a subgraph of Γ wich strictly contains T. Since every vertex is in T, there must be an edge e with $e \in \Delta$ and $e \notin T$. Then Δ contains two distinct irreducible paths from $e\sigma$ to $e\tau$, one consisting of the edge e only, the other being a path in T. Hence Δ is not a tree.//

Lemma 10 *Let T be a finite tree with more than one vertex. Then T has (at least) two vertices each of which is the start of exactly one edge (and the end of exactly one edge).*

Proof Since T is finite, there is a bound on the lengths of paths all of whose vertices are distinct. Let $e_1,\ldots,e_n$ be the longest such irreducible path. Suppose that there is an edge $e \neq \bar{e}_n$ with $e\sigma = e_n\tau$. If $e\tau = e_r\sigma$ for some σ we would have an irreducible loop $e_r,\ldots,e_n,e$. This is impossible as T is a tree. It follows that $e_1,\ldots,e_n,e$ is also an irreducible path with distinct vertices, contradicting our choice of n. Hence e_n and $\bar{e}_n$ are the only edges of which $e_n\tau$ is a vertex, so that $e_n\tau$ is one of the required vertices. Similarly $e_1\sigma$ is the other vertex required.//

Proposition 11 *Let* Γ *be a finite connected graph with n vertices and m edge-pairs. Then* Γ *is a tree iff* $n = m+1$.

Proof Let Γ be a tree. If $n = 1$ then $m = 0$ (since any edge would be a circle). Otherwise let v be a vertex which is the start of exactly one edge, which we call e. Let Δ be the graph obtained from Γ by deleting the edges e and $\bar{e}$ and the vertex v. (This is a graph, since no other edge has v as a vertex.) Then Δ is a forest. If it is a tree, the result will follow by induction.

Take any vertices u and w of Δ. They can be joined by an irreducible path $e_1,\ldots,e_n$ in Γ. Suppose that there is some r with $e_r\tau = v$. Then $e_{r+1}\sigma = v$. Since e is the only edge starting at v, we must have $e_{r+1} = e = \bar{e}_r$. The path is then reducible, contrary to hypothesis. Hence the path does not go through the vertex v, and so it is a path in Δ, as required.

Now let Γ be a finite connected graph with $m = n-1$. Let T be a maximal tree in Γ. Since T contains every vertex of Γ, it has n vertices. By what we have just proved, it must contain $n-1$ edge-pairs. By hypothesis, these will be all the edge-pairs of Γ, and so $T = \Gamma$.//

Let T_λ, for λ in the index set Λ, be subtrees of the graph Γ such that $T_\lambda \cap T_\mu = \emptyset$ for $\lambda \neq \mu$. Assume that every vertex is in some T_λ (if this did not hold we would just add further trees, each consisting of one vertex). Construct a graph Δ as follows. The vertex set of Δ is to be Λ, and the edge set of Δ is to consist of those edges of Γ which are not in $\cup T_\lambda$. The map $^-$ on Γ evidently provides a map $^-$ on Δ. When e is an edge of Δ whose start, regarded as an edge in Γ, lies in T_λ we define the start of e as an edge of Δ to be λ. We say that Δ is obtained from Γ by *contracting the trees* T_λ.

Lemma 12 *Let Δ be obtained from Γ by contracting the subtrees T_λ. Then Δ is connected iff Γ is connected, and Δ is a forest iff Γ is a forest.*

Proof Let $e_1,\ldots,e_n$ be a path in Γ. Suppose that, for some $r<s$, we have $e_r \notin \cup T_\lambda$ and $e_s \notin \cup T_\lambda$ but $e_i \in \cup T_\lambda$ for $r<i<s$. If $e_i \in T_\lambda$ and $e_{i+1} \in T_\mu$ then $e_i\tau = e_{i+1}\sigma$ is in $T_\lambda \cap T_\mu$. Hence $\lambda = \mu$, by our requirements on the subtrees. It follows that there is some ν such that T_ν contains $e_r\tau$ and $e_s\sigma$. We see from this that when we omit from the path $e_1 \ldots,e_n$ all the edges lying in $\cup T_\lambda$ the result is a path in Δ.

Suppose that the path in Γ is irreducible, and keep the same notation. We cannot have $e_s = \bar{e}_r$ with $s = r+1$, since the path in Γ is irreducible. And we cannot have $e_s = \bar{e}_r$ with $s > r+1$, as then $e_{r+1},\ldots,e_{s-1}$ would be an irreducible loop in T_ν. We deduce that the path in Δ is also irreducible.

Now let us start with an irreducible path $e_1,\ldots,e_n$ in Δ. Then for each i there is some μ with $e_i\tau$ and $e_{i+1}\sigma$, as vertices of Γ, both lying in T_μ. Hence there is an irreducible path p_i in T_μ from $e_i\tau$ to $e_{i+1}\sigma$. Then the path $q = e_1.p_1.e_2. \ldots .p_{n-1}.e_n$ is an irreducible path in Γ. More generally, let $e_1\sigma$ and $e_n\tau$ in Δ be α and β, and let v and w be vertices of Γ in T_α and T_β, respectively. Then there are paths p_0 and p_n in T_α and T_β such that $p_0.q.p_n$ is an irreducible path in Γ from v to w.

Let Γ be connected, and let α and β be vertices of Δ. Take vertices v and w of Γ in T_α and T_β. Then a path in Γ from v to w gives rise, by the above, to a path in Δ from α to β. Hence Δ is connected.

Let Δ be connected, and let v and w be vertices of Γ. Take α and β with $v \in T_\alpha$ and $w \in T_\beta$. Then there is a path in Δ from α to β, which gives rise, by the above, to a path in Γ from v to w.

If Γ is not a forest it contains an irreducible loop. The corresponding path in Δ will be an irreducible loop, and so Δ will not be a forest.

If Δ is not a forest there are vertices α and β such that there are two distinct irreducible paths in Δ from α to β. Take vertices v and w of Γ in T_α and T_β. These paths give rise to two distinct irreducible paths in Γ from v to w, and so Γ is not a forest.//

The remaining property of trees will be used only in the proof of Grushko's Theorem in Chapter 8. It is the beginning of homology theory.

Let Γ be a graph. Let C_0 be the free abelian group with basis the vertices of Γ. Let C_1 be obtained from the free abelian group with basis the edges of Γ by factoring out the subgroup generated by all $e+\bar{e}$. This may also be described as the free abelian group with one basis element for each edge-pair.

Let $\varepsilon: C_0 \to \mathbb{Z}$ be the homomorphism given by $v\varepsilon = 1$ for each vertex v. Let $\partial: C_1 \to C_0$ be the homomorphism given by $e\partial = e\tau - e\sigma$ for each edge e (it is easy to check that ∂ is defined on C_1). Plainly $\partial\varepsilon = 0$, and so $C_1\partial \subseteq \ker\varepsilon$.

Lemma 13 *Γ is connected iff $C_1\partial = \ker\varepsilon$, and Γ is a forest iff $\ker\partial = 0$.*

Proof Let Γ be connected, and let v and w be vertices. Take a path $e_1,\ldots,e_n$ from v to w. Then $e_1+\ldots+e_n$ is an element of C_1 which maps by ∂ to $w-v$. It is easy to check that $\ker\varepsilon$ is generated by the elements $w-v$ of C_0 for all vertices v and w. Hence $\ker\varepsilon = C_1\partial$,

Let Γ be disconnected, and let Δ be a component of Γ. It is easy to check that $\partial\varepsilon' = 0$, where ε' is the homomorphism given by $v\varepsilon' = 1$ if $v \in \Delta$ and $v\varepsilon' = 0$ otherwise. So $w-u$ is not in $C_1\partial$ if u is in Δ and w is not in Δ.

Suppose that Γ is not a forest. Then Γ contains a circuit $e_1,\ldots,e_n$. By the definition of a circuit, we cannot have $e_i = \bar{e}_j$. Hence $e_1+\ldots+e_n$ is a non-zero element of C_1 which maps by ∂ to 0.

Suppose that there is a non-zero element x of C_1 with $x\partial = 0$. Write x as $\Sigma m_i e_i$, with each $m_i \in \mathbb{Z}$. We may assume the e_i are all distinct and that each $m_i \neq 0$. We may also assume that we never have $e_i = \bar{e}_j$. For suppose that $e_i = \bar{e}_j$. We need only rewrite $m_i e_i + m_j e_j$ as $(m_i - m_j)e_i$ (note that the definition of a graph ensures that $i \neq j$), and similarly for other pairs, to get rid of this possibility. Finally, we can assume that each m_i is positive, replacing $m_i e_i$ by $(-m_i)\bar{e}_i$ if necessary.

Suppose this done. Let Δ be the finite subgraph of Γ consisting of all e_i and $\bar{e}_i$ and their vertices. Because $x\partial = 0$, it is easy to see, using the conditions we have required on x, that for each i there must be j and k such that $e_i\sigma = e_j\tau$ and $e_i\tau = e_k\sigma$. By Lemma 10, this tells us that Δ is not a forest. Hence Γ is not a forest.//

Exercise 1 Let e be an edge in a connected graph Γ. Show that $\Gamma - \{e,\bar{e}\}$ is connected if e is in some circuit and consists of exactly two components if e is not in any circuit.

5.2 COMPLEXES AND THEIR FUNDAMENTAL GROUPS

Let f be a loop in a graph. The set F consisting of f and all its cyclic permutations is called a *cyclic loop.* The cyclic loop $\bar{F}$ consisting of $\bar{f}$ and all its cyclic permutations is called the *inverse* of F.

A *complex* K consists of three disjoint sets, V, E, and C (the *vertices, edges,* and *cells* of K) with the following properties;

(i) the vertices and edges form a graph;

(ii) there is a map $^-$ from C to C such that, for all $c \in C$, $\bar{c} \neq c$ and $\bar{\bar{c}} = c$;

(iii) to each $c \in C$ there is a cyclic loop $c\partial$ such that $\bar{c}\partial = \overline{c\partial}$.

We will refer to an edge of $c\partial$ or $\bar{c}\partial$ as an edge of c, and to a vertex of such an edge as a vertex of c. For convenience of notation, we will usually write $c\partial$ to denote some loop belonging to the cyclic loop, rather than the cyclic loop itself. We call $c\partial$ the *boundary* of c.

As with graphs, readers are left to decide for themselves in each situation whether the empty complex is permitted or not.

The graph consisting of the vertices and edges of the complex K is called the 1-*skeleton* K^1 of K. By a path (loop, circuit, and so on) in K we will mean a path in K^1.

A complex as defined above is the combinatorial analogue of the topological (and partly combinatorial) notion of a 2-dimensional CW-complex. (We have no need of this concept and shall not discuss it further, but readers with some familiarity with the notion will find that some of the concepts of this chapter are suggested by results in that theory.) This seems to me to be the most natural setting in which to work. Fortunately we do not require analogues of CW-complexes of higher dimension, which would be messy to define. Any 2-dimensional simplicial complex gives a complex in the above sense (once we have added inverses of edges and cells).

A graph is just a complex with no cells.

As already remarked about graphs, the use of the map $^-$ on cells and edges is just a technical convenience. When giving a picture of a complex we usually only refer to one edge or cell in each pair.

A *subcomplex* K_1 of a complex K consists of sets $V_1 \subseteq V$, $E_1 \subseteq E$, and $C_1 \subseteq C$ such that if $c \in C_1$ then $\bar{c} \in C_1$ and every edge of c is in E_1 and if $e \in E_1$ then $\bar{e} \in E_1$ and $e\sigma \in V_1$. It then follows that K_1 is a complex, the operations being those of K. The intersection and union of subcomplexes of K is obviously a subcomplex of K.

The *full subcomplex* of K corresponding to the subset V_1 of V is the subcomplex whose vertex set is V_1, whose edge set consists of all edges whose vertices are in V_1, and whose cell set consists of all cells whose vertices are in V_1.

We say that the complex K is *connected* if any two vertices can be joined by a path. Thus K is connected iff K^1 is connected. More generally, we can obtain an equivalence relation on the vertices by making v related to w iff there is a path from v to w. Then K is connected iff there is only one equivalence class. A full subcomplex whose vertices form an equivalence class is a *component* of K. If L is a component of K then L^1 is a component of K^1.

A *map* or *morphism* φ from a graph Γ to a graph Δ consists of two maps, both denoted by φ, one from $V(\Gamma)$ to $V(\Delta)$ and the other from $E(\Gamma)$ to $E(\Delta)$ such that, for all $e \in E(\Gamma)$, $\bar{e}\varphi = \overline{e\varphi}$ and $e\varphi\sigma = e\sigma\varphi$. It immediately follows that $e\varphi\tau = e\tau\varphi$.

Let f be the path $e_1,\ldots,e_n$ in Γ. It is easy to check that $e_1\varphi,\ldots,e_n\varphi$ is a path in Δ. We denote this path by $f\varphi$. If f is a loop then $f\varphi$ is a loop. If F is a cyclic loop containing the loop f then the cyclic loop containing $f\varphi$ depends only on F, and not on the choice of f. We denote it by $F\varphi$.

A *map* or *morphism* φ from the complex K to the complex L consists of three maps, all denoted by φ, one from $V(K)$ to $V(L)$, one from $E(K)$ to $E(L)$, and one from $C(K)$ to $C(L)$ such that, for $e \in e(K)$ and $c \in C(K)$, $\bar{e}\varphi = \overline{e\varphi}$ and $e\varphi\sigma = e\sigma\varphi$, and $\bar{c}\varphi = \overline{c\varphi}$ and $c\varphi\partial = c\partial\varphi$. We call φ an *isomorphism* if there is a map $\psi: L \to K$ such that $\varphi\psi$ and $\psi\varphi$ are identity maps.

If $V_1 \subseteq V$, $E_1 \subseteq E$, and $C_1 \subseteq C$ then (V_1,E_1,C_1) is a subcomplex of K iff the inclusion is a map of complexes.

Evidently, the identity map on a complex K is a map of complexes in the above sense. Also, if $\varphi: K \to L$ and $\psi: L \to M$ are maps of complexes then $\varphi\psi$ is a map of complexes. In category language, we have a category of complexes and maps (and also a category of graphs and maps).

Let $\varphi: K \to L$ be a map of complexes, and let K_1 and L_1 be subcomplexes of K and L. Then $K_1\varphi$ and $L_1\varphi^{-1}$ are easily seen to be subcomplexes of L and K, respectively.

Let $\varphi: K \to L$ be a map of complexes, and let K be connected. Then it is easy to check that $K\varphi$ is also connected.

Note that the maps in our category are not the analogue of cellular maps in CW-complexes; they correspond to a special kind of cellular map.

Let f be the path $e_1, \ldots, e_n$. We say that the path g is obtained from f *by elementary reduction* if

either (i) for some i, $e_{i+1} = \bar{e}_i$ and g is $e_1, \ldots, e_{i-1}, e_{i+2}, \ldots, e_n$

or (ii) for some i and j with $i \leq j$ and some cell c, the path $e_i, \ldots, e_j$ is a loop belonging to the cyclic loop $c\partial$, and g is the path $e_1, \ldots, e_{i-1}, e_{j+1}, \ldots, e_n$.

It is easy to check that the deletion of the stated edges does leave a path in either case if not all edges have been deleted. If the deletion would remove every edge we regard g as being the path consisting of the vertex $e_1\sigma$.

We call the path f *homotopic to* the path g, and we write $f \simeq g$, if there is a sequence of paths f_r for $1 \leq r \leq m$ (some m) such that f_1 is f, f_m is g, and, for all $r < m$, either f_{r+1} is an elementary reduction of f_r or f_r is an elementary reduction of f_{r+1}. This is obviously an equivalence relation. We shall use the phrase *edge-homotopic* when we need to distinguish this from the topological notion of homotopy.

It is immediate that if f is a path from v to w and $f \simeq g$ then g is also a path from v to w, and $\bar{f} \simeq \bar{g}$. Further, for any path f' starting at w, if f_r is a sequence of paths which show that $f \simeq g$ then the sequence $f_r.f'$ shows that $f.f' \simeq g.f'$. Similarly, if f'' is a path ending at v then $f''.f \simeq f''.g$. It follows that f' is a path starting at w and $f' \simeq g'$ then $f.f' \simeq g.f' \simeq g.g'$.

The homotopy classes of paths therefore form a groupoid, which we call the *fundamental groupoid* (or *edge-groupoid*) of K, written $\gamma(K)$. This groupoid has vertex set $V(K)$. The vertex group of the groupoid at the vertex v (that is, the set of homotopy classes of loops starting at v) is called the *fundamental group* of K with base-point v, written $\pi_1(K,v)$. When K is connected this does not depend on v, and we often just write $\pi_1(K)$. Any path from v to w induces an isomorphism from $\pi_1(K,v)$ to $\pi_1(K,w)$.

We will frequently find it convenient to write f^{-1} to denote the inverse of the path f, rather than $\bar{f}$, to emphasise that its class is the inverse of the class of f.

Let $\varphi: K \to L$ be a map of complexes. Let the path g be obtained from the path f of K by elementary reduction. Then $g\varphi$ is obtained from $f\varphi$ by elementary reduction (for instance, if we remove $c\partial$ from f to get g then we need only remove $c\partial\varphi$ from $f\varphi$ to get $g\varphi$, and $c\partial\varphi = c\varphi\partial$ is the boundary of $c\varphi$). It follows that if $f \simeq g$ then $f\varphi \simeq g\varphi$. Hence φ gives rise to a homomorphism $\varphi^*: \gamma(K) \to \gamma(L)$, and also to a homomorphism φ^* from $\pi_1(K,v)$ to $\pi_1(L,v\varphi)$.

If $\psi: L \to M$ is a map of complexes then $(\varphi\psi)^* = \varphi^*\psi^*$. If φ is the identity map on K then φ^* is the identity homomorphism on $\gamma(K)$. Hence, if φ is an isomorphism of complexes then φ^* is an isomorphism of group(oid)s. In category language, we have a covariant functor from the category of complexes to the category of group(oid)s.

Examples Let K be the complex with two vertices v and w and one edge-pair $\{e,\bar{e}\}$ with $e\sigma = v$ and $e\tau = w$. Then K is connected and $\gamma(K)$ is the tree groupoid on $\{v,w\}$, so that $\pi_1(K,v)$ is trivial. For any non-trivial path must have its edges alternately e and $\bar{e}$, and so it must be homotopic to one of v, w, e, or $\bar{e}$, and no two of these can be homotopic, since no two have the same start and finish.

Let K be a graph with one vertex v and one edge-pair $\{e,\bar{e}\}$; we must have $e\sigma = v = e\tau$ Then $\pi_1(K,v)$ is infinite cyclic, generated by the class of e. For any path is a loop, and any path is plainly homotopic either to v or to e^n or to $\bar{e}^n$ for some $n > 0$. If a path has r occurrences of e and s occurrences of $\bar{e}$ then $r-s$ in unchanged by elementary reduction. Hence $r-s$ is constant on homotopy classes, which shows that no two of the above loops are homotopic.

The theorems on fundamental group(oids) of spaces have analogues for the fundamental group(oid)s of complexes, which are usually much easier to prove.

Theorem 14 (van Kampen's Theorem) *Let K_1 and K_2 be subcomplexes of the complex K, and let K_0 be $K_1 \cap K_2$. Then the inclusions induce a pushout of groupoids*

$$\begin{array}{ccc} \gamma(K_0) & \to & \gamma(K_1) \\ \downarrow & & \downarrow \\ \gamma(K_2) & \to & \gamma(K) . \end{array}$$

Corollary 1 *Suppose further that S is a subset of $V(K)$ such that, for $i = 0,1,2$, S meets every component of K_i, and let S_i be $S \cap V(K_i)$. Then the inclusions induce a pushout of groupoids*

$$\begin{array}{ccc} \gamma(K_0)_{S_0} & \to & \gamma(K_1)_{S_1} \\ \downarrow & & \downarrow \\ \gamma(K_2)_{S_2} & \to & \gamma(K)_S . \end{array}$$

Corollary 2 *Suppose further that K_0, K_1, and K_2 are connected (and then K will also be connected). Let $v \in V(K_0)$. Then the inclusions induce a pushout of groups*

$$\begin{array}{ccc} \pi_1(K_0,v) & \to & \pi_1(K_1,v) \\ \downarrow & & \downarrow \\ \pi_1(K_2,v) & \to & \pi_1(K,v) \; . \end{array}$$

Corollary 3 *Let K_1 and K_2 be connected, and let $K_1 \cap K_2 = \{v\}$. Then $\pi_1(K,v) = \pi_1(K_1,v) * \pi_1(K_2,v)$.*

Proof Corollary 3 is a special case of Corollary 2, which in turn is a special case of Corollary 1. Corollary 1 follows from the theorem and Example 3 on pushouts of groupoids.

The proof of the theorem is similar to the topological case, but simpler.

We take homomorphisms φ_i from $\gamma(K_i)$ to a groupoid G for $i = 1,2$, such that the two homomorphisms from $\gamma(K_0)$ to G are the same.

We may extend the φ_i to multiplicative maps ψ_i from the set of paths of K_i into G. Our assumption tells us that $\psi_1 = \psi_2$ on paths of K_0. In particular, since each edge or vertex of K may be regarded as a path and lies in either K_1 or K_2, these maps give a map ψ from the edges and vertices of K into G. Further, for any vertex v, $v\psi$ will be an identity in G, since v is an identity on the paths of K.

Let e and e' be edges such that $e\tau = e'\sigma$. If e and e' are both in K_1 (or in K_2) we know that $(e\psi)(e'\psi)$ is defined (and equals $(e.e')\psi$), since ψ_1 is multiplicative. If one of them is in K_1 and the other in K_2 then $(e\psi)(e'\psi)$ is still defined, because $(e\tau)\psi$ is a right identity for $e\psi$ and a left identity for $e\,\psi$.

Now let f be any path, say the path $e_1,\dots,e_n$. By the previous paragraph, the product $(e_1\psi)\dots(e_n\psi)$ exists in G. We define $f\psi$ to be this product. Evidently ψ is a multiplicative map which extends ψ_1 and ψ_2.

Let f' be obtained from f by elementary reduction. If we have $e_{r+1} = \bar{e}_r$ then $e_{r+1}\psi = (e_r\psi)^{-1}$, and so $f'\psi = f\psi$. Otherwise, f' comes from f by deleting the sequence of edges $e_r,\dots,e_s$ which form the boundary of a cell c. Now c is in K_1 or in K_2. Hence the path $e_r,\dots,e_s$ is homotopic in K_1 (or in K_2) to a trivial path. Since ψ_i arises from φ_i, which is defined on homotopy classes, it follows that $(e_r\psi)\dots(e_s\psi)$ is an identity in G. Again we find that $f'\psi = f\psi$.

Since ψ is unchanged by an elementary reduction, it is constant on homotopy classes. Hence it gives rise to a homomorphism $\varphi:\gamma(K) \to G$, as required. There is only one possibility for such a map φ, since its value on the class of an edge is given to us, and any element is the product of such classes.//

By a similar simplification of the topological case, we obtain the following results on direct limits.

Theorem 15 *Let Λ be a directed set, and let K_λ (for $\lambda \in \Lambda$) be subcomplexes of the complex K such that $K = \cup K_\lambda$ and $K_\lambda \subseteq K_\mu$ for $\lambda < \mu$. Then $\gamma(K)$ is the direct limit of the $\gamma(K_\lambda)$ (the maps being induced by the inclusions.*

Suppose further that each K_λ is connected, and that $v \in \cap K_\lambda$. Then $\pi_1(K,v)$ is the direct limit of the $\pi_1(K_\lambda)$.//

Corollary 1 *For any complex K, $\gamma(K) = \varinjlim \gamma(L)$, where L runs over all finite subcomplexes of K.//*

Corollary 2 *Let K be a connected complex. Then $\pi_1(K,v) = \varinjlim \pi_1(L,v)$, where L runs over all connected subcomplexes of K which contain v.//*

Example Let Γ be a graph with only one vertex v and a collection of edge-pairs $\{e_i, \bar{e}_i\}$. Then $\pi_1(\Gamma)$ is free with basis $\{e_i\}$. For we have seen that this holds when there is only one edge-pair. Inductively, using van Kampen's Theorem, it holds when there are only finitely many edge-pairs. It holds in general by Corollary 2.

We now obtain a presentation for $\pi_1(K)$, where K is a connected complex. It turns out to be inconvenient to use van Kampen's Theorem, because the boundary of a cell need not be a circuit.

Any path in K can be regarded as an element of $F(E)$, the free group with basis the set E of edges of K, regarding the path $e_1, \ldots, e_n$ as the product $e_1 \ldots e_n$ in $F(E)$. We shall use the same symbol for a path and the corresponding element of $F(E)$.

Let $R \subseteq F(E)$ be $\{e\bar{e}$; all $e \in E\} \cup \{c\partial$; all $c \in C\}$. We call R the set of *relators of the first kind.*

Choose a vertex a of K. For each vertex v choose a path f_v from a to v, and let f_a be the trivial path. Let $S \subseteq F(E)$ be $\{f_v$; all $v \in V\}$. We call S the set of *relators of the second kind.*

Theorem 16 *$\langle E;\ R \cup S\rangle$ is a presentation of $\pi_1(K,a)$. In this presentation the edge e from v to w maps to the homotopy class of the path $f_v.e.f_w^{-1}$.*

Remarks We can replace the loop $c\partial$ by another loop giving the same cyclic loop. This just replaces an element of R by a conjugate. We plainly need only take $c\partial$ for one cell c from each cell-pair. We could take only one generator e from each edge-pair, replacing the corresponding $\bar{e}$ by e^{-1} in the relators; we will always do this when we have specific examples.

Proof Let G be $\langle E;\ R \cup S\rangle$. Let N be $\langle R \cup S\rangle^{F(E)}$, so that wN is the element of G corresponding to $w \in F(E)$. Let $[f]$ be the homotopy class of the path f.

Take the homomorphism from $F(E)$ to $\pi_1(K,a)$ which sends e to $[f_v.e.f_w^{-1}]$, where e starts at v and ends at w. It is easy to check (inductively) that, for any path f from v to w, this sends f to $[f_v.f.f_w^{-1}]$.

In particular, this homomorphism sends $e\bar{e}$ to the identity. Also, if the loop $c\partial$ starts at v, it sends $c\partial$ to $[f_v.c\partial.f_v^{-1}]$, which is the identity. Finally, it sends f_v to $[f_a.f_v.f_v^{-1}]$, which is also the identity. Hence it induces a homomorphism $\varphi: G \to \pi_1(K,a)$.

To each path f we have a corresponding element fN of G. This is plainly a multiplicative map. Also $e\bar{e}$ and $c\partial$ go to the identity. Hence this map is constant on homotopy classes. It therefore defines a homomorphism ψ from $\gamma(K)$ to G, which restricts to a homomorphism $\psi: \pi_1(K,a) \to G$.

In particular, f_v maps to 1. Hence, when e begins at v and ends at w, $(eN)\varphi\psi = [f_v.e.f_w^{-1}]\psi = [f_v]\psi[e]\psi([f_w]\psi)^{-1} = eN$.

Also, for any edge e from v to w, we have $[e]\psi\varphi = (eN)\varphi = [f_v.e.f_w^{-1}]$. Inductively, for any path f from v to w, $[f]\psi\varphi = [f_v.f.f_w^{-1}]$. In particular, since f_a is trivial, if f is a loop based at a then $[f]\psi\varphi = [f]$.

It follows that the homomorphisms φ and ψ are inverses of each other, and so φ is an isomorphism.//

This theorem will rarely be used, as the special case of it given below is more useful. We will refer to the presentation below as the presentation *using the maximal tree T*.

Theorem 17 *Let T be a maximal tree in K. Then $\pi_1(K)$ is presented by $\langle E;\ R \cup \{e;\ e \in T\}\rangle$.*

Proof For each vertex v, let f_v be the unique irreducible path in T from a to v. Evidently the elements f_v of $F(E)$ are consequences of $\{e;\ e \in T\}$.

Conversely, let e be an edge in T from v to w. If f_v does not end in $\bar{e}$ then $f_v.e$ will be an irreducible path in T from a to w; that is, $f_v.e = f_w$. If f_v does end in $\bar{e}$ then removal of this edge gives an irreducible path in T from a to w; that is, $f_v = f_w.\bar{e}$. In either case, e is a consequence of $\{f_v;\ \text{all } v \in V\} \cup \{e\bar{e};\ \text{all } e \in E\}$. Hence both presentations give the same group.//

Corollary 1 *Let* Γ *be a connected graph, and let* T *be any maximal tree in* Γ. *Then* $\pi_1(\Gamma)$ *is free, with one basis element for each edge-pair not in* T. *Further, the basis element corresponding to a chosen edge* e *is the class of* $f_v.e.f_w^{-1}$, *where* e *starts at* v *and ends at* w, *and* f_v *is the unique irreducible path in* T *from the base-point to* v.//

Corollary 2 *A connected graph is a tree iff its fundamental group is trivial.*//

Corollary 3 *Let* Γ *be a connected graph with* n *vertices and* m *edge-pairs. Then the rank of* $\pi_1(\Gamma)$ *is* $m-n+1$, *provided* n *is finite.*

Proof We know there is one basis element for each edge-pair not in a chosen maximal tree. Now the maximal tree has exactly n vertices, so by Proposition 11 it has $n-1$ edge-pairs. Since there are m edge-pairs altogether, we see that the number of basis elements is $m-n+1$.//

Corollary 4 *Let* K *be a connected complex, and let* L *be* $K \cup \{c, \bar{c}\}$, *where* c *is a cell not in* K *and* $c\partial$ *is a cyclic loop in* K. *Then* $\pi_1(L)$ *is obtained from* $\pi_1(K)$ *by cancelling the class of the loop* $c\partial$.//

Remark More precisely, when we specify base-points, $\pi_1(L,a)$ is obtained from $\pi_1(K,a)$ by cancelling the class of $f.c\partial.f^{-1}$, where f is any path from a to the start of $c\partial$.

We can extend Corollary 4 inductively to the case where L is obtained from K by adding a finite number of cells. Then Theorem 15 allows us to extend to the case where we add an arbitrary number of cells. Thus we have the following result.

Corollary 5 *Let K be a connected complex, and let L be $K \cup \{c_\lambda, \bar{c}_\lambda\ (\lambda \in \Lambda)\}$, where c_λ is a cell not in K and $c_\lambda \partial$ is a cyclic loop in K. Then $\pi_1(L)$ is obtained from $\pi_1(K)$ by cancelling the classes of the loops $c_\lambda \partial$.*

Theorem 18 *For any group G there is a complex K such that $\pi_1(K)$ is (isomorphic to) G.*

Proof Let $\langle X;R \rangle$ be a presentation of G. Let K^1 be the graph with only one vertex and with one edge-pair for each element of X. Then $\pi_1(K^1)$ is the free group on X.

Hence any element of R can be represented as a loop in K^1. We obtain a complex K from K^1 by adding cells. Let K have one cell-pair $\{c_r, \bar{c}_r\}$ for each $r \in R$, with $c_r \partial$ being the cyclic loop containing the loop corresponding to r. Evidently $\pi_1(K)$ is isomorphic to G.//

Theorem 19 *Let L be a connected complex, and let φ be a homomorphism from $\pi_1(L)$ to the group G. Then there is a complex K containing L, such that $\pi_1(K)$ is G and the inclusion of L in K induces φ.*

Proof Let M be a complex with $\pi_1(M) = G$, and let N be the union of L and M with one common vertex. Then $\pi_1(N) = \pi_1(L) * G$, by Corollary 3 to Theorem 14.

We obtain K from N by adding cells. Let Y generate $\pi_1(L)$. For each $y \in Y$ take a loop f_y in N whose class is $(y\varphi)y^{-1}$. Let K be obtained from N by adding cell-pairs $\{c_y, \bar{c}_y\}$ for each $y \in Y$, where $c_y \partial$ is the cyclic loop containing f_y. Then, by Corollary 5 to Theorem 17, $\pi_1(K)$ is obtained from $\pi_1(N)$ by cancelling all the elements $(y\varphi)y^{-1}$. This gives the result.//

Corollary *Let L_i be connected complexes, let G be a group, and let $\varphi_i : \pi_1(L_i) \to G$ be homomorphisms. (i) There is a complex K containing (copies of) the L_i and a vertex v_0 such that $L_i \cap L_j = \{v_0\}$ for $i \neq j$ and such that $\pi_1(K)$ is G and the inclusion of L_i in K induces φ_i. (ii) There is a complex M containing (copies of) the L_i such that $L_i \cap L_j = \emptyset$ for $i \neq j$ and such that $\pi_1(M)$ is G and the homomorphism induced by inclusion of L_i in K is φ_i followed by a conjugation.*

Remark In (ii) we have to use different base-points for the different L_i, and in order to refer each to a fixed base-point for K we must use the isomorphism induced by a path. A different path would provide another isomorphism, related to the first by conjugation. In fact our construction will give a preferred path to use.

Proof (i) Let L be obtained from the disjoint union of the L_i by identifying one vertex from each, so that $L_i \cap L_j = \{v_0\}$ for $i \neq j$. Either from Theorem 17 and Lemma 6 or from van Kampen's Theorem and the direct limit argument, we see that $\pi_1(L)$ is $*\pi_1(L_i)$. Now apply the previous theorem to L and the homomorphism $*\varphi_i$.

(ii) We may assume that the L_i are disjoint. Choose a vertex v_i of L_i. Let M_i be obtained from L_i by adding a new vertex w (the same for each i) and an edge-pair $\{e_i, \bar{e}_i\}$ with $e_i\sigma = w$ and $e_i\tau = v_i$. Then $\pi_1(M_i, w)$ is isomorphic to $\pi_1(L_i, v_i)$, and we need only apply part (i) to the M_i with common vertex w.//

Let K be a connected complex, and let K_0 and K_1 be connected subcomplexes of K such that $K_0 \cap K_1 = \{v\}$ and such that there is an isomorphism $\vartheta: K_0 \to K_1$ with $v\vartheta = v$. .We may obtain a complex L from K by identifying vertices, edges, and cells of K_0 with the corresponding elements of K_1 under ϑ.

Let i_0 and i_1 be the inclusions of K_0 and K_1 into K.

Proposition 20 *With this notation,* $\pi_1(L, v)$ *is obtained from* $\pi_1(K, v)$ *by cancelling the elements* $(gi_0*\vartheta*)^{-1}(gi_1*)$ *for all* $g \in \pi_1(K_0, v)$.

Proof To begin with we show that L is a tree if K is a tree.

Let K be a tree. Then K_0 and K_1 are both trees, and so is $K_0 \cup K_1$, since $K_0 \cap K_1$ is a vertex. By Lemma 12, the graph obtained from K by contracting $K_0 \cup K_1$ is also a tree. This graph is plainly also the graph obtained from L by contracting the tree K_0. By Lemma 12 again, L is a tree..

We now return to the general case. Let T_0 be a maximal tree in K_0, and let T_1 be $T_0\vartheta$, which is a maximal tree in K_1. Then $T_0 \cup T_1$ is a tree, and so (by the corollary to Proposition 8) it is contained in a maximal tree T of K. By the previous paragraph, the image S in L of T is a tree, which is a maximal tree, since it contains every vertex of L.

Now look at the presentations of $\pi_1(K,v)$ and $\pi_1(L,v)$ given by Theorem 17, using the maximal trees T and S. It is clear that the presentation of $\pi_1(L,v)$ comes from the presentation of $\pi_1(K,v)$ by first adding new relations $e = e\vartheta$ for all edges e of K_0 not in T_0 and then using Tietze transformations to eliminate the generators $e\vartheta$.

Since $\{e;\ e \in K_0, e \notin T_0\}$ is a set of generators for $\pi_1(K_0,v)$, we obtain the result.//

Now suppose that K is a connected complex, that K_0 and K_1 are connected subcomplexes with $K_0 \cap K_1 = \emptyset$, and that $\vartheta: K_0 \to K_1$ is an isomorphism. Let L be obtained from K by identifying K_0 with K_1 using ϑ, as before. Let i_0 and i_1 be the inclusions of K_0 and K_1 in K. Let v_0 be a vertex of K_0 and let v_1 be $v_0\vartheta$. Let f be a path in K from v_1 to v_0.

Proposition 21 *With this notation,* $\pi_1(L)$ *is the pseudo-HNN extension* $\langle \pi_1(K,v_0), t;\ (\pi_1(K_0,v_0)i_0{*})^t = \pi_1(K_1,v_1)i_1{*}f{\#} \rangle$.

Remark By (ii) of the corollary above, every pseudo-HNN extension can be obtained in this way.

Proof Let M be obtained from K by identifying v_0 with v_1. Then L comes from M by identifying the images of K_0 and K_1. So we may apply the previous proposition, once we have determined $\pi_1(M,v_0)$ and the homomorphisms induced by the inclusions j_0 and j_1 of K_0 and K_1 in M.

Let $N = K \cup \{e,\bar{e}\}$, where e is a new edge with $e\sigma = v_0$ and $e\tau = v_1$. Then M is obtained from N by contracting the edge e. Take the presentation of $\pi_1(N,v_0)$ corresponding to a maximal tree containing e and the presentation of $\pi_1(M,v_0)$ corresponding to the image of this tree (which will be a maximal tree in M). We see that the two groups are isomorphic, since the only difference between the presentations is that the former has an extra generator e and also an extra relator e.

By the groupoid version of van Kampen's Theorem and Example 1 on pushouts of groupoids, $\pi_1(N,v_0) = \pi_1(K,v_0) * \langle t \rangle$, where t is the class of the path $e.f$. If we regard $\pi_1(M,v_0)$ as the same as $\pi_1(N,v_0)$ then $j_0{*}$ is plainly the same as $i_0{*}$. Also $j_1{*}$ will be $i_1{*}\bar{e}_1{\#}$. Since t is the class of $e.f$, it follows that, for $g \in \pi_1(K_1,v_1)$, $gj_1{*} = t(gi_1{*}f{\#})t^{-1}$. The result now follows.//

Exercise 2 Let $\varphi: K \to L$ be a map. Let K_1 and L_1 be subcomplexes of K and L. Shows that $K_1\varphi$ and $L_1\varphi^{-1}$ are subcomplexes of L and K.

Exercise 3 Prove Theorem 15.

5.3 *FREE GROUPS AND THEIR AUTOMORPHISMS*

In Chapter 7 we show that any subgroup of a free group is free. Three proofs are given in that chapter, all very closely related, and another related proof is given in section 8.1. The first proof in Chapter 7 requires the theory of covering complexes, which is developed in Chapter 6. The other proofs, however, could be read at this point. In this section we look ar automorphisms of free groups.

Let α be an automorphism of a finitely generated free group F, and let Fixα denote the subgroup $\{g \in F;\ g\alpha = g\}$ For some years it was an open question whether Fixα was always finitely generated, and some intertesting partial results were obtained. Then, almost simultaneously, several different prooofs appeared. We shall follow the proof due to Goldstein and Turner (1986), which is remarkably simple, and which shows that Fixη is finitely generated for any endomorphism η of F.

We begin with a lemma about graphs. An *orientation* of a graph Γ is a subset A of E such that $A \cap \bar{A} = \emptyset$ and $A \cup \bar{A} = E$. The *outdegree* $v\omega$ of the vertex v (with respect to A) is $|\{e \in A;\ e\sigma = v\}|$.

Lemma 22 *Let Γ be a connected graph. Then $\pi_1(\Gamma)$ has finite rank iff there is an orientation A such that $v\omega$ is finite for all v and $v\omega \leq 1$ for all but finitely many v.*

Proof Suppose that $\pi_1(\Gamma)$ has finite rank. Let T be a maximal tree in Γ. By Corollary 1 to Theorem 17, there are only finitely many edges not in T.

Take any vertex v_0. Let A_0 consist of those edges of T such that the irreducible path in T from $e\sigma$ to v_0 begins with e. It is easy to see that A_0 is an orientation of T such that $v_0\omega = 0$ and $v\omega = 1$ for $v \neq v_0$ (this orientation, called the orientation of T *inwards towards* v_0, will also be used in section 8.5). Enlarge A_0 to an orientation A of Γ by orienting the edges not in T arbitrarily. Clearly A satisfies the required conditions.

Now suppose that there is an orientation A of Γ satisfying the conditions. Removing finitely many edge-pairs, we obtain a graph Δ such that $v\omega \leq 1$ for every vertex v (in the orientation $A \cap \Delta$ of Δ). Since Γ is connected, and we remove only finitely many edge-pairs, it follows from Exercise 1 that Δ has only finitely many components.

It is enough to show that each component of Δ has finitely generated fundamental group. For we may then take a maximal tree in each component which, by Corollary 1 to Theorem 17, contains all but finitely many edges in that component. The union of these is a forest in Δ containing all but finitely many edges of Δ. This forest can be enlarged to a maximal tree of Γ, which will contain all but finitely many edges of Γ. Then $\pi_1(\Gamma)$ will be finitely generated, using Corollary 1 to Theorem 17 again.

So let Δ_1 be a component of Δ. If Δ_1 is not a tree it contains a circuit $e_1,\ldots,e_n$. We may assume that $e_n \in A$ (if not, replace the circuit by the circuit $\bar{e}_{n-1},\ldots,\bar{e}_1,\bar{e}_n$). Since $e_n\sigma$ has outdegree at most 1, we see that $\bar{e}_{n-1} \notin A$, and so $e_{n-1} \in A$. By induction on $n-r$, we find that $e_r \in A$ for all r.

Now suppose that Δ_1 has a vertex v such that $v\omega = 0$. Let w be any vertex, and let $x_1,\ldots,x_m$ be the irreducible path from w to v. Since $v\omega = 0$, we cannot have $\bar{x}_m \in A$, and so $x_m \in A$. As in the previous paragraph, we find that $x_i \in A$ for all i. We also find, since no vertex has outdegree >1, that the only irreducible paths starting at w in which all edges are in A are the paths $x_1,\ldots,x_k$ for $0 \le k \le m$; in particular, there is a bound to the lengths of such paths.

It follows that if Δ_1 contains a vertex of outdegree 0 then it cannot contain any circuit, since a circuit $e_1,\ldots,e_n$ with all $e_i \in A$ gives rise to arbitrarily long irreducible paths starting at $e_1\sigma$ and with all edges in A; namely, the paths $(e_1,\ldots,e_n)^p$ for all integers p.

Thus if Δ_1 contains a vertex of outdegree 0 then it is a tree. Also, if Δ_1 contains a circuit, then we may remove an edge-pair occurring in the circuit, and we get a connected graph with a vertex whose outdegree is now 0; that is, Δ_1 is obtained from a tree by adding just one edge-pair, and so $\pi_1(\Delta_1)$ is infinite cyclic in this case.//

Let F be a finitely generated free group with basis S, and let η be an endomorphism of F. We construct a graph Γ, together with a function from $E(\Gamma)$ to $S \cup S^{-1}$ (called the *labelling* of the edges) as follows. $V(\Gamma)$ is to be F. There is to be an edge from v to w with label $x \in S \cup S^{-1}$ iff $w = x^{-1}v(x\eta)$. When this happens, we also have an edge labelled x^{-1} from w to v (since $v = (x^{-1})^{-1}w(x^{-1}\eta)$), and this is to be the inverse of the first edge. Γ need not be connected, and we let Γ_1 be the component of 1.

Lemma 23 *With the above notation,* $Fix\eta$ *is isomorphic to* $\pi_1(\Gamma_1)$.

Proof First observe that there is a homomorphism from the fundamental groupoid of Γ_1 into F which sends each edge into its label. This induces a homomorphism from $\pi_1(\Gamma_1,1)$ into F.

Notice that there is only one edge starting (or ending) at a given vertex and with a given label. Also, if e has label x then $\bar{e}$ has label x^{-1}. It follows that the image of an irreducible path is an irreducible word of the same length. In particular, the homomorphism from $\pi_1(\Gamma_1,1)$ to F is a monomorphism.

To any $x_1,\ldots,x_n$ there is a unique path starting at 1 whose edges have labels $x_1,\ldots,x_n$ respectively. This path ends at v, where $v = w^{-1}(w\eta)$ and $w = x_1\ldots x_n$. In particular, this path is a loop iff $w = w\eta$. Thus the image of $\pi_1(\Gamma_1,1)$ is Fixη.//

Theorem 24 *Let η be an endomorphism of a finitely generated free group F. Then Fixη is finitely generated.*

Proof With the above notation, it is enough to find an orientation A of Γ satisfying the hypotheses of Lemma 22.

Let e be an edge, and let e start at v, end at w, and have label x, so that $w = x^{-1}v(x\eta)$. Suppose that v does not cancel completely in forming the product $v(x\eta)$ and that w does not cancel completely in forming the product $w(x^{-1}\eta)$. Then we put e in A iff v begins with x. For the moment we do not say whether or not e is in A if the conditions on cancellation are not satisfied.

It is clear that there is at most one edge in A beginning at a given vertex v. Since $w(x^{-1}\eta) = x^{-1}v$, and we are assuming that w does not cancel completely in forming $w(x^{-1}\eta)$, the first letter of w will be the first letter of $x^{-1}v$. That is, it will be x^{-1} iff v does not begin with x. Hence, when the conditions on cancellation are satisfied, exactly one of e and $\bar{e}$ is in A.

Now, to each x, there are only finitely many choices of v such that v cancels completely in $v(x\eta)$, namely those v such that v^{-1} is an initial segment of $x\eta$. Similarly, there are only finitely many choices of w such that w cancels completely in $w(x^{-1}\eta)$. Since F is finitely generated, it follows that there are finitely many edge-pairs not previously considered. We make an arbitrary choice of which edge of the pair to put in A for these pairs.

Plainly, if v is not one of the finitely many vertices where the cancellation condition is not satisfied then $v\omega \leq 1$, as required.//

We can extend this theorem, but we need a result from Chapter 8.

Let H be a (finitely generated) subgroup of a free group F. All mention of words, cancellation, reduction, and so on, wil be with respect to a fixed basis of F, which we do not need to name. It is shown in section 8.2 that H is free, with a basis S satisfying the following condition. To any $x \in S \cup S^{-1}$

there are words $\lambda(x)$, $\mu(x)$, and $\rho(x)$ such that $\mu(x)$ has length 1, $\lambda(x^{-1}) = \rho(x)^{-1}$, $\mu(x^{-1}) = \mu(x)^{-1}$, and $\rho(x^{-1}) = \lambda(x)^{-1}$, and $x = \lambda(x)\mu(x)\rho(x)$, reduced as written, and such that, for any $y \in S \cup S^{-1}$ with $y \neq x$, when cancellation is performed on $y^{-1}x$ it does not reach $\mu(x)$ or $\mu(y)^{-1}$.

Proposition 25 *Let H be a finitely generated subgroup of a free group F. Let $\varphi: H \to F$ be a homomorphism. Then $\{w \in H;\ w\varphi = w\}$ is finitely generated.*

Corollary *Let F be a finitely generated free group, let ϑ and η be endomorphisms of F such that μ is one-one. Then $\{w \in F;\ w\vartheta = w\eta\}$ is finitely generated.*

Proof The corollary follows from applying the proposition to the homomorphism $\eta^{-1}\vartheta$ on $F\eta$.

To prove the proposition, we proceed much as before. This time we take S to be a basis of H with the special property described above. As before, we take a graph Γ whose vertex set is F, and whose edges have labels in $S \cup S^{-1}$, there being an edge from v to w with label x iff $w = x^{-1}v(x\varphi)$. As before, $\{w;\ w\varphi = w\}$ is isomorphic to $\pi_1(\Gamma_1)$, and we look for an orientation of Γ satisfying the conditions of Lemma 22.

If v cancels completely in forming $x^{-1}v(x\varphi)$ then there are words p and q such that $v = pq$, reduced as written, with p being a final segment of x and q^{-1} being an initial segment of $x\varphi$. Thus there are only finitely many such v.

Let e be an edge from v to w with label x. Suppose that v does not cancel completely in $x^{-1}v(x\varphi)$ and w does not cancel completely in $xw(x^{-1}\varphi)$. Then we put e in A iff in $x^{-1}v$ at least $(\mu(x)\lambda(x))^{-1}$ cancels from x^{-1}, or, equivalently, if v begins with $\lambda(x)\mu(x)$.

If $y \neq x$, by our choice of S, cancellation in $y^{-1}x$ does not reach $\mu(y)^{-1}$ or $\mu(x)$. Since both x and v begin with $\lambda(x)\mu(x)$, it follows that cancellation in $y^{-1}v$ does not reach $\mu(y)^{-1}$. We see that at most one edge starting at v can lie in A.

We complete the proof as before, once we have shown that, when v and w satisfy the conditions on cancellation, exactly one of e and $\bar{e}$ is in A.

Suppose that $e \in A$, so that $v = \lambda(x)\mu(x)q$, reduced as written, for some word q. Since v does not cancel completely in $x^{-1}v(x\varphi)$, it follows that q cannot cancel completely in $v(x\varphi)$, and so the reduced form of $xw = v(x\varphi)$ must begin with $\lambda(x)\mu(x)$, and the reduced form of $\lambda(x)^{-1}xw$ must begin with $\mu(x)$. If we could write w as $\rho(x)^{-1}\mu(x)^{-1}u$, reduced as written, then u could not begin with $\mu(x)$. Since u is the reduced form of $\lambda(x)^{-1}xw$, we get a contradiction. Thus w does not begin with $\rho(x)^{-1}\mu(x)^{-1}$, and so $\bar{e}$ is not in A.

Suppose that $e \notin A$. Then the reduced form of $x^{-1}v$ will begin with $\rho(x)^{-1}\mu(x)^{-1}$ (since the cancellation cannot go further than $\lambda(x)^{-1}$). Since v does not cancel completely in $x^{-1}v(x\varphi)$, it follows that the reduced form of $w = x^{-1}v(x\varphi)$ also begins with $\rho(x)^{-1}\mu(x)^{-1}$, which tells us that $\bar{e} \in A$.//

It has recently been proved by Bestvina (unpublished at present) that if α is an automorphism of a free group F of finite rank then Fixα has rank at most the rank of F. We shall not prove this. But we will give a bound for the rank of Fixη which can be calculated easily from the endomorphism η, and which sometimes gives a better bound.

We take an arbitrary set B of edges of a graph Γ. As before, we define the outdegree $v\omega$ of a vertex v to be $|\{e \in B;\ e\sigma = v\}|$. We call an edge-pair $\{e,\bar{e}\}$ *repelling* if neither e nor $\bar{e}$ is in B, *ordinary* if exactly one of e and $\bar{e}$ is in B, and *attracting* if both e and $\bar{e}$ are in B. (To explain the wording, consider a picture for Γ, in which as usual one edge in the picture corresponds to an edge-pair. Place an arrow near the vertex $e\sigma$ of the edge, pointing away from $e\sigma$ if $e \in B$ and towards $e\sigma$ if $e \notin B$. There will be another arrow near $e\tau = \bar{e}\sigma$. Then an attracting edge has its arrows $\rightarrow\!\!-\!\!-\!\!\leftarrow$, an ordinary edge has arrows $\rightarrow\!\!-\!\!-\!\!\rightarrow$ or $\leftarrow\!\!-\!\!-\!\!\leftarrow$, and a repelling edge has its arrows $\leftarrow\!\!-\!\!-\!\!\rightarrow$.) A *ray* is defined to be an infinite sequence of edges $e_1, e_2, \ldots$ all in B such that $e_i\tau = e_{i+1}\sigma$ for all i, and $e_i\sigma \neq e_j\sigma$ for $i \neq j$. Two rays are called equivalent if they become the same after deleting a finite number of edges from each.

Let r and a denote the number of repelling and attracting edge-pairs, let c be the number of components, and let n be the number of equivalence classes of rays. Let $d = \Sigma d_i$, where d_i is the rank of $\pi_1(\Gamma_i)$, where the Γ_i are the components of Γ. Notice that if $v\omega$ is finite for all v and $v\omega \leq 1$ for all but finitely many v then the sum $\Sigma(v\omega - 1)$ is well-defined (It may be $-\infty$, but cannot be ∞).

Lemma 26 *Suppose that $v\omega$ is finite for all v and $v\omega \leq 1$ for all but finitely many v. Suppose also that r and c are finite. Then*

$$r + c + \Sigma(v\omega - 1) = d + a + n.$$

In particular, if Γ is connected and $v\omega \leq 1$ for all v and $v_1\omega = 0$ for some v_1 then $d \leq r$. More generally, if $v\omega \leq 1$ for all v and $v_1\omega = 0$ for some v_1, and Γ_1 is the component of v_1 then $\pi_1(\Gamma_1)$ has rank at most r.

Proof The particular case is immediate from the formula, since we then have $c = 1$ and $\Sigma(v\omega - 1) \leq -1$. The more general case then follows, since we $\pi_1(\Gamma_1)$ will have rank at most the number of repelling edge-pairs in Γ_1, which is certainly at most r.

Let us remove an edge-pair $\{e,\bar{e}\}$ from Γ. Then $\Sigma(v\omega - 1)$ is decreased by 0, 1, or 2 when the edge-pair is repelling, ordinary, or attracting, respectively. Also n remains unchanged.

If the edge-pair lies in a circuit of a component Δ of Γ, then, by Exercise 1, $\Delta - \{e,\bar{e}\}$ is connected, so c remains unchanged. Any maximal tree of $\Delta - \{e,\bar{e}\}$ will be a maximal tree of Δ, and Corollary 1 to Theorem 17 shows us that d is decreased by 1. If the edge-pair is in Δ but does not lie in a circuit of Δ, then Exercise 1 tells us that $\Delta - \{e,\bar{e}\}$ has two components Δ' and Δ'', and so c increases by 1. If $T'$and T'' are maximal trees in Δ' and Δ'' then $T' \cup T'' \cup \{e,\bar{e}\}$ is a maximal tree in Δ (using Lemma 6), and Corollary 1 to Theorem 17 shows us that d remains unchanged.

It follows that removal of an edge-pair does not alter the difference between the two sides. We may therefore remove the finitely many repelling pairs, and the finitely many edges at those vertices v with $v\omega > 1$, and prove the result for the graph obtained. That is, we may assume that Γ has no repelling edges, and that $v\omega \leq 1$ for all v.

We now show that under these circumstances there can be at most one attracting edge-pair in each component. For if there were two we could take an irreducible path $e_1, \ldots, e_n$ joining a vertex of one attracting edge-pair to a vertex of the other. If e_1 were not in the attracting edge-pair, we could add one edge of the pair at the beginning of this path, and similarly for the end. Thus we may assum that e_1 and e_n are attracting. Then e_n is the only edge in B starting at $e_n\sigma$, since every vertex has outdegree ≤ 1. Since there are no repelling edges and $\bar{e}_{n-1} \notin B$, we find that $e_{n-1} \in B$. Inductively, we see that $e_{n-i} \in B$ for all i. In particular, $e_2 \in B$. We get a contradiction, since both e_2 and $\bar{e}_1$ are in B and both begin at $e_2\sigma$.

It follows that there are only finitely many attracting edge-pairs. As before, we can remove them all, so we may assume that all the edge-pairs are ordinary.

It is now enough to show that, when all edge-pairs are ordinary and $v\omega \leq 1$ for all v, if Γ is connected then there is either exactly one circuit or exactly one vertex of outdegree 0 or exactly one equivalence class of rays, and that no two of these can hold simultaneously. The proof of Lemma 22 shows that if there is a vertex of outdegree 0 then there can be no circuits and no rays.

If there is a circuit, the proof of Lemma 22 shows that we may remove an edge-pair, and reduce to the case where there is a vertex of outdegree 0.

If v is a vertex of outdegree 0, the proof of Lemma 22 shows that, for any vertex w, the irreducible path from w to v has all its edges in B. In particular, if $w \neq v$ then w has outdegree ≥ 1. So there is at most one vertex of outdegree 0.

Now suppose that there are no circuits, and no vertices of outdegree 0, and so all vertices have outdegree 1. It follows immediately that any path $e_1, \ldots, e_n$ all of whose edges are in B can be extended by a further edge e_{n+1}. Since there are no circuits, this gives rise to a ray, $e_1, e_2, \ldots$. Let w be any vertex, and let $x_1, \ldots, x_m$ be the shortest irreducible path from w to a point of the ray. Let this path end at $e_n\sigma$. Since e_n is the only edge in B beginning at $e_n\sigma$, and all edges are ordinary, we see that $x_m \in B$, and then, inductively, that $x_j \in B$ for all j. It is then clear, since all vertices have outdegree 1, that the only ray beginning at w is $x_1, \ldots, x_m, e_n, e_{n+1}, \ldots$, and so all rays are equivalent.//

We take the same graph as in Theorem 24. Let e be an edge from v to w with label x. Then e is to be in B iff v begins with x. Thus 1 has outdegree 0, and all other vertices have outdegree 1. So the lemma applies, and we have only to find a way of calculating the number of repelling edge-pairs.

Take any edge-pair. Let the edge in the pair whose label is $s \in S$ (rather than s^{-1}) start at v and end at w. If v begins with s then the edge-pair is attracting or ordinary.

Suppose that v does not begin with s. Then $w = s^{-1}v(s\eta)$. Since v does not begin with s, either w begins with s^{-1}, and so the edge-pair is ordinary, or else $s^{-1}v$ cancels completely in reducing $s^{-1}v(s\eta)$. In the latter case we find that $s\eta = v^{-1}sw$, reduced as written, and so w does not begin with s^{-1}. In this case the edge-pair is repelling. Conversely, if there are words v and w such that $s\eta = v^{-1}sw$, reduced as written, then there is a repelling edge-pair joining v and w.

Thus the number of repelling edge-pairs, which is an upper bound for the rank of Fixη, equals the total number of occurrences for all $s \in S$ of s (not s^{-1}) in $s\eta$.

Examples In $F(a,b)$, let $a\eta = ba$ and $b\eta = bba$. Then a occurs once in $a\eta$, and b occurs twice in $b\eta$. Thus Fixη has rank at most $1+2$.

In $F(a,b,c)$, let $a\varphi = baca^{-1}b^{-1}ac^{-1}$, $b\varphi = a^{-1}ba$, $c\varphi = baca^{-1}b^{-1}aca^{-1}bac^{-1}a^{-1}b^{-1}$. Then a occurs twice in $a\varphi$, b occurs once in $b\varphi$, and c occurs twice in $c\varphi$. Thus Fixφ has rank at most $2+1+2$. We shall analyse this example more closely in a moment, and we will show that Fix$\varphi = \{1\}$.

We do not need to count all the repelling edge-pairs in Γ, but only those in Γ_1. These correspond to the expressions of $s\eta$ as $v^{-1}sw$ with $v \in \Gamma_1$. As we have already seen, $v \in \Gamma_1$ iff there is some u such that $v = u^{-1}(u\eta)$.

We cannot expect in general to find out whether or not a given v satisfies this condition. But it may well happen that we know enough about η to find a necessary condition for this to occur.

For instance, η always induces an endomorphism of the free abelian group F/F'. If this induced endomorphism is the identity then $v = u^{-1}(u\eta)$ only if $v = 1$ in F/F', which we can check. In particular, in our second example φ does induce the identity on F/F', and none of the v which begin a repelling edge are in Γ_1. Hence Fixφ is trivial.

Exercise 4 Show that the attracting edge-pairs correspond to occurences of s^{-1} in $s\eta$, and that if $s\eta = ps^{-1}q$, reduced as written, then the corresponding edge-pair is in Γ_1 iff $sp^{-1} \in \Gamma_1$.

5.4 *COVERINGS OF COMPLEXES*

The notion of a covering of a complex is remarkably tricky to define. The definition in several books is wrong; that is, the authors claim that certain properties follow from their definitions, but there are examples showing that their claims are false. Even knowing this, the definition I gave in an earlier version of this book was also wrong. I think that I have dealt with all the problems in the definitions I give below.

The difficulties arise because the boundary of a cell is a loop which may be a proper power of some other loop. Because of this, we define some special classes of complexes.

The complex K is called *powerless* if there is no cell c such that $c\partial$ can be written as a proper power $(e_1, \ldots, e_r)^k$ with $k > 1$. We say the complex *K has simple boundaries* if, for every cell c, the boundary $c\partial$ is a circuit.

The complex K is a *simplicial complex* if it satisfies the following conditions: (i) no edge is a circle, (ii) for any two vertices v and w there is at most one edge e with $e\sigma = v$ and $e\tau = w$, (iii) the boundary of each cell is a loop e_1, e_2, e_3 with three distinct edges, and (iv) for any three distinct edges e_1, e_2, and e_3 such that e_1, e_2, e_3 is a loop there is at most one cell with this loop as boundary.

Plainly a simplicial complex has simple boundaries, and a complex with simple boundaries is powerless.

Let $\varphi: K \to L$ be a map. Then K will be powerless if L is powerless, since if $c\partial$ were $(e_1, \ldots, e_r)^k$ with $k > 1$ we would have $(c\varphi)\partial = c\partial\varphi = (x_1, \ldots, x_r)^k$, where x_i is $e_i\varphi$. Also K will have simple boundaries if L has simple boundaries. If L is a simplicial complex then K has no circles, and each $c\partial$ is of the required form, but the uniqueness properties required for a simplicial complex need not hold for K.

In Theorem 18 we constructed for each presentation of a group G a complex whose fundamental group is G. This complex will be powerless iff no relator is a proper power. To any presentation $\langle X;R \rangle$ of G there is another presentation which has this property, namely $\langle X, y\ ;\ \{yr \text{ for all } r \in R\}, y \rangle$. Hence, for any group G there is a powerless complex whose fundamental group is G. Further, if L is a powerless complex then the complex containing L which is constructed in Theorem 19 can also be taken to be powerless.

For any group G there is a complex with simple boundary, and even a simplicial complex, whose fundamental group is G. However, this result is harder than the previous ones. It will be shown in the next section that there is a process of subdivision of a complex which yields a simplicial complex with the same fundamental group.

We can now define coverings. We start with an easy case, and become progressively more general.

Definition Let K and $\tilde{K}$ be connected graphs, and let $p: \tilde{K} \to K$ be a map. Then p is called a *covering map* and $\tilde{K}$ is called a *covering graph* of K if, for every vertex $\tilde{v}$ of $\tilde{K}$, p induces a bijection from the set of edges of $\tilde{K}$ starting at $\tilde{v}$ to the set of edges of K starting at $\tilde{v}p$.

Examples Let K be the complex with one vertex v and one edge-pair $\{e, \bar{e}\}$. Let $\tilde{K}$ have one vertex v_n and one edge-pair $\{e_n, \bar{e}_n\}$ for each $n \in \mathbf{Z}$, with $e_n\sigma = v_n$ and $e_n\tau = e_{n+1}$. The obvious map is a covering map. Note that in K both e and $\bar{e}$ start at v, while in $\tilde{K}$ the edges starting at v_n are e_n and $\bar{e}_{n-1}$.

Another covering L_k of K is given by the complex with vertices v_n and edge-pairs $\{e_n, \bar{e}_n\}$ for $1 \leq n \leq k$. Here $e_n\sigma = v_n$ for all n, while $e_n\tau = v_{n+1}$ for $n < k$ and $e_k\tau = v_1$.

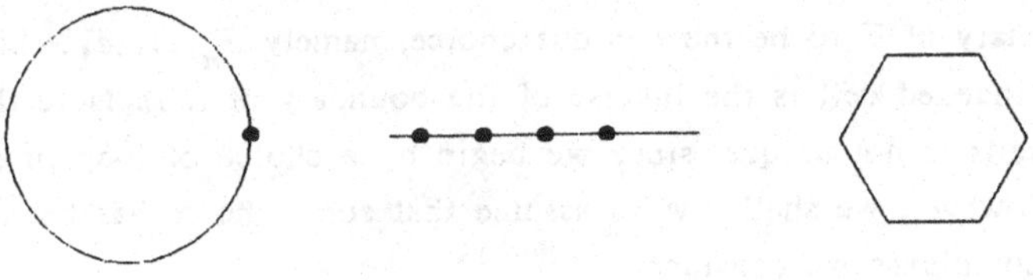

K $\tilde{K}$ L_6

Figure 1

Lemma 27 *Let* $p:\tilde{K}\to K$ *be a covering of graphs. Then* $\tilde{K}p = K$.

Proof The condition ensures that p maps $E(\tilde{K})$ onto $E(K)$ provided that it maps $V(\tilde{K})$ onto $V(K)$. By Lemma 1 it is enough to show that if $v\in V(\tilde{K})p$ and e is an edge with $e\sigma = v$ then $e\tau\in V(\tilde{K})p$. We need only take some $\tilde{v}$ with $\tilde{v}p = v$, and then take an edge $\tilde{e}$ with $\tilde{e}\sigma = \tilde{v}$ and $\tilde{e}p = e$.//

Warning The examples above show that a covering p need not be one-one on the set of edges having $\tilde{v}$ as a vertex. This situation can occur when K has a circle with vertex $\tilde{v}p$.

Definition Let K and $\tilde{K}$ be connected complexes with simple boundaries. Then the map $p:\tilde{K}\to K$ is a *covering map* and $\tilde{K}$ is a *covering complex* of K if, for each vertex $\tilde{v}$ of $\tilde{K}$, p induces a bijection from the set of edges of $\tilde{K}$ starting at $\tilde{v}$ to the set of edges of K starting at $\tilde{v}p$ and also induces a bijection from the set of cells of $\tilde{K}$ with $\tilde{v}$ as a vertex to the set of cells of K with $\tilde{v}p$ as a vertex.

Plainly, if K is a graph then this definition coincides with the previous one. It is also easy to check that if K is a simplicial complex then any covering complex is a simplicial complex. Lemma 27 and the hypothesis on cells immediately tell us that a covering map is onto.

When K does not have simple boundaries we need to look at a slightly more delicate notion.

Let c be any cell of a complex K, and choose a specific loop $e_1,\dots,e_n$ of the cyclic loop $c\partial$. With the cell c we associate n objects which we call *indexed cells*, which we denote by $(c,1),\dots,(c,n)$. The boundary of the indexed cell (c,i) is to be the loop (not the cyclic loop) $e_i,\dots,e_n,e_1,\dots,e_{i-1}$, and we define the *starting vertex* of (c,i) to be $e_i\sigma$. The inverse of the indexed cell (c,i) is to be the indexed cell $(\bar{c},n+2-i)$, where we regard $n+1$ as 1. When we take the boundary of $\bar{c}$ to be the obvious choice, namely $\bar{e}_n,\ldots,\bar{e}_1$, the boundary of this indexed cell is the inverse of the boundary of (c,i). Note that the indexing of cells is not unique, since we begin by a choice of loop in each cyclic loop $c\partial$. However, we shall always assume that some choice has been made in all the complexes we consider.

When K has simple boundaries the indexed cell (c,i) is determined by specifying the cell c and the vertex which is the starting vertex of the indexed cell. When K does not have simple boundaries there will always be a cell c and two distinct indexed cells (c,i) and (c,j) with the same starting vertex. When K is powerless the indexed cell (c,i) is determined by specifying the cell c and the boundary of (c,i), but this is not enough in general. For let us write $c\partial$ as $(e_1,\ldots,e_r)^k$ with k as large as possible. Then it is easy to check that the loops $e_i,\ldots,e_{i-1}$ and $e_j,\ldots,e_{j-1}$, which are the boundaries of (c,i) and (c,j), are the same iff $j-i$ is a multiple of r.

Let $\varphi:L\to K$ be a map of powerless complexes. Let c be a cell of L, and let the indexed cell $(c,1)$ have boundary $e_1,\ldots,e_n$. Then the loop $e_1\varphi,\ldots,e_n\varphi$ of K is the boundary of some indexed cell corresponding to $c\varphi$. Suppose this indexed cell is (c,j). Then we have an induced map of indexed cells sending (c,i) to $(c\varphi,i+j-1)$ (where a number greater than n is to be reduced mod n), with the boundary of the image being the image of the boundary. This induced map will send an indexed cell and its inverse to an indexed cell and its inverse.

Definition Let $p:\tilde{K}\to K$ be a map of connected powerless complexes. Then p is a *covering map* and $\tilde{K}$ is a *covering complex* of K if, for each vertex $\tilde{v}$ of $\tilde{K}$, p induces a bijection from the set of edges of $\tilde{K}$ starting at $\tilde{v}$ to the set of edges of K starting at $\tilde{v}p$ and also induces a bijection from the set of indexed cells of $\tilde{K}$ starting at $\tilde{v}$ to the set of indexed cells of K starting at $\tilde{v}p$.

Again, coverings are onto. Also, this notion coincides with the previous one when the complex K has simple boundary, in which case we know that $\tilde{K}$ also has simple boundary.

Example Let K have one vertex v, two edge-pairs $\{x,\bar{x}\}$ and $\{y,\bar{y}\}$, and one cell-pair $\{c,\bar{c}\}$ with $c\partial$ being x,y. Thus K has two indexed cells associated with c, namely $(c,1)$ with boundary x,y and $(c,2)$ with boundary y,x; both these cells have starting vertex v.

Let $\tilde{K}$ have vertices v_n, edge-pairs $\{x_n,\bar{x}_n\}$ and $\{y_n,\bar{y}_n\}$, and cell-pairs $\{c_n,\bar{c}_n\}$ for all $n\in\mathbb{Z}$. Let $x_n\sigma = v_n = y_n\tau$ and $x_n\tau = v_{n+1} = y_n\sigma$, and let $c_n\partial$ be x_n,y_n. Let p be the obvious map. Then p does not give a bijection of cells with vertex v_n to cells with vertex v. However, the indexed cells corresponding to c_n are $(c_n,1)$ with starting vertex v_n and $(c_n,2)$ with starting vertex v_{n+1}. Thus p maps the indexed cells $(c_n,1)$ and $(c_{n-1},2)$, which both start at v_n, onto the indexed cells $(c,1)$ and $(c,2)$. We see that p is a covering map.

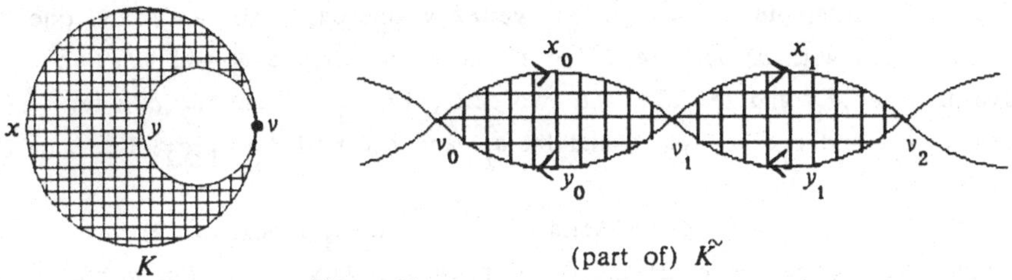

Figure 2

Finally, we must look at general complexes. We see that a map $\varphi: L \to K$ does not determine a unique map on indexed cells. Suppose that the boundary of $c\varphi$ can be written as $(x_1,\ldots,x_r)^k$ with $k>1$, and that the image under φ of the boundary of the indexed cell $(c,1)$ is the boundary of the indexed cell (c,j). Then we can map the indexed cells (c,i) to the indexed cells $(c\varphi, j+i-1)$, as before. But we are also permitted to take any integer m, and then map (c,i) to $(c\varphi, mr+j+i-1)$.

Consequently, we have to make a choice, which we call $\varphi_!$, of maps of indexed cells corresponding to φ. We may make the choice so that $\varphi_!$ sends an indexed cell and its inverse to an indexed cell and its inverse. Note that if $\psi: M \to L$ is another map then $\psi_!\varphi_!$ is a possible choice for $(\psi\varphi)_!$; because of the non-uniqueness, we cannot just say that $(\psi\varphi)_!$ is $\psi_!\varphi_!$. However, if our complexes are powerless, so that only one choice is possible (once we have specified the indexing of the cells), then we do know that $\psi_!\varphi_! = (\psi\varphi)_!$.

Definition Let $p: \tilde{K} \to K$ be a map of connected complexes. Then p is a *covering map* and $\tilde{K}$ is a *covering complex* of K if there is a choice of $p_!$ such that, for every vertex $\tilde{v}$ of $\tilde{K}$, p induces a bijection of the set of edges of $\tilde{K}$ starting at $\tilde{v}$ to the set of edges of K starting at $\tilde{v}p$ and $p_!$ induces a bijection of the set of indexed cells of $\tilde{K}$ starting at $\tilde{v}$ to the set of indexed cells of K starting at $\tilde{v}p$.

This coincides with the previous definition when K has simple boundaries, in which case $\tilde{K}$ also has simple boundaries.

As before, covering maps are onto.

It would be more precise to refer to the pair $(p, p_!)$ as a covering map. The example below shows that one choice of $p_!$ may satisfy the required property, while another choice may not. However, we shall not need to pay much attention to $p_!$, so the simpler notation is preferable.

Example Let K have one vertex v, one edge-pair $\{e,\bar{e}\}$, and one cell-pair $\{c,\bar{c}\}$ with $c\partial$ being e, e. Let $\tilde{K}$ have two vertices v_1 and v_2, two edge-pairs $\{e_1,\bar{e}_1\}$ and $\{e_2,\bar{e}_2\}$, and two cell pairs $\{c_1,\bar{c}_1\}$ and $\{c_2,\bar{c}_2\}$; let $e_1\sigma = v_1 = e_2\tau$ and $e_1\tau = v_2 = e_2\sigma$, and let $c_1\partial$ and $c_2\partial$ both be e_1, e_2. Let p be the obvious map.

Define the indexed cells of $\tilde{K}$ so that $(c_1,1)$ and $(c_2,2)$ have boundary e_1, e_2, while $(c_1,2)$ and $(c_2,1)$ have boundary e_2, e_1. Let $p_!$ map $(c_1,1)$ to $(c,1)$. We may choose $p_!$ either to map $(c_2,1)$ to $(c,1)$ or to map $(c_2,1)$ to $(c,2)$; in the latter case it maps $(c_2,2)$ to $(c,1)$. It is easy to check that the first choice of $p_!$ gives a covering map, while the second does not.

Let L be a complex with two vertices w_1 and w_2, two edge-pairs $\{x_1,\bar{x}_1\}$ and $\{x_2,\bar{x}_2\}$, and one cell-pair $\{d,\bar{d}\}$; let $x_1\sigma = w_1 = x_2\tau$ and $x_1\tau = w_2 = x_2\sigma$, and let $d\partial$ be x_1, x_2. With K and $\tilde{K}$ as above, there are two obvious maps from L to $\tilde{K}$, the first, φ, sending d to c_1 and the second, ψ, sending d to c_2. Plainly $\varphi p = \psi p$. However, with $(d,1)$ being the indexed cell starting at w_1, we must define $\varphi_!$ to map $(d,1)$ to $(c_1,1)$, This explains the conditions in the next proposition. It is examples like this one which need the extreme care in our definitions, and which show the inadequacies of many other suggested definitions.

Proposition 28 *Let $p:\tilde{K} \to K$ be a covering. Let L be a connected complex, and let $\varphi,\psi:L \to \tilde{K}$ be maps such that $\varphi p = \psi p$, and such that there is a vertex v with $v\varphi = v\psi$. Then (i) $\varphi = \psi$ on the 1-skeleton L^1; (ii) if the complexes are powerless then $\varphi = \psi$; (iii) if we can choose $\varphi_!$ and $\psi_!$ so that $\varphi_! p_! = \psi_! p_!$ then $\varphi_! = \psi_!$, and so $\varphi = \psi$.*

Proof (i) We first show that $\varphi = \psi$ on $V(L)$. By Lemma 1, it is enough to show that if e is an edge of L such that $e\sigma\varphi = e\sigma\psi$ then $e\tau\varphi = e\tau\psi$. Now $e\varphi$ and $e\psi$ both begin at the same vertex, since $(e\varphi)\sigma = (e\sigma)\varphi$ and $(e\psi)\sigma = (e\sigma)\psi$, and they have the same image under p. By the definition of a covering, this implies that $e\varphi = e\psi$. It follows that $e\tau\varphi = e\tau\psi$ (because $e\tau\varphi = e\varphi\tau$, etc.), as required. The same argument now tells us that $e\varphi = e\psi$ for any edge e.

(ii) This can be proved directly, but we may as well use part (iii). Because our complexes are powerless, we know that $\varphi_! p_! = (\varphi p)_! = (\psi p)_! = \psi_! p_!$.

(iii) Let (c,i) be an indexed cell of L, and let its starting vertex be v. Then $(c,i)\varphi_!$ and $(c,i)\psi_!$ have the same image under $p_!$. Because $\varphi = \psi$ on the vertices, they both have the starting vertex $v\varphi$. The definition of a covering now tells us that $(c,i)\varphi_! = (c,i)\psi_!$, as required.//

Proposition 29 *Let $p:\tilde{K} \to K$ be a covering. Let f be a path of K starting at v. Then, for any vertex $\tilde{v}$ of $\tilde{K}$ such that $\tilde{v}p = v$, there is a unique path $\tilde{f}$ of $\tilde{K}$ starting at $\tilde{v}$ and such that $\tilde{f}p = f$.*

Proof Let f be $e_1, \ldots, e_n$. Suppose that for some r we have found uniquely edges $\tilde{e}_1, \ldots, \tilde{e}_r$ of $\tilde{K}$ such that $\tilde{e}_1, \ldots, \tilde{e}_r$ is a path of $\tilde{K}$ starting at $\tilde{v}$ and such that $\tilde{e}_i p$ is e_i for $i \leq r$. Then $\tilde{e}_r \tau p = e_r \tau$, and so there is a unique edge $\tilde{e}_{r+1}$ starting at $\tilde{e}_r \tau$ and with $\tilde{e}_{r+1} p = e_{r+1}$. The result follows by induction.//

Proposition 30 *Let $p:\tilde{K} \to K$ be a covering. Let $\tilde{f}$ and $\tilde{g}$ be paths of $\tilde{K}$ starting at the same point and such that $\tilde{f}p \simeq \tilde{g}p$. Then $\tilde{f} \simeq \tilde{g}$.*

Proof It is enough to show that if $\tilde{g}p$ comes from $\tilde{f}p$ by an elementary reduction then $\tilde{g}$ comes from $\tilde{f}$ by elementary reduction. Let $\tilde{f}$ be $\tilde{e}_1, \ldots, \tilde{e}_n$, and let e_r be $\tilde{e}_r p$ for $r \leq n$.

Suppose that $e_{i+1} = \bar{e}_i$, and that $\tilde{g}p$ comes from $\tilde{f}p$ by deletion of e_i, e_{i+1}. Then $\tilde{e}_{i+1} = \bar{\tilde{e}}_i$, since both start at $\tilde{e}_i \tau$ and they have the same images under p. Then removal of $\tilde{e}_i, \tilde{e}_{i+1}$ is an elementary reduction, and the path obtained must be $\tilde{g}$, by the previous proposition, since both it and $\tilde{g}$ start at the start of $\tilde{f}$, and they evidently have the same image under p.

Suppose that $\tilde{g}p$ comes from $\tilde{f}p$ by removal of the loop $e_i, \ldots, e_j$, where this loop is the boundary of a cell c. For convenience, assume that c is indexed so that this loop is the boundary of the indexed cell $(c,1)$. Then there is a cell $\tilde{c}$ of $\tilde{K}$ one of whose indexed cells, say $(\tilde{c}, k)$, starts at $\tilde{e}_i \sigma$ and maps to $(c,1)$. Then $(\tilde{c}, k)\partial p = (\tilde{c}, k)p_! \partial = (c,1)\partial$ is the loop $e_i, \ldots, e_j$. Since the path $\tilde{e}_i, \ldots, \tilde{e}_j$ also starts at $\tilde{e}_i \sigma$ and maps to this loop by p, the previous proposition tells us that $(\tilde{c}, k)\partial$ is $\tilde{e}_i, \ldots, \tilde{e}_j$. Hence deletion of these edges is an elementary reduction of $\tilde{f}$. The result of this deletion must be $\tilde{g}$, since they both start at the start of $\tilde{f}$ and they have the same image under p.//

5.5 *SUBDIVISIONS*

In this section we will show how, given a complex K, we can construct a simplicial complex related to K with the same fundamental group as K. In the course of the construction we will obtain other complexes with the same fundamental group as K, one of which is powerless and the other of which has simple boundaries. This material is not needed for the later development of the theory. But it shows that we may use the simpler approaches to coverings of complexes if we wish, without running into any problems in the applications.

Let K be a connected complex. Its *edge subdivision* K^e is the complex defined as follows. The vertices of K^e are the vertices of K together with one new vertex for each edge-pair $\{e.\bar{e}\}$ of K. This new vertex will be denoted by v_e or by $v_{\bar{e}}$. There are two edges of K^e to each edge e of K. These are denoted by e_- and e_+. The edge e_- starts at $e\sigma$ and ends at v_e, while the edge e_+ starts at v_e and ends at $e\tau$. We define $\overline{e_-}$ and $\overline{e_+}$ to be $\bar{e}_+$ and $\bar{e}_-$, resepctively. The cells of K^e are the cells of K. The boundary in K^e of a cell is obtained from its boundary in K by replacing each edge e by the two edges e_-, e_+.

The *cell-subdivision* K^c of K is defined as follows. The vertices of K^c are the vertices of K together with one new vertex for each cell-pair. It is most convenient to take a set C_0 of cells which contains exactly one cell from each pair, and to let the new vertices be v_c for all $c \in C_0$. The edges of K^c are the edges of K together with new edges. Let $c \in C_0$ have as boundary the loop $e_1, \ldots, e_n$. Then there are to be n new edges f_{ci} for $1 \le i \le n$-with $f_{ci}\sigma = v_c$ and $f_{ci}\tau = e_i\sigma$, together with the inverses $\bar{f}_{ci}$ of these. The cells of K^c are all to be new cells. Corresponding to the cell $c \in C_0$ there are n new cell-pairs $\{d_{ci}, \bar{d}_{ci}\}$ for $1 \le i \le n$. The boundary of d_{ci} is to be $f_{ci}, e_i, \bar{f}_{c,i+1}$.

Example In Figure 3 below, K has four vertices, four edge-pairs, and one cell-pair. Then K^e has eight vertices, eight edge-pairs, and one cell-pair, while K^c has five vertices, eight edge-pairs, and four cell-pairs. Finally $(K^e)^c$ has nine vertices, sixteen edge-pairs, and eight cell-pairs.

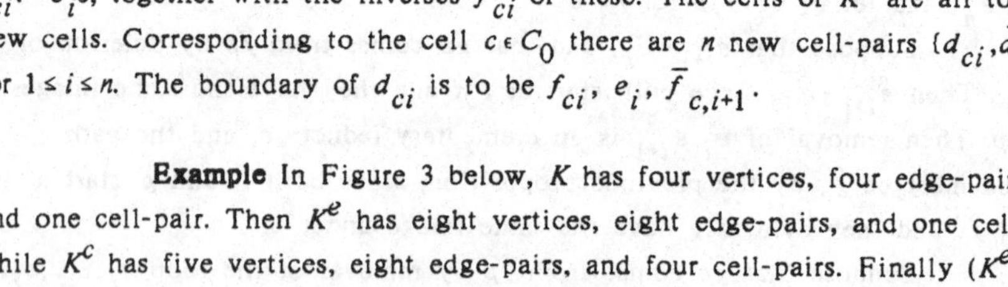

K K^e K^c $(K^e)^c$

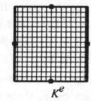

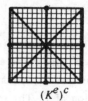

Figure 3.

It is easy to check the following facts. K^e has no circles. If K has no circles then K^e has at most one edge with a given start and finish. If K is powerless then K^e is powerless. If K has simple boundaries then K^e has simple boundaries.

Also, if K has no circles then K^c has no circles. If K has no circles then K^c has simple boundaries. If K has at most one edge with a given start and finish and K has simple boundaries then K^c has at most one edge with a given start and finish.

The *barycentric subdivision* K' of K is defined to be $(K^e)^c$. Then K' has simple boundaries, and K'' is a simplicial complex.

Proposition 31 *$\pi_1(K^e)$ is isomorphic to $\pi_1(K)$ for any connected K.*

Proof Let T be a maximal tree in K, and let $E_0 \subseteq E$ consist of exactly one edge from each edge-pair not in T. Let S be the subgraph of K^e consisting of all vertices of K^e, and whose edges are e_- and e_+ for all $e \in T$ and e_- and $\overline{e_-}$ for all $e \in E_0$. Note that if $e \in T$ then $\bar{e} \in T$, and so $\overline{e_-}$ and $\overline{e_+}$, being $\bar{e}_+$ and $\bar{e}_-$ respectively, are in S.

We shall show that S is a tree. Choose some vertex a of K. Let v be any vertex of K. Then there is a path in T from a to v. Replacing each edge e by the pair of edges e_-, e_+ in S, we get a path in S from a to v. Now take a vertex v_e of K^e. We may assume that $e \in E_0 \cup T$. Then there is a path in S from a to v_e which first goes from a to $e\sigma$, and follows this with the edge e_-. Hence S is connected.

In an irreducible path in K^e, the edge (if any) following an edge e_- must be the corresponding e_+. It follows that two irreducible paths in S from a to v must be the same, since they both come from irreducible paths in T by splitting edges into two, and these two paths in T are the same, as T is a tree. Now consider two irreducible paths f and g in S from a to the vertex v_e. If they both end in e_- or both end in $\bar{e}_-$ then, deleting the last edge, we get two irreducible paths in S from a to a vertex of K. We have just seen that these must be the same. Suppose that f ends in e_- and g ends in $\bar{e}_-$. Then if we remove $\bar{e}_-$ from g we get an irreducible path in S from a to $e\tau$ which does not end in e_+ (as $\bar{e}_- = \overline{e_+}$). If we add e_+ to f we get an irreducible path in S from a to $e\tau$ which does end in e_+. Thus we get two distinct irreducible paths in S from a to $e\tau$. Hence this case cannot happen.

It follows that S is a tree in K^e, and it must be a maximal tree, since it contains every vertex.

Take the presentation of $\pi_1(K^e,a)$ using the maximal tree S and the presentation of $\pi_1(K,a)$ using the maximal tree T. In both presentations, the relations say that certain generators are trivial. If we delete these generators, the resulting presentations of the two groups become identical.//

Proposition 32 *$\pi_1(K^c)$ is isomorphic to $\pi_1(K)$ for any connected K.*

Proof Let T be a maximal tree of K. Let S be the subgraph of K^c conisting of all the vertices of K, all the edges of T, and all the edges f_{c1} and $\bar{f}_{c1}$ for $c \in C_0$. Because each vertex v_c is a vertex of exactly one edge-pair in S, it is easy to check that S is a tree. As S contains every vertex, it is a maximal tree in K^c.

The presentation of $\pi_1(K^c)$ using this tree is the following. It has generators e for all $e \in E$ together with the f_{ci} for $c \in C_0$ (we do not need $\bar{f}_{ci}$, which we just regard as f_{ci}^{-1}). The relations are $e\bar{e} = 1$ for all $e \in E$ and $e = 1$ for all $e \in T$, together with further relations for each $c \in C_0$. If $c\partial$ is $e_1, \ldots, e_n$ then there are $n+1$ relations corresponding to this c. They are $f_{c1} = 1$, $f_{ci}e_i = f_{c,i+1}$ for $0 < i < n$, and $f_{cn}e_n = f_{c1}$. Tietze transformations replace these with $f_{c1} = 1$, $f_{ci} = e_1 \ldots e_{i-1}$ for $1 < i < n$, and $f_{cn}e_n = f_{c1}$, and a further Tietze transformation lets us replace the last of these with the relation $e_1 \ldots e_n = 1$. Now Tietze transformations let us delete every f_{ci}. The result is just the presentation of $\pi_1(K)$ using the tree T.//

Let $\varphi: K \to L$ be a map. We can define a map $\varphi^e: K^e \to L^e$ such that φ^e corresponds to φ under the isomorphism of Proposition 31. We can also define a map $\varphi^c: K^c \to L^c$ such that φ^c corresponds to φ under the isomorphism of proposition 32. While φ^e is uniquely given by φ, there are choices for φ^c, but it is uniquely given by φ and $\varphi_!$. The details are left to the reader.

The constructions can easily be modified so that we do not subdivide all edges or cells, but just choose some of them to subdivide. In particular, suppose that L is a subcomplex of K, and that L is a simplicial complex. Then, by subdividing the remaining edges and cells (twice) we obtain a simplicial complex M which contains L (rather than containing a subdivision of L) and such that the inclusion of L in M and the inclusion of L in K induce the same homomorphisms of fundamental groups.

5.6 *GEOMETRIC REALISATIONS*

In this section, we will associate a space $|K|$ with each complex K in such a way that $\pi_1(|K|)$ is isomorphic to $\pi_1(K)$. When K is the complex associated with the group G by Theorem 18 then $|K|$ will be the space associated with G by Theorem 4.12. The results of this section are not needed elsewhere. But I feel it is worth noting that the analogy between the theory of fundamental groups of complexes and the theory of fundamental groups of spaces is more than merely formal.

We begin with a graph Γ, and take a set E_0 of edges which contains exactly one member of each edge-pair. We can regard the set V of vertices as a discrete space. For each $e \in E_0$ take a copy I_e of the unit interval I, these copies being disjoint. We attach $\cup I_e$ to V by the map which sends 0 and 1 in I_e to $e\sigma$ and $e\tau$ in V. The resulting space is defined to be the *geometric realisation* $|\Gamma|$ of Γ.

Let f be the edge-path $e_1, \ldots, e_n$ of Γ. We associate with it a standard path $|f|$ in $|\Gamma|$ as follows. If $e_i \in E_0$ then $|f|$ maps the interval $[(i-1)/n, i/n]$ linearly onto the image of I_{e_i} in $|\Gamma|$, sending $(i-1)/n$ to 0 and i/n to 1. If $e_i \notin E_0$ then $\bar{e}_i \in E_0$; In this case $|f|$ maps $[(i-1)/n, i/n]$ linearly to the image of $I_{\bar{e}_i}$, this time sending $(i-1)/n$ to 1 and i/n to 0. If f is an edge-loop then $|f|$ is a standard loop. In this case we can, as usual, regard $|f|$ as a map from S^1 to Γ.

Now let K be a complex. Take a set C_0 of cells, containing exactly one cell-from each pair. We have already constructed the space $|K^1|$ corresponding to the 1-skeleton K^1 of K. For any cell $c \in C_0$ choose an edge-loop (not just a cyclic loop) as its boundary. As we have seen, this gives rise to a map ϑ_c from S^1 to $|K^1|$. Take a copy D_c of the disc D^2 for each $c \in C_0$. We now define $|K|$, the *geometric realisation* of K, to be the space obtained from $|K^1|$ by attaching $\cup D_c$ using the maps ϑ_c. A different choice of the edge loop in the cyclic loop $c\partial$ would provide a homeomorphic space.

Let K be the complex associated with the group G by Theorem 18. It is easy to check that $|K|$ is the space associated with G by Theorem 4.12.

Proposition 33 $\pi_1(|K|)$ *is isomorphic to* $\pi_1(K)$ *for any connected complex* K.

Proof The direct limit arguments of Theorem 15 and Proposition 4.10 tell us that the result is true in general if it is true when K is a finite complex.

Let K be a finite complex. Let c be a cell of K, and let L be the complex $K - \{c, \bar{c}\}$. Then Corollary 4 to Theorem 17 and the corollary to Proposition 4.11 ensure that $\pi_1(|K|$ is isomorphic to $\pi_1(K)$ provided that $\pi_1(|L|)$ is isomorphic to $\pi_1(L)$. Thus, inductively, it is enough to prove the result when K is a finite graph.

Let Γ be a finite graph, and let T be a maximal tree in Γ. Suppose Γ has an edge e not in T. Let Δ be the graph $\Gamma - \{e, \bar{e}\}$. Then Δ is also a connected graph. By Corollary 1 to Theorem 17 and the remark after Proposition 4.9, $\pi_1(|\Gamma|)$ is isomorphic to $\pi_1(\Gamma)$ provided that $\pi_1(|\Delta|)$ is isomorphic to $\pi_1(\Delta)$. Thus, inductively, it is enough to prove the result when K is a finite tree T.

We know from Lemma 10 that any finite tree T has a vertex v belonging to exactly one edge-pair $\{e, \bar{e}\}$. Let S be obtained from T by deleting the edge-pair $\{e, \bar{e}\}$ and the vertex v. Then S is a tree. It is easy to check that $|T|$ is obtained from $|S|$ by attaching an interval at one end-point. So, by the example preceding Proposition 4.9, $|S|$ is a strong deformation retract of $|T|$. It follows that $|T|$ is contractible if $|S|$ is contractible. Inductively, we find that $|T|$ is contractible, and so $\pi_1(|T|)$ is trivial. Since we know that $\pi_1(T)$ is trivial, the result is proved.//

Given a map $\varphi: K \to L$ we can obtain a map $|\varphi|: |K| \to |L|$. This map is not unique, but is uniquely defined by φ and $\varphi_!$. The homomorphism induced by φ and $|\varphi|$ correspond under the isomorphisms of the proposition. Proofs are left to the reader.

It needs some care to establish the topological conditions needed to show that $|p|$ is a covering of spaces when p is a covering of complexes, but this can be done. In our examples this fact will usually be clear.

For instance, let Γ be the graph with one vertex and one edge-pair. Then $|\Gamma|$ is S^1. Our two examples of coverings of Γ give rise to the covering of S^1 by $\mathbb{R}$ and the covering of S^1 by itself in which the map sends the complex number z to z^k.

Let K be the complex with one vertex, one edge-pair and one cell-pair, with $c\partial$ being e,e, and let $p: \tilde{K} \to K$ be the covering discussed. Then $|\tilde{K}|$ is S^2, and $|K|$ is the real projective space $\mathbb{R}P^2$.

6 COVERINGS OF SPACES AND COMPLEXES

In this chapter we will develop the theory of covering spaces and covering complexes, and in the next chapter we will use this theory to obtain properties of groups.

Most of our results will be true both for spaces and for complexes. The proofs for the two cases will be closely related; the proof for complexes is often just a simplification of the proof for spaces, and I omit the proof for complexes when this happens. When proving results for complexes, I will give the general case using the map $p_!$. For all the results that we need, however, it would be sufficient to consider powerless complexes only. Thus we need never consider $p_!$, and readers may make this simplification if they wish.

All spaces we consider will be assumed path-connected and locally path-connected, and all complexes will be assumed connected, unless otherwise stated. The symbols $\tilde{X}$ and $\tilde{K}$ will always denote coverings of the space X and the complex K, and p will always denote a covering map (from $\tilde{X}$ to X or from $\tilde{K}$ to K). Also $\tilde{v}$ will denote a point of $\tilde{X}$ (or vertex of $\tilde{K}$) such that $\tilde{v}p = v$, where v is a point of X (or vertex of K). We will also say that $\tilde{v}$ *lies above* v.

Recall (Propositions 4.15 and 5.29) that given any path f (in X or K) starting at v and some $\tilde{v}$ lying above v then there is a unique path $\tilde{f}$ in $\tilde{X}$ (or $\tilde{K}$) starting at $\tilde{v}$ such that $\tilde{f}p = f$. We call $\tilde{f}$ the *lift* of f starting at $\tilde{v}$. We also showed (Propositions 4.16 and 5.30) that homotopic paths starting at v have homotopic lifts starting at $\tilde{v}$; in particular, these lifts end at the same point.

We showed further (Lemma 4.14) that if two maps φ and ψ from (Y, y_0) to $(\tilde{X}, \tilde{v})$ are such that $\varphi p = \psi p$ then $\varphi = \psi$. The same holds (by Proposition 5.28) for maps from (L, w) to $(\tilde{K}, \tilde{v})$ when the complexes are powerless. For general complexes we need to assume that $\varphi_! p_! = \psi_! p_!$, and we can then deduce that $\varphi_! = \psi_!$, and so $\varphi = \psi$.

We will find it convenient to denote the inverse of a path f by f^{-1}, rather than $\bar{f}$. Plainly, if $\tilde{f}$ is a lift of f then $\tilde{f}^{-1}$ is a lift of f^{-1}.

Lemma 1 *A loop at* v *lifts to a loop at* $\tilde{v}$ *iff it represents an element of* $\pi_1(\tilde{X},\tilde{v})p*$ *(or of* $\pi_1(\tilde{K},\tilde{v})p*$ *for the case of complexes).*

Proof If $f = \tilde{f}p$, where $\tilde{f}$ is a loop based at $\tilde{v}$, then plainly f represents an element of $\pi_1(\tilde{X}.\tilde{v})p*$.

Conversely, let f represent at element of $\pi_1(\tilde{X},\tilde{v})p*$. Then there is a loop $\tilde{g}$ based at $\tilde{v}$ such that $f \simeq \tilde{g}p$. Since $\tilde{g}p$ lifts to $\tilde{g}$, which begins and ends at $\tilde{v}$, the lift $\tilde{f}$ of f starting at $\tilde{v}$ must also end at $\tilde{v}$.//

Corollary 1 *Let* f *and* g *be paths from* v *to* w, *with lifts* $\tilde{f}$ *and* $\tilde{g}$ *starting at* $\tilde{v}$. *Then* $\tilde{f}$ *and* $\tilde{g}$ *end at the same point iff* $f.g^{-1}$ *represents an element of* $\pi_1(\tilde{X},\tilde{v})p*$ *(or* $\pi_1(\tilde{K},\tilde{v})p*$ *for the case of complexes).*

Proof Suppose that $\tilde{f}$ and $\tilde{g}$ end at the same point. Then we can form $\tilde{f}.\tilde{g}^{-1}$, and this is a loop. The loop $f.g^{-1}$ is the image of this loop under p, and so $f.g^{-1}$ represents an element of $\pi_1(\tilde{X},\tilde{v})p*$.

Conversely, let $f.g^{-1}$ represent an element of $\pi_1(\tilde{X},\tilde{v})p*$. By the lemma, $f.g^{-1}$ lifts to a loop $\tilde{h}$ based at $\tilde{v}$. Then $(f.g^{-1}).g$ will lift to $\tilde{h}.\tilde{g}$. In particular, the lift $\tilde{h}.\tilde{g}$ of $(f.g^{-1}).g$ starting at $\tilde{v}$ has the same end-point as $\tilde{g}$. Since $f \simeq (f.g^{-1}).g$, we know that the end-points of $\tilde{f}$ and $\tilde{h}.\tilde{g}$ are the same.//

Corollary 2 *The cardinality of* xp^{-1} *is independent of* x, *where* x *is a point of* X *(a vertex of* K*).*

Remark This cardinality is called the *number of sheets* of the covering.

Proof Take a path f from v to x. To each $\tilde{v}$ above v there is a unique lift $\tilde{f}$ of f starting at $\tilde{v}$. The end-point of $\tilde{f}$ will be a point above x. Thus f defines a function from vp^{-1} to xp^{-1}. Similarly, the path f^{-1} defines a function from xp^{-1} to vp^{-1}. Let $\tilde{f}$ start at $\tilde{v}$ and end at $\tilde{x}$. Then $\tilde{f}^{-1}$ is a lift of f^{-1} which starts at $\tilde{x}$ and ends at $\tilde{v}$. Thus the composite of the two maps (in either order) is the identity (on vp^{-1} or xp^{-1}, respectively).//

Corollary 3 *There is a bijection from* vp^{-1} *to the set of cosets of* $\pi_1(\tilde{X},\tilde{v})p*$ *in* $\pi_1(X,v)$.

Proof To each point $\tilde{w}$ in vp^{-1} there is at least one path from $\tilde{v}$ to $\tilde{w}$, and the image of this path under p represents an element of $\pi_1(X,v)$. Two different paths from $\tilde{v}$ to $\tilde{w}$ give rise to elements of $\pi_1(X,v)$ in the same coset of $\pi_1(\tilde{X},\tilde{v})p^*$, by Corollary 1. Conversely, any element of $\pi_1(X,v)$ corresponds to some $\tilde{w} \in vp^{-1}$ in this way, and two elements in the same coset correspond to the same point.//

Theorem 2 *p^* is a monomorphism for any covering p.*

Proof Let $\tilde{f}$ be a loop at $\tilde{v}$ which represents an element α with αp^* the identity element. Then $\tilde{f}p$ is homotopic to the trivial loop at v in X (or K). Hence $\tilde{f}$ is homotopic to the trivial loop at $\tilde{v}$, and so α is the identity.//

Proposition 3 *Let $\tilde{v}_1$ be another point in vp^{-1}. Then $\pi_1(\tilde{X},\tilde{v}_1)p^*$ is conjugate in $\pi_1(X,v)$ to $\pi_1(\tilde{X},\tilde{v})$. Further, any conjugate of $\pi_1(\tilde{X},\tilde{v})p^*$ is of the form $\pi_1(\tilde{X},\tilde{v}_1)p^*$ for a suitable choice of $\tilde{v}_1$ in vp^{-1}. The same holds for complexes.*

Proof Let $\tilde{f}$ be a path in $\tilde{X}$ from $\tilde{v}$ to $\tilde{v}_1$, and let $f = \tilde{f}p$. Then f is a loop at v, which represents an element α of $\pi_1(X,v)$. We know that $\pi_1(\tilde{X},\tilde{v}_1) = \pi_1(\tilde{X},\tilde{v})\tilde{f}^{\#}$. Hence $\pi_1(\tilde{X},\tilde{v}_1)p^* = (\pi_1(\tilde{X},\tilde{v})p^*)(\tilde{f}p)^{\#}$, and this is the conjugate of $\pi_1(\tilde{X},\tilde{v})p^*$ by α.

Conversely, given α in $\pi_1(X,v)$, we represent it by a loop f at v. This loop lifts to a path in $\tilde{X}$ starting at $\tilde{v}$. The end of this path will be a point $\tilde{v}_1$ in vp^{-1}. The argument of the previous paragraph then shows that $\pi_1(\tilde{X},\tilde{v}_1)p^*$ is the conjugate of $\pi_1(\tilde{X},\tilde{v})p^*$ by α.//

We say that the covering $p:\tilde{X} \to X$ (of spaces or complexes) *belongs to* the subgroup $\pi_1(\tilde{X},\tilde{v})p^*$. Evidently, the subgroup of $\pi_1(X,v)$ to which a given covering belongs is determined only up to conjugacy. We shall see later that there is essentially at most one covering belonging to a given subgroup. Also, for sufficiently complicated spaces, some subgroups have no coverings belonging to them, but there is always a covering of a complex belonging to a given subgroup. We now investigate under what conditions there is a covering of a space X belonging to the subgroup H of $\pi_1(X,v)$.

We call H *sufficiently large* if each $x \in X$ has a path-connected open neighbourhood U such that, for any path f from v to x and any loop g in U based at x, the loop $f.g.f^{-1}$ represents an element of H. Such a neighbourhood

U will be called a *basic neighbourhood* of x.

Let U be a basic neighbourhood of x and let $y \in U$. Then U is also a basic neighbourhood of y. For we may choose a path h in U from y to x. Then, given any path f in X from v to y, and any loop g in U based at y, the loop $(f.h).(h^{-1}.g.h).(f.h)^{-1}$ represents, by definition of a basic neighbourhood of x, an element of H. Since this loop is homotopic to $f.g.f^{-1}$, we see that U satisfies the required condition with respect to y. Also, any connected open neighbourhood of x contained in U will itself be a basic neighbourhood.

Plainly the trivial subgroup is sufficiently large (and hence every subgroup is sufficiently large) iff each $x \in X$ has an open neighbourhood U such that $\pi_1(U,x)i*$ is trivial, where $i:U \to X$ is the inclusion. A space with this property is sometimes called *semi-locally simply connected* (if $\pi_1(U,x)$ were itself trivial, we would refer to local simple connectedness). In particular, X is semi-locally simply connected if it is locally contractible; that is, if each point has a contractible neighbourhood.

The geometric realisation $|K|$ of a complex K is locally contractible. This follows from the next lemma, going from the discrete space $V(K)$ to the space $|K^1|$, and then to $|K|$.

Lemma 4 *Let X be a locally contractible space, and let Y be obtained from X by attaching disks of various dimensions. Then Y is locally contractible.*

Proof Let the disks D_i be attached by maps f_i. Any point of $Y - X$ lies in the interior of some D_i, and so has a contractible neighbourhood.

Let $x \in X$, and let U be a contractible neighbourhood of x in X. For each i, the set Uf_i^{-1} is an open subset of the boundary $(n-1)$-sphere S_i of the n-disk D_i.

We can regard D_i as consisting of all points ta, where $a \in S_i$ and $t \in [0,1]$. Let V_i be $\{ta;\ t \in (0,1]$ and $a \in Uf_i^{-1}\}$. Then V_i has a strong deformation retraction onto Uf_i^{-1}. Now the set $W = U \cup \bigcup\{V_i;$ all $i\}$ is an open neighbourhood of x in Y. By Proposition 2.4 $W \times I$ is an identification space of the disjoint union of $U \times I$ and the sets $V_i \times I$. Thus the maps from $V_i \times I$ to Uf_i^{-1} which show the strong deformation retraction combine to give a continuous function from $W \times I$ to U. Then W is contractible, since it has a strong deformation retraction onto the contractible set U.//

Corollary *With the same hypotheses, X is locally path-connected.*//

Theorem 5 *A path-connected and locally path-connected space X has a covering belonging to the subgroup H of $\pi_1(X,v)$ iff H is sufficiently large.*

Theorem 6 *A complex K has a covering belonging to any subgroup H.*

Proof of Theorem 5 Let $p:\tilde{X}\to X$ be a covering, and let H be $\pi_1(\tilde{X},v)p*$. Take $x\in X$, and let U be an elementary neighbourhood of x. Let f be a path from v to x. Then f lifts to a path $\tilde{f}$ starting at $\tilde{v}$. Let $\tilde{x}$ be the end of $\tilde{f}$. Then $\tilde{x}$ has a neighbourhood $\tilde{U}$ which p maps homeomorphically onto U. Let g be any loop in U based at x.

Then g lifts to a loop $\tilde{g}$ in $\tilde{U}$ based at $\tilde{x}$. So the loop $f.g.f^{-1}$ is the image under p of the loop $\tilde{f}.\tilde{g}.\tilde{f}^{-1}$, and therefore represents an element of H. Hence H is sufficiently large.

Now let H be sufficiently large. If there is a covering belonging to H then we know that every path starting at v determines a point of the covering space (the end of the lift starting at $\tilde{v}$ of the path), that every point of the covering space comes from such a path, and we know (by Corollary 1 to Lemma 1) when two paths determine the same point.

This leads to the following construction. We consider all paths in X starting at v, and we write $f\equiv g$ if $f.g^{-1}$ represents an element of H. This is easily seen to be an equivalence relation, and we denote the equivalence class of f by $[f]$. We define $\tilde{X}$ to be the set of equivalence classes, with $\tilde{v}$ being the class of the trivial path. We define $p:\tilde{X}\to X$ by requiring $[f]p$ to be the end-point of f.

Take $\tilde{x}\in\tilde{X}$, and let x be $\tilde{x}p$. Choose f with $\tilde{x}=[f]$. Let U be a basic neighbourhood of x. Define $(\tilde{x},U)$ to be the subset of $\tilde{X}$ consisting of all points $[f.g]$ with g a path in U starting at x. Notice that this set does not depend on the choice of f, since $f'.g\equiv f.g$ if $f'\equiv f$. We shall show that the sets $(\tilde{x},U)$ form a basis for a topology.

We first show that if $\tilde{y}\in(\tilde{x},U)$ then $(\tilde{y},U)=(\tilde{x},U)$. For there must be a path g in U from x to y such that $\tilde{y}=[f.g]$. Thus $(\tilde{y},U)=\{[f.g.h];\ h$ is a path in U starting at $y\}$. Hence $(\tilde{y},U)\subseteq(\tilde{x},U)$. Also, $\tilde{x}=[f.g.g^{-1}]$, and so $\tilde{x}\in(\tilde{y},U)$. For the same reason, this gives $(\tilde{x},U)\subseteq(\tilde{y},U)$.

Now let $\tilde{z}\in(\tilde{x},U)\cap(\tilde{y},V)$. Let W be the component of $U\cap V$ containing z. Then W is also a basic neighbourhood (it is open because X is locally path-connected). We see that $(\tilde{z},W)\subseteq(\tilde{z},U)\cap(\tilde{z},V)=(\tilde{x},U)\cap(\tilde{y},V)$. It follows that the sets $(\tilde{x},U)$ form a basis for a topology on $\tilde{X}$, which is the topology we use.

We now show that p is continuous. It is enough to show that Up^{-1} is open for every basic neighbourhood U. Since Up^{-1} is the union, over all $\tilde{x} \in xp^{-1}$ and all $x \in U$, of the sets $(\tilde{x},U)$, this is immediate.

The next step is to show that p is a bijection from $(\tilde{x},U)$ onto U. Take any $y \in U$. Since basic neighbourhoods are path-connected, there is a path g in U from x to y. Then $[f.g]$ is in $(\tilde{x},U)$ and $[f.g]p = y$, and so p is onto. Now suppose that $[f.g]$ and $[f.h]$ are elements of $(\tilde{x},U)$ with the same image under p. Then g and h are both paths in U from x to some point y, and so $g.h^{-1}$ is a loop in U. Since H is sufficiently large, the loop $(f.g).(f.h)^{-1}$ represents an element of H. So, by definition, $[f.g] = [f.h]$, and therefore p is one-one.

Since p maps each member $(\tilde{x},U)$ of a basis for $\tilde{X}$ onto an open set U, p maps every open set of $\tilde{X}$ to an open set of X. In particular, since p is a bijection on $(\tilde{x},U)$, p is a homeomorphism from $(\tilde{x},U)$ onto U. It follows further that $\tilde{X}$ is locally path-connected, since each point $\tilde{x}$ has a neighbourhood $(\tilde{x},U)$ which is locally path-connected, being homeomorphic to U.

The next step is to show that Up^{-1} is the disjoint union of the sets $(\tilde{x},U)$ for all $\tilde{x} \in xp^{-1}$. First take any $\tilde{y} \in Up^{-1}$. Then $(\tilde{y},U)p = U$, so that we can find $\tilde{x}_1 \in xp^{-1}$ with $\tilde{x}_1 \in (\tilde{y},U)$. We have already seen that this gives $(\tilde{x}_1,U) = (\tilde{y},U)$, and so $\tilde{y} \in (\tilde{x}_1,U)$. That is, Up^{-1} is the union of the sets $(\tilde{x},U)$.

Now suppose that $(\tilde{x}_1,U) \cap (\tilde{x}_2,U) \neq \emptyset$, where $\tilde{x}_1 p = x = \tilde{x}_2 p$. Let $\tilde{y}$ be in this intersection. Then $(\tilde{x}_1,U) = (\tilde{y},U) = (\tilde{x}_2,U)$. Since p is a bijection on $(\tilde{y},U)$, we see that $\tilde{x}_1 = \tilde{x}_2$, and so Up^{-1} is the disjoint union of the sets $(\tilde{x},U)$.

Since these sets are open and connected, p will be a covering provided that $\tilde{X}$ is path-connected.

Let $\tilde{x} \in \tilde{X}$ be $[f]$, where $f:[0,r] \to X$ is a path starting at v. Let f_t be the path defined on $[0,t]$ such that $sf_t = sf$ for $0 \leq s \leq t$. Let $\tilde{f}$ be the function from from $[0,r]$ to $\tilde{X}$ defined by $t\tilde{f} = [f_t]$. If $\tilde{f}$ is continuous, it is plainly the lift of f starting at v, and it ends at $\tilde{x}$. Hence $\tilde{X}$ is path-connected if $\tilde{f}$ is continuous.

Take $s \in [0,r]$. Let U be a basic neighbourhood of sf. Choose $\delta > 0$ such that $Jf \subseteq U$, where $J = [0,r] \cap \{t;|t-s| < \delta\}$. Let $\tilde{U}$ be $([f_s],U)$. If $t \in J$ then either $f_t = f_s . g$ or $f_s = f_t . g$ (depending on whether $t \geq s$ or $t \leq s$), where g is a path in U. In either case, we see that $[f_t] \in \tilde{U}$. Hence $J\tilde{f} \subseteq \tilde{U}$. Since $\tilde{f}p = f$ and p is a homeomorphism from $\tilde{U}$ to U, we see that $\tilde{f}$ is continuous at s.

We have now shown that p is a covering, but we still have to identify the subgroup it belongs to. So we take a loop f based at v. This lifts to a path $\tilde{f}$ in $\tilde{X}$ starting at $\tilde{v}$. The discussion above shows that $\tilde{f}$ ends at $[f]$. Thus $\tilde{f}$ is a loop iff $[f] = \tilde{v}$; that is, iff f represents an element of H. Hence, by Lemma 1, $\pi_1(\tilde{X},\tilde{v})p* = H$, as required.//

Proof of Theorem 6 Now let K be a complex, and let H be a subgroup of $\pi_1(K,v)$. Much as before, we define an equivalence relation on the edge-paths starting at v by $f \equiv g$ iff f and g have the same end-point and $f.g^{-1}$ represents an element of H, and we let $\tilde{v}$ be the path of zero length at v. This time the set of equivalence classes forms the vertex set $V(\tilde{K})$. As before, we define $[f]p$ to be the end-point of f.

We define the set of edges $E(\tilde{K})$ to consist of all pairs $(\tilde{x},e)$ where $\tilde{x} \in V(\tilde{K})$, $e \in E(K)$, and $\tilde{x}p = e\sigma$. Let $e\sigma$ be x and let $e\tau$ be y, and let $\tilde{x}$ be $[f]$. We define $(\tilde{x},e)\sigma$ to be $\tilde{x}$, and $\overline{(\tilde{x},e)}$ to be $(\tilde{y},\bar{e})$, where $\tilde{y}$ is $[f.e]$. Since $[f.e.\bar{e}] = [f]$, we see that $^{-}$ satisifes the required conditions for the vertices and edges of $\tilde{K}$ to be a graph. We define $(\tilde{x},e)p$ to be e. Then p is clearly a map of graphs, which induces a bijection from the set of edges of $\tilde{K}$ starting at $\tilde{x}$ to the set of edges of K starting at x.

Let g be the path $e_1 \ldots . e_n$, and let $e_i\sigma$ be x_i. Let $\tilde{x}_1 \in x_1p^{-1}$ be $[f]$. Define $\tilde{x}_i$ to be $[f.e_1 \ldots, e_{i-1}]$, and let $\tilde{e}_i$ be $(\tilde{x}_i,e_i)$. It is easy to check that $\tilde{e}_1 \ldots . \tilde{e}_n$ is the unique path in $\tilde{K}$ which starts at $\tilde{x}_1$ and is a lift of g. In particular, if g starts at v then this path starts at $\tilde{v}$ and ends at $[g]$. Hence K is connected.

Take the lift $\tilde{g}$ starting at $\tilde{v}$ of a loop g based at v. From the previous paragraph, it is easy to see that $\tilde{g}$ ends at $[g]$. In particular, $\tilde{g}$ is a loop iff g is equivalent to the trivial loop; that is, iff g represents an element of H. Hence, by Lemma 1, once we have defined the cells of $\tilde{K}$ so as to make it a covering of K, the covering belongs to the subgroup H.

We see also that, for an arbitrary path g, the lift of g starting at $[f]$ ends at $[f.g]$. In particular, when g is the boundary $c\partial$ of a cell c then the lift of g starting at $[f]$ is a loop, since in this case $f.g.f^{-1} \simeq f.f^{-1}$ which is homotopic to the trivial loop.

We now proceed to define the cells of $\tilde{K}$. It is most convenient to choose a set $C_+(K)$ consisting of one cell c from each cell-pair $\{c,\bar{c}\}$ of K. We shall assume that an indexing of the cells of K has been made. Let $C_+(\tilde{K})$ consist of the pairs $(\tilde{x},c)$ for each chosen cell c and vertex $\tilde{x} \in xp^{-1}$, where x is the vertex of the indexed cell $(c,1)$. Let $C_-(\tilde{K})$ be bijective with $C_+(\tilde{K})$ and disjoint from it, with the bijection being denoted by $^{-}$. The set $C(\tilde{K})$ of cells of $\tilde{K}$ is to be the union of these two sets. When $\tilde{c}$ is $(\tilde{x},c)$ we define $\tilde{c}\partial$ to be the lift starting at $\tilde{x}$ of the boundary $(c,1)\partial$ of the indexed cell $(c,1)$. We saw in the previous paragraph that this lift is a loop. We define p by $(\tilde{x},c)p = c$ and $\overline{(\tilde{x},c)}p = \bar{c}$. Plainly p is a map of complexes.

Let $\tilde{c}$ be $(\tilde{x},c)$. We define the corresponding indexed cells so that $(\tilde{c},1)$ has vertex $\tilde{x}$ and boundary the lift $\tilde{e}_1,\ldots,\tilde{e}_n$ of the boundary $e_1,\ldots,e_n$ of $(c,1)\partial$. We then define $p_!$ to send $(\tilde{c},j)$ to (c,j), and we extend the definition to the indexed cells corresponding to the cells of $C_-(\tilde{K})$ so that $p_!$ sends an indexed cell and its inverse to an indexed cell and its inverse.

Let the indexed cell (c,i) with $c \in C_+(K)$ have vertex y. Let $\tilde{y}$ be in yp^{-1}, and take a path h with $\tilde{y} = [h]$. Let g be the path $e_1\ldots\ldots e_{i-1}$ from x to y. Let $\tilde{x}$ be $[h.g^{-1}]$, and let $\tilde{g}$ be the lift of g starting at $\tilde{x}$. Then the end-point of $\tilde{g}$ is $[h.g^{-1}.g] = [h] = \tilde{y}$. By definition of the indexing of cells, it follows that $\tilde{y}$ is the vertex of the indexed cell $(\tilde{c},i)$, where $\tilde{c}$ is $(\tilde{x},c)$. Hence (c,i) is in the image under $p_!$ of the set of indexed cells with vertex $\tilde{y}$.

Now suppose that $(\tilde{c},i)$ and $(\tilde{d},j)$ are two indexed cells with vertex $\tilde{y}$ with the same image under $p_!$ and both in $C_+(\tilde{K})$. Let $\tilde{c}$ be $(\tilde{x},c)$ and let $\tilde{d}$ be $(\tilde{z},d)$. Plainly we must have $c = d$, and then $x = z$, as both are the vertex of $(c,1)$. Also $i = j$, since $(c,i) = (\tilde{c},i)p_! = (\tilde{d},j)p_! = (c,j)$. It then follows, with g as before, that the lifts of g starting at $\tilde{x}$ and $\tilde{z}$ both end at $\tilde{y}$. Hence $\tilde{x} = \tilde{z}$, both being the end of that lift of g^{-1} which starts at $\tilde{y}$. Thus $(\tilde{c},i) = (\tilde{d},j)$.

Now a cell in $C_-(\tilde{K})$ cannot have the same image under p as a cell in $C_+(\tilde{K})$, since the latter image is in $C_+(K)$ while the former is not. It follows easily from this and the previous two paragraphs that $p_!$ is a bijection from the set of indexed cells with vertex $\tilde{y}$ to the set of indexed cells with vertex y. Hence p (with the associated $p_!$) is a covering.//

Theorem 7 (*i*) *Let* $\varphi:(Y,y_0) \to (X,v)$ *be a map such that* $\pi_1(Y,y_0)\varphi_* \subseteq \pi_1(\tilde{X},\tilde{v})p_*$. *Then there is a unique map* $\tilde{\varphi}:(Y,y_0) \to (\tilde{X},\tilde{v})$ *such that* $\tilde{\varphi}p = \varphi$. (*ii*) *Under the similar conditions for complexes, and with a given choice of* $\varphi_!$, *there is a unique* $\tilde{\varphi}$ *such that* $\tilde{\varphi}p = \varphi$ *and* $\tilde{\varphi}_!p_! = \varphi_!$.

Remark We call $\tilde{\varphi}$ a *lift* of φ.

Proof We already know there can be at most one such map.

(i) Take any point $y \in Y$. Then there is a path f in Y from y_0 to y. The path $f\varphi$ in X lifts to a path in $\tilde{X}$ starting at $\tilde{v}$. We show first that the end of this path depends only on y, and not on the choice of f.

So we take another path g from y_0 to y. Then $f.g^{-1}$ is a loop at y_0. By hypothesis, the path $(f\varphi).(g\varphi)^{-1} = (f.g^{-1})\varphi$ represents an element of $\pi_1(\tilde{X},\tilde{v})p*$. It follows, by Corollary 1 to Lemma 1, that the lifts of $f\varphi$ and $g\varphi$ end at the same point.

Thus we have defined a function $\tilde{\varphi}:Y\to\tilde{X}$ which clearly satisfies $\tilde{\varphi}p=\varphi$ and $y_0\tilde{\varphi}=\tilde{v}$. We still have to show that $\tilde{\varphi}$ is continuous.

Now let $\tilde{w}$ be $y\tilde{\varphi}$, so that $\tilde{w}p = y\varphi$ and let $\tilde{W}$ be a neighbourhood of $\tilde{w}$. Let W be $\tilde{W}p$. We may assume (replacing $\tilde{W}$ by a smaller set if necessary, using the remarks about elementary neighbourhoods following Lemma 4.17) that W is an elementary neighbourhood of $\tilde{w}p$.

Since Y is assumed locally pathwise-connected, there is a pathwise connected open neighbourhood U of y such that $U\varphi\subseteq W$. Choose a path f in Y from y_0 to y. Given $z\in U$, we may take a path g in U from y to z. Now $y\tilde{\varphi}$ is the end of that lift $\tilde{h}$ of $f\varphi$ which begins at $\tilde{v}$, while $z\tilde{\varphi}$ is the end of that lift of $(f.g)\varphi$ which begins at $\tilde{v}$. Equivalently, $z\tilde{\varphi}$ is the end of that lift of $g\varphi$ which begins at $\tilde{w}$. Now $g\varphi$ is a path in W, and p is a homeomorphism from $\tilde{W}$ onto W. Thus the lift of $g\varphi$ starting at $\tilde{w}$ lies completely in $\tilde{W}$. In particular $z\tilde{\varphi}$ is in $\tilde{W}$. As this holds for all $z\in U$, we see that $\tilde{\varphi}$ is continuous.

(ii) Now let $\varphi:(L,y_0)\to(K,v)$ be a map of complexes such that $\pi_1(L,y_0)\varphi*\subseteq\pi_1(\tilde{K},\tilde{v})p*$. Let y be any vertex of L. Then there is a path f in L from y_0 to y. The path $f\varphi$ in K lifts to a path in $\tilde{K}$ starting at $\tilde{v}$. As in (i), the end of this path does not depend on the choice of f, and we denote it by $y\tilde{\varphi}$. Plainly, $y\tilde{\varphi}p=y\varphi$ and $y_0\tilde{\varphi}=\tilde{v}$.

We next have to define $\tilde{\varphi}$ on the edges. Let e be an edge of L from y to z. Then $e\varphi$ is an edge of K starting at $y\varphi$. Since p is a covering, there is a unique edge of $\tilde{K}$ starting at $y\tilde{\varphi}$ whose image under p is $e\varphi$. Let $e\tilde{\varphi}$ be this edge. By construction $e\tilde{\varphi}\sigma=e\sigma\tilde{\varphi}$. With f as before, we may construct $z\tilde{\varphi}$ using the path $f.e$. Thus $z\tilde{\varphi}$ is the end of that lift of $(f\varphi).(e\varphi)$ which starts at $\tilde{v}$. Since the lift of $f\varphi$ ends at $y\tilde{\varphi}$, it follows that $z\tilde{\varphi}$ is the end of that edge above $e\varphi$ which starts at $y\tilde{\varphi}$. This edge is $e\tilde{\varphi}$, and so $e\tilde{\varphi}\tau=e\tau\tilde{\varphi}$. Further, $\bar{e}\tilde{\varphi}=\overline{e\tilde{\varphi}}$, since both edges start at $z\tilde{\varphi}$ and both have $\bar{e}\varphi$ as their image under p.

Finally we have to define $\tilde{\varphi}$ and $\tilde{\varphi}_!$ on the cells and indexed cells. Let $(c,\bar{c})$ be a cell-pair in L. Let the indexed cell $(c,1)$ have vertex y and boundary $e_1,\ldots,e_n$. Let $(c,1)\varphi_!$ be $(c\varphi,j)$. This indexed cell has vertex $y\varphi$, and there is a unique indexed cell $(\tilde{d},k)$ with vertex $y\tilde{\varphi}$ such that $(\tilde{d},k)p_!=(c\varphi,j)$. We define $c\tilde{\varphi}$ to be $\tilde{d}$, with $(\tilde{c},1)\tilde{\varphi}_!$ being defined as $(\tilde{d},k)$. We then define $\bar{c}\tilde{\varphi}$ to be $\bar{\tilde{d}}$.

Now $e_1\tilde{\varphi},\ldots,e_n\tilde{\varphi}$ is a loop in $\tilde{K}$ starting at $y\tilde{\varphi}$ which maps by p to the loop $e_1\varphi,\ldots,e_n\varphi$. This loop is $(c\varphi, j)\partial$, and so it is the image under p of the loop $(\tilde{d}, k)\partial$, which also starts at $y\tilde{\varphi}$. Hence these two loops are the same; that is, $(c,1)\partial\tilde{\varphi} = (c,1)\tilde{\varphi}_!\partial$, as required. We can now define $\tilde{\varphi}_!$ on the remaining indexed cells corresponding to c, and on the indexed cells corresponding to $\bar{c}$, in exactly one consistent way.//

Definition Let $p_1:\tilde{X}_1 \to X$ and $p_2:\tilde{X}_2 \to X$ be coverings of spaces. A map $\varphi:\tilde{X}_1 \to \tilde{X}_2$ is a *homomorphism* if $\varphi p_2 = p_1$, and it is an *isomorphism* if it is a bijection and both φ and φ^{-1} are homomorphisms. For a homomorphism of complexes we require that $\varphi p_2 = p_1$ as before, but we also require that the chosen $\varphi_!$ is such that $\varphi_! p_{2!} = p_{1!}$.

Proposition 8 *Let $p_1:\tilde{X}_1 \to X$ and $p_2:\tilde{X}_2 \to X$ be coverings (of spaces or complexes), and let $\tilde{v}_1$ and $\tilde{v}_2$ be points such that $\tilde{v}_1 p_1 = \tilde{v}_2 p_2$. Then there is a homomorphism from $\tilde{X}_1$ to $\tilde{X}_2$ sending $\tilde{v}_1$ to $\tilde{v}_2$ iff $\pi_1(\tilde{X}_1,\tilde{v}_1)p_{1*} \subseteq \pi_1(\tilde{X}_2,\tilde{v}_2)p_{2*}$, and an isomorphism from $(\tilde{X}_1,\tilde{v}_1)$ to $(\tilde{X}_2,\tilde{v}_2)$ iff $\pi_1(\tilde{X}_1,\tilde{v}_1)p_{1*} = \pi_1(\tilde{X}_2,\tilde{v}_2)p_{2*}$.*

Proof The result for homomorphisms is immediate from Theorem 4.

Suppose that $\pi_1(\tilde{X}_1,\tilde{v}_1)p_{1*} = \pi_1(\tilde{X}_2,\tilde{v}_2)p_{2*}$. Then there will be homomorphisms $\varphi:(\tilde{X}_1,\tilde{v}_1) \to (\tilde{X}_2,\tilde{v}_2)$ and $\psi:(\tilde{X}_2,\tilde{v}_2) \to (\tilde{X}_1,\tilde{v}_1)$. Then $\varphi\psi p_1 = p_1$ (and, for complexes, $\varphi_!\psi_! p_{1!} = p_{1!}$ also). Thus uniqueness of liftings shows that $\varphi\psi$ (and, similarly, $\psi\varphi$) is the identity; in the case of complexes, the extra condition needed also holds.//

Corollary *The coverings $\tilde{X}_1$ and $\tilde{X}_2$ of X are isomorphic iff $\pi_1(\tilde{X}_1,\tilde{v}_1)p_{1*}$ and $\pi_1(\tilde{X}_2,\tilde{v}_2)p_{2*}$ are conjugate in $\pi_1(X,v)$, where $\tilde{v}_1 p_1 = v = \tilde{v}_2 p_2$. This holds for spaces or complexes.*

Proof Let φ be an isomorphism. Then $\pi_1(\tilde{X}_1,\tilde{v}_1)p_{1*} = \pi_1(\tilde{X}_2,\tilde{v}_1\varphi)p_{2*}$, and this is conjugate to $\pi_1(\tilde{X}_2,\tilde{v}_2)p_{2*}$, since $\tilde{v}_1\varphi p_2 = \tilde{v}_2 p_2$.

Conversely, if the groups are conjugate we can find a point $\tilde{w}$ of $\tilde{X}_2$ such that $\tilde{w}p_2 = v$ and $\pi_1(\tilde{X}_1,\tilde{v}_1)p_{1*} = \pi_1(\tilde{X}_2,\tilde{w})p_{2*}$. By the proposition, this gives us an isomorphism from $(\tilde{X}_1,\tilde{v}_1)$ to $(\tilde{X}_2,\tilde{w})$.//

By Proposition 8 and its corollary, the (conjugacy class of the) subgroup to which a covering belongs determines the covering up to isomorphism.

Proposition 9 *Let $\varphi:\tilde{X}_1 \to \tilde{X}_2$ be a homomorphism of coverings of X. Then φ is a covering. The same result holds for complexes.*

Proof Let U_1 and U_2 be elementary neighbourhoods of $x \in X$ for p_1 and p_2. Then U, the component of $U_1 \cap U_2$ containing x, is an elementary neighbourhood for both p_1 and p_2.

Take any point $\tilde{x}_2$ of $\tilde{X}_2$, and let x be $\tilde{x}_2 p_2$. Let $\tilde{U}$ be the component of Up_2^{-1} containing $\tilde{x}_2$. Then Up_2^{-1} is the disjoint union of $\tilde{U}$ and another open set (since the components of open sets are open). Hence $Up_2^{-1}\varphi^{-1}$ is the disjoint union of $\tilde{U}\varphi^{-1}$ and another open set. It follows that each component of $\tilde{U}\varphi^{-1}$ is a component of $Up_2^{-1}\varphi^{-1} = Up_1^{-1}$. Now p_2 is a homeomorphism from $\tilde{U}$ onto U, while p_1 is a homeomorphism from each component of Up_1^{-1} onto U. Hence φ is a homeomorphism from each component of $\tilde{U}\varphi^{-1}$ onto $\tilde{U}$, as required.

The result for complexes is left as an exercise.//

With a change of notation, this result says that if $p:\tilde{X} \to X$ and $q:\tilde{\tilde{X}} \to \tilde{X}$ are maps such that p and qp are coverings then q is a covering.

For complexes, it is easy to check that if both p and q are coverings then qp is also a covering, with $(qp)_!$ being defined as $q_! p_!$. The following example shows that in the case of spaces qp need not be a covering. The proposition below shows that qp will be a covering if it possibly can be, in a sense made precise in the proposition.

Example Let X be the Hawaiian earring. That is, $X \subseteq \mathbb{R}^2$ is $\cup X_n$, where X_n is the circle $\{(x,y);\ x^2+y^2 = 2x/n\}$ with centre $(1/n,0)$ and radius $1/n$ (for every positive integer n). Let $\tilde{X}$ be the covering of X obtained by unwrapping the first circle; that is, $\tilde{X}$ is obtained from $\mathbb{R}$ by attaching at each integer point a copy of those X_n with $n > 1$. We can obtain a covering $\tilde{\tilde{X}}$ of $\tilde{X}$ by unwrapping, for all $n > 1$, the copy of X_n which is attached at the integer point n. It is easy to check that this is a covering.

Now any neighbourhood U of $(0,0)$ contains X_n for n large enough. Consequently, some component of the counter-image of U in $\tilde{\tilde{X}}$ will contain an unwrapping of X_n, and so will not map homeomorphically onto U. Thus $\tilde{\tilde{X}}$ is not a covering of X.

It is clear that the trivial subgroup is not sufficiently large. Additionally, any sufficiently large subgroup must contain, for all large enough n, an element which maps to the generator of $\pi_1(X_n)$ under the homomorphism induced by the retraction of X to X_n. Further, the discussion of the Hawaiian earring in section 4.5 shows that any sufficiently large subgroup must be uncountable.

Proposition 10 *Let* $p:\tilde{X} \to X$ *and* $q:\tilde{\tilde{X}} \to \tilde{X}$ *be coverings of spaces. Then* qp *is a covering. iff* X *has a covering belonging to the subgroup* $\pi_1(\tilde{\tilde{X}})q*p*$.

Proof The condition is clearly necessary.

Suppose that the condition holds, and let $r:Y \to X$ be the covering belonging to this subgroup. By Theorem 7 there is a map $\varphi:Y \to \tilde{X}$ such that $r = \varphi p$, and φ is a covering by Proposition 9. Since $p*$ is a monomorphism and $\pi_1(\tilde{\tilde{X}})q*p* = \pi_1(Y)\varphi*p*$, we see that $\pi_1(\tilde{\tilde{X}})q* = \pi_1(Y)\varphi*$. By Proposition 8, the coverings $\tilde{\tilde{X}}$ and Y of $\tilde{X}$ are isomorphic. Under this isomorphism qp corresponds to φp. As the latter is a covering, so is the former.//

From Proposition 8 we see at once that if X is simply connected then (up to isomorphism) the only covering of X is the identity.

Suppose that X has a covering $\tilde{X}$ such that $\tilde{X}$ is simply connected. We then call $\tilde{X}$ the *universal covering* of X. By Theorem 7, for any covering Y of X there is a homomorphism from $\tilde{X}$ to Y, and by Proposition 9 this homomorphism is a covering. Thus $\tilde{X}$ is a covering of every covering of X, which explains its name. These properties hold both for spaces and complexes.

Example Let Y_+ be the cone on the Hawaiian earring. Let Y_- be another copy of Y_+, and let Y be the join of Y_+ and Y_-, the common point v being (0,0) in the earrings. It was shown in section 4.5 that $\pi_1(Y)$ is uncountable. We will show that Y has (up to isomorphism) no covering other than the identity. Thus, for complicated spaces, there may be a covering which is not simply connected but which still covers every covering.

Take a covering $p:(\tilde{Y},\tilde{v}) \to (Y,v)$. Let $\tilde{Y}_+$ and $\tilde{Y}_-$ be the components of Y_+p^{-1} and Y_-p^{-1} containing $\tilde{v}$. By Proposition 12 below, $p:\tilde{Y}_+ \to Y_+$ is a covering. Since Y_+ is simply connected, it follows that p is a homeomorphism from $\tilde{Y}_+$ to Y_+ and also from $\tilde{Y}_-$ to Y_-.

This implies that p is a homeomorphism from $\tilde{Y}_+ \cup \tilde{Y}_-$ onto Y. For let $\tilde{U}$ be any open subset of this union. Then $\tilde{U}p$ is the union of the sets $(\tilde{U} \cap \tilde{Y}_+)p$ and $(\tilde{U} \cap \tilde{Y}_-)p$, which are open in Y_+ and Y_- respectively, and if v is in one of these two sets it is also in the other. Hence $\tilde{U}p$ is open in Y.

Now take any point $\tilde{y}$ of $\tilde{Y}$. It is the end of some path starting at $\tilde{v}$. This path is the unique lift of some path in Y starting at v. Since, by the previous paragraph, every path in Y starting at v lifts to a path in $\tilde{Y}_+ \cup \tilde{Y}_-$ starting at $\tilde{v}$, we see that $\tilde{Y}$ must be $\tilde{Y}_+ \cup \tilde{Y}_-$.

Proposition 11 *Let $p:(\tilde{X},\tilde{v}) \to (X,v)$ be a covering of spaces (or complexes). Let A be a path-connected subset (a connected subcomplex) of X with $v \in A$. Let H be the subgroup $\pi_1(\tilde{X},\tilde{v})p*$ of $\pi_1(X,v)$, and let $i:A \to X$ be the inclusion. Then Ap^{-1} is a path-connected space (a connected complex) iff $\pi_1(A,v)i*$ meets every right coset of H.*

Proof Let Ap^{-1} be path-connected. Let $\alpha \in \pi_1(X,v)$ be represented by the loop f at v. Lift f to a path $\tilde{f}$ starting at $\tilde{v}$, and let $\tilde{f}$ end at $\tilde{w}$. Then $\tilde{w} \in vp^{-1} \subseteq Ap^{-1}$. Hence there is a path $\tilde{g}$ in Ap^{-1} from $\tilde{v}$ to $\tilde{w}$. Now $\tilde{g}p$ is a loop in A, so it represents in $\pi_1(X,v)$ an element β of $\pi_1(A,v)i*$. Also $\beta\alpha^{-1}$ is represented by $(\tilde{g}p).f^{-1} = (\tilde{g}.\tilde{f}^{-1})p$. Since $\tilde{g}.\tilde{f}^{-1}$ is a loop in $\tilde{X}$, it follows from Lemma 1 that $\beta\alpha^{-1} \in H$. Hence $\pi_1(A,v)i*$ meets the coset $H\alpha$.

Conversely, suppose that $\pi_1(A,v)i*$ meets every coset of H. Take $\tilde{x} \in Ap^{-1}$. There is a path in A from $\tilde{x}p$ to v, and this path lifts to a path in Ap^{-1} from $\tilde{x}$ to some point $\tilde{w} \in vp^{-1}$. So we need only find a path in Ap^{-1} from $\tilde{v}$ to $\tilde{w}$.

Take any path $\tilde{f}$ in $\tilde{X}$ from $\tilde{v}$ to $\tilde{w}$. Let f be $\tilde{f}p$. Then f is a loop in X based at v. By the hypothesis, there is a loop g in A based at v such that $f.g^{-1}$ represents an element of H. Then g will lift to a path $\tilde{g}$ in Ap^{-1} starting at $\tilde{v}$. By Corollary 1 to Lemma 1, the paths $\tilde{f}$ and $\tilde{g}$ have the same end-point, as required.//

Proposition 12 *Let $p:(\tilde{X},\tilde{v}) \to (X,v)$ be a covering (of spaces or complexes) belonging to the subgroup H of $\pi_1(X,v)$. Let A be a path-connected and locally path-connected subspace of the space X (or a connected subcomplex of the complex X) with $v \in A$. Let $\tilde{A}$ be the component of Ap^{-1} containing $\tilde{v}$. Then $p:(\tilde{A},\tilde{v}) \to (A,v)$ is a covering belonging to the subgroup $Hi*^{-1}$, where $i:A \to X$ is the inclusion.*

Proof First suppose that X is a space. Take any $a \in A$. Let U be an elementary neighbourhood of a. Since A is locally path-connected, the component of $A \cap U$ containing a is an open subset of A. Hence it is $A \cap V$, for some open subset V of X. Replacing U by the component of $U \cap V$ containing a, it follows that we can assume that the elementary neighbourhood U of a is such that $A \cap U$ is connected.

Take any $\tilde{a} \in \tilde{A}$ with $\tilde{a}p = a$, and let $\tilde{U}$ be the component of Up^{-1} containing $\tilde{a}$. Then the component of $(A \cap U)p^{-1}$ containing $\tilde{a}$ is $\tilde{A} \cap \tilde{U}$, and it maps homeomorphically onto $A \cap U$ by p. Hence $\tilde{A}$ is a covering space of A.

Now suppose that X is a complex. Take any vertex a of A, and any vertex $\tilde{a}$ of $\tilde{A}$ with $\tilde{a}p = a$. Then any edge or cell of $\tilde{X}$ which has $\tilde{a}$ as a vertex and maps into A by p must itself lie in $\tilde{A}$. It is then easy to check from the definition that the restriction to $\tilde{A}$ of p (together with the corresponding $p_!$) is a covering.

In either case, let K be $\pi_1(\tilde{A},\tilde{v})p*$, and let $\tilde{i}:\tilde{A} \to \tilde{X}$ be the inclusion. Then $Ki* = \pi_1(\tilde{A},\tilde{v})p*i* = \pi_1(\tilde{A},\tilde{v})\tilde{i}*p* \subseteq H$. Conversely, let $\alpha \in Hi*^{-1}$. Then α is represented by a loop f in A starting at v and such that f lifts to a loop $\tilde{f}$ in $\tilde{X}$ starting at $\tilde{v}$. On the other hand, since $\tilde{A}$ is a covering of A, f lifts to a path in $\tilde{A}$ starting at $\tilde{v}$. Uniqueness of lifts now tells us that $\tilde{f}$ must lie in $\tilde{A}$, and so α is in K, being represented by f.//

Let $\tilde{X}$ be a covering of X. An isomorphism of $\tilde{X}$ with itself is called a *covering transformation* of $\tilde{X}$. They are clearly of importance when looking at a covering.

Proposition 13 *Let $\tilde{X}$ be a covering of X belonging to the subgroup H. Then the group of covering transformations of $\tilde{X}$ is isomorphic to $N(H)/H$, where $N(H)$ is the normaliser of H in $\pi_1(X,v)$.*

Proof Let φ be a covering transformation. By Corollary 3 to Lemma 1, there is a coset of H corresponding to $\tilde{v}\varphi^{-1} \in vp^{-1}$. One element of this coset is the homotopy class α of $\tilde{f}p$, where $\tilde{f}$ is a path from $\tilde{v}$ to $\tilde{v}\varphi^{-1}$.

By Proposition 3, $\pi_1(\tilde{X},\tilde{v}\varphi^{-1})p* = \alpha^{-1}H\alpha$. Since φ is an isomorphism, $\pi_1(\tilde{X}, \tilde{v}\varphi^{-1})p* = H$. Hence α is in $N(H)$.

Conversely, given $\alpha \in N(H)$ let $\tilde{f}$ be a path starting at $\tilde{v}$ such that $\tilde{f}p$ represents α. Then the end-point $\tilde{w}$ of $\tilde{f}$ lies in vp^{-1}, and, by Corollary 3 to Lemma 1, it depends only on the coset $H\alpha$. Since $\pi_1(\tilde{X},\tilde{w})p* = \alpha^{-1}H\alpha = H$, by Proposition 8 there is an isomorphism φ of $\tilde{X}$ sending $\tilde{w}$ to $\tilde{v}$. Thus we have found to every covering transformation a coset of H in $N(H)$ and to every coset a covering transformation. It is easy to check that these functions from the group of covering transformations to $N(H)/H$ and vice versa are inverses of each other. That is, we have a bijection from the group of covering transformations to $N(H)/H$.

Let ψ be another isomorphism, and let $\tilde{g}$ be a path from $\tilde{v}$ to $\tilde{v}\psi^{-1}$. Let β be the class of $\tilde{g}p$. Since $\tilde{g}\varphi^{-1}$ is a path from $\tilde{v}\varphi^{-1}$ to $\tilde{v}\psi^{-1}\varphi^{-1}$, the path $\tilde{f}.(\tilde{g}\varphi^{-1})$ is a path from $\tilde{v}$ to $\tilde{v}(\varphi\psi)^{-1}$. The image of this path under p represents $\alpha\beta$, and so the image of $\varphi\psi$ is the coset of $\alpha\beta$. Hence the bijection is a homomorphism.//

A covering is called *regular* if for every $\tilde{w} \in vp^{-1}$ there is an isomorphism sending $\tilde{v}$ to $\tilde{w}$. The proof of Proposition 13 shows that this holds iff $N(H) = \pi_1(X,v)$; that is, iff H is normal in $\pi_1(X,v)$.

If a covering is regular then any loop in X which lifts to a loop starting at $\tilde{v}$ will also lift to a loop starting at $\tilde{w}$ for every $\tilde{w} \in vp^{-1}$ (by applying to the loop starting at $\tilde{v}$ the isomorphism sending $\tilde{v}$ to $\tilde{w}$). Conversely, suppose this condition is satisfied. Then every element of H lies in $\pi_1(\tilde{X},\tilde{w})p*$ for every $\tilde{w} \in vp^{-1}$. By Proposition 3, this tells us that $H \subseteq \alpha^{-1}H\alpha$ for every α in $\pi_1(X,v)$. Hence H is normal in $\pi_1(X,v)$.

Let Y be a space, and let G be a group of homeomorphisms of Y. We say that the action of G is *properly discontinuous* if each $y \in Y$ has a neighbourhood U such that $Ug \cap Uh = \emptyset$ for $g \neq h$.

Let G act properly discontinuously on Y, and let X be the quotient space Y/G (whose points are the orbits of G, and whose topology is the identification topology). It is easy to see that Y is a regular covering space of X, whose group of covering transformations is G. Conversely, let $\tilde{X}$ be a covering space of X, and let G be the group of covering transformations. Then G acts properly discontinuously on $\tilde{X}$, the relevant neighbourhoods being the components of Up^{-1}, where U is an elementary neighbourhood.

When Y is simply connected it will be the universal covering of X. In this case the group of covering transformations is $\pi_1(X)$, by Proposition 13. Hence $\pi_1(X) = G$. In general, Proposition 13 tells us that $\pi_1(X)$ is an extension of $\pi_1(Y)$ by G.

Similar results hold when G is a group of automorphisms of a complex K, provided that G acts freely and without inversions; that is, provided that if $g \neq 1$ then g fixes no vertex, edge, or cell, and $eg \neq \bar{e}$ and $cg \neq \bar{c}$ for every edge e and cell c.

These results can be extended to situations where G has fixed points, and the extended results can be derived from similar results about fundamental groupoids.

Exercise 1 Prove Proposition 9 for complexes.

Exercise 2 Let $p:L \to K$ and $q:M \to L$ be coverings of complexes. Show that qp is a covering.

Exercise 3 In Proposition 12 for complexes, check that the restriction of p to $\tilde{A}$ is a covering.

Exercise 4 Find $\pi_1(\mathbb{R}P^2)$, both by writing it as the quotient of a simply connected space by a properly discontinuous group action, and by the method used to obtain π_1 of the torus and Klein bottle.

Exercise 5 With the notation of Proposition 11, show that there is a bijection between the components of Ap^{-1} and the double cosets of $(H, \pi_1(A,v)i*)$. Use this to deduce Proposition 11.

7 COVERINGS AND GROUP THEORY

Now that we have developed the theory of covering complexes and covering spaces we will use this theory to obtain major results in combinatorial group theory. We will give three proofs of Schreier's theorem which states that all subgroups of free groups are themselves free. The first proof uses covering complexes, while the third is an algebraic formulation of the first. The second is essentially the same as the first, but is redesigned so that no explicit mention of coverings is needed. Yet another proof, which is also essentially the same as the others, is given in section 8.1, while section 8.2 contains the very different proof due to Nielsen.

Theorem 1 (weak form of Schreier's Theorem) *A subgroup of a free group is free. Further, if H is a subgroup of finite index r in a free group of rank n then H has rank* $r(n-1)+1$.

Proof Let F be free with basis X. Take a graph Γ with one vertex v and one edge-pair for each element of X. By Corollary 1 to Theorem 5.17, $\pi_1(\Gamma)$ is isomorphic to F. There is a covering $\tilde{\Gamma}$ of Γ belonging to the subgroup H. Then H is isomorphic to $\pi_1(\tilde{\Gamma})$, which is free, by Corollary 1 to Theorem 5.17 again, since $\tilde{\Gamma}$ is a graph.

Since Γ has only one vertex, the set of vertices of $\tilde{\Gamma}$ is vp^{-1}, and this is bijective with the set of cosets of H. That is, $\tilde{\Gamma}$ has r vertices. Since Γ has $2n$ edges, all of which start at v, $\tilde{\Gamma}$ has $2n$ edges starting at each vertex of $\tilde{\Gamma}$, by the definition of a covering. Hence $\tilde{\Gamma}$ has $2rn$ edges in total, and so it has rn edge-pairs. By Corollary 3 to Theorem 5.17, when r is finite the rank of $\pi_1(\tilde{\Gamma})$ is $rn-r+1$.//

This version of the theorem does not tell us a basis for H, which is the reason it is called the weak form of Schreier's Theorem. Theorem 4 gives the more precise version which does tell us a basis.

When r is infinite nothing can be said about the rank of H. For instance, the free group of rank 2 contains a free subgroup of infinite rank which has infinite index. Contained in this subgroup there are free subgroups of every finite rank, all of which have infinite index.

We now give a variant of the above proof which does not require knowledge of coverings.

We construct a graph Δ whose vertices are the cosets Hg of H in F. In particular, the coset H is a vertex which we take as our base-point. There is to be a function λ from the set of edges of Δ into $X \cup X^{-1}$, which we call the *labelling* of the edges. Starting at each vertex there is to be exactly one edge with a given label. The edge with label x^{ε} starting at Hg is to end at Hgx^{ε}, and the inverse of the edge with label x from Hg to Hgx is to be the edge with label x^{-1} from Hgx to Hg.

Using Corollary 3 to Lemma 6.1, we can show that Δ is isomorphic to $\tilde{\Gamma}$, so that $\pi_1(\Delta)$ is isomorphic to H. We shall show directly, without using coverings, that this is true. Then, as above but using Δ instead of $\tilde{\Gamma}$, we see that H is free, and we find its rank when its index is finite. The strong form of Schreier's Theorem given in Theorem 4 can also be proved using Δ instead of $\tilde{\Gamma}$.

The map λ obviously extends to a multiplicative map from paths of Δ to words on X. If the path f' comes from the path f by elementary reduction then the word corresponding to f' comes from that corresponding to f by elementary reduction. It follows that λ induces a homomorphism from $\gamma(\Delta)$, and also from $\pi_1(\Delta, H)$, to F.

In particular, an irreducible path maps to a reduced word of the same length, and so the homomorphism from $\pi_1(\Delta, H)$ to F has trivial kernel.

An easy induction on the length of the word shows that for any word w and coset Hg there is exactly one path in Δ starting at Hg whose corresponding word is w, and this path ends at Hgw. Thus an element w of F is in the image of $\pi_1(\Delta, H)$ iff $Hw = H$. That is, the image is H.

The next proposition can be regarded as a partial converse to the numerical part of Theorem 1. It can be proved for subnormal subgroups, not just for normal subgroups.

Proposition 2 *Let H be a non-trivial finitely generated normal subgroup of a free group F. Then H has finite index and F has finite rank.*

Proof There is a finitely generated free factor of F which contains H. This is the subgroup of F generated by those elements of the basis X of F which occur in a chosen finite set of generators of H. Since (Exercise 1.18) a proper free factor of a group does not contain any non-trivial normal subgroup, it follows that F must be finitely generated.

Take Γ and $\tilde{\Gamma}$ as before. Since H is non-trivial, there is a non-trivial irreducible loop in $\tilde{\Gamma}$ based at $\tilde{v}$. Let this loop have length k. Because H is a normal subgroup of F, the covering is regular. Hence there is an isomorphism of $\tilde{\Gamma}$ taking $\tilde{v}$ onto any vertex. In particular, every vertex of $\tilde{\Gamma}$ lies in some irreducible loop of length k.

Let T be a maximal tree in $\tilde{\Gamma}$. Because H is finitely generated, there are only finitely many edges of $\tilde{\Gamma}$ not in T. Each irreducible loop must contain one of these edges. Hence $\tilde{\Gamma}$ has a finite subset S of vertices such that every vertex can be joined to a vertex of S by a path of length less than k. Since F is finitely generated, there are only finitely many edges of $\tilde{\Gamma}$ starting at a given vertex, and so only finitely many paths of a given length with a given start.

It follows that $\tilde{\Gamma}$ has only finitely many vertices, and so H has finite index.//

The set of edge-pairs of Γ is bijective with the basis X of F. Hence we may label the edges of Γ by the elements of $X \cup X^{-1}$. We may then give the edges of $\tilde{\Gamma}$ the same label as the corresponding edge of Γ. When this is done, the paths in $\tilde{\Gamma}$ starting at $\tilde{v}$ (and all paths in Γ, which are all loops) may be identified with the words on $X \cup X^{-1}$, and the irreducible paths may be identified with the reduced words; that is, with elements of $F(X)$. The vertices of $\tilde{\Gamma}$ may be regarded as cosets of H, and a collection of irreducible paths starting at $\tilde{v}$ with exactly one ending at each vertex may be regarded as a transversal of H in $F(X)$.

In particular, let T be a maximal tree in $\tilde{\Gamma}$, and let S be the transversal corresponding to the set of irreducible paths in T starting at $\tilde{v}$. If we delete the last edge from such a path the result is another such path. It follows that if $x_{i_1}^{\varepsilon_1} \dots x_{i_n}^{\varepsilon_n} \in S$ then $x_{i_1}^{\varepsilon_1} \dots x_{i_{n-1}}^{\varepsilon_{n-1}} \in S$ (we shall assume unless otherwise stated that $x_{i_r} \in X$ and $\varepsilon_r = \pm 1$ fo $r \leq n$, and that $\varepsilon_{r+1} \neq -\varepsilon_r$ for $r < n$). Such a transversal is called a *Schreier transversal* of H in the free group F with basis X (or, more simply, of H in $F(X)$; but note that the concept depends on the basis X and not just on the group). Plainly we have $1 \in S$ for any Schreier transversal S.

Conversely, let S be a Schreier transversal. We regard S, as already discussed, as giving a set of irreducible paths in $\tilde{\Gamma}$. Let T be the graph whose vertices are the vertices of these paths, and whose edges are the edges in these paths and their inverses. The lemma below tells us that T is a tree, and that the irreducible paths in T are exactly the paths we started with. In particular, S is the Schreier transversal associated with the tree T.

Lemma 3 *Let Δ be a graph. Let P be a set of irreducible paths starting at v such that there is exactly one path in P ending at each vertex of Δ. Suppose that whenever a path is in P then so is the path obtained by deleting its last edge. Let T be the subgraph whose vertices are the vertices of the paths in P and whose edges are the edges of the paths in P and their inverses. Then T is a maximal tree in Δ. Also the paths in P are exactly the irreducible paths in T starting at v.*

Proof It is enough to show that each irreducible path in T starting at v is in P. For our hypotheses then tell us that there is only one irreducible path in T from v to a given vertex. Hence T is a forest. Since each path in P is plainly an irreducible path in T, we see that T is a tree. As T contains every vertex, it will be a maximal tree, and the lemma will be proved.

The hypotheses ensure that the trivial path is in P. We will show by induction on n that an irreducible path $e_1, \ldots, e_n$ in T starting at v belongs to P.

By the definition of T, either e_n or $\bar{e}_n$ belongs to a path in P. By the hypotheses, this edge will be the last edge of a path belonging to P. By induction, the path $e_1, \ldots, e_{n-1}$ belongs to P, and it is then the only path in P ending at $e_{n-1}\tau$. Since $e_{n-1} \neq \bar{e}_n$ but $e_{n-1}\tau = \bar{e}_n\tau$, no path in P ends with $\bar{e}_n$.

Hence there is a path in P whose last edge is e_n. If we delete this edge we get another path in P. This path ends at $e_{n-1}\tau$, and so must be $e_1, \ldots, e_{n-1}$. Hence the path in P with last edge e_n must be $e_1, \ldots, e_n$, as required.//

Theorem 4 (Schreier's Subgroup Theorem) *Let H be a subgroup of a free group with basis X. Let S be a Schreier transversal for H. For all $s \in S$ and $x \in X$, let y_{sx} be sxt^{-1}, where t is the element of S such that $Ht = Hsx$. Then H is free with basis conisting of those y_{sx} which are non-trivial.*

Proof Take Γ and $\tilde{\Gamma}$ as before, and let T be the maximal tree in $\tilde{\Gamma}$ corresponding to S. Theorem 5.17 gives a presentation of $\pi_1(\tilde{\Gamma})$, and tells us explicitly a set of loops which represent the generators. Under the isomorphism of $\pi_1(\tilde{\Gamma})$ with H, we see that H is free with the stated basis.//

We can look at this theorem and Theorem 1 somewhat differently. Let $\tilde{\tilde{\Gamma}}$ be the universal covering of Γ. Then it is a tree, being a connected graph with trivial fundamental group. The group F acts freely on $\tilde{\tilde{\Gamma}}$ without inversions (that is, it fixes no vertex or edge, and sends no edge to its inverse), being the group of covering transformations. Hence the subgroup H also acts freely without inversions. $\tilde{\tilde{\Gamma}}$ is also a covering of $\tilde{\Gamma}$. The group of covering transformations of this covering is H, and $\tilde{\Gamma} = \tilde{\tilde{\Gamma}}/H$.

More generally, let T be a tree on which a group H acts freely without inversions. The remarks after Proposition 6.13 then tell us the following facts. We may form the quotient graph T/H (whose vertices and edges are the orbits of H in T), and the natural map from T to T/H is a covering. The group of covering transformations is H. Also, T is simply connected, so the group of covering transformations must be $\pi_1(T/H)$. Hence H is isomorphic to $\pi_1(T/H)$, and this group is free, being the fundamental group of a graph.

Thus we have proved the following theorem, from which the weak form of Schreier's Theorem follows at once. This theorem will be obtained again from another viewpoint in Chapter 8.

Theorem 5 *A group is free iff it acts freely and without inversions on some tree.//*

We now give a purely algebraic proof of Theorem 4. This proof, like most good proofs of similar results, is closely modelled on the covering complex approach, and we look at their relationship.

Lemma 6 *Let G be any group and let X generate G. Let $\varphi: G \to G$ be a homomorphism such that $\varphi^2 = \varphi$. Then*

(*i*) $G = (\ker\varphi)(\operatorname{im}\varphi)$,

(*ii*) $\ker\varphi \cap \operatorname{im}\varphi = \{1\}$,

(*iii*) $\operatorname{im}\varphi = \langle X\varphi \rangle$,

(*iv*) $\ker\varphi = \langle x x^{-\varphi};\ x \in X \rangle$.

Proof Note that $g^{-\varphi}$ is defined to be $(g^{-1})\varphi = (g\varphi)^{-1}$. Now

$g = (gg^{-\varphi})(g\varphi)$. Since $\varphi^2 = \varphi$, we have $gg^{-\varphi} \in \ker\varphi$, and so (i) holds. Also, (ii) is easy and (iii) is trivial.

Let $N = \langle xx^{-\varphi};\ x \in X\rangle^G$. Plainly $N \subseteq \ker\varphi$. Since $x \equiv x\varphi \bmod N$ for every $x \in X$, and $G = \langle X\rangle$, we see easily that $g \equiv g\varphi \bmod N$ for every $g \in G$. So if $g \in \ker\varphi$ we have $g \in N$.//

Let G be free with basis X, and let H be a subgroup of G. Let T be a transversal for H with $1 \in T$. Denote the cosets of H by H_i (where i runs over some index set I), and let $\overline{H_i}$ denote that member of T which is in H_i.

Let Y be the set of symbols y_{ix}, where $i \in I$ and $x \in X$, and let $F(Y)$ be the free group with basis Y. Define a homomorphism $\vartheta: F(Y) \to H$ by $y_{ix}\vartheta = \overline{H_i}\, x\, (\overline{H_i x})^{-1}$.

We define inductively for each coset H_i a function from W, the set of not necessarily reduced words on $X \cup X^{-1}$, to $F(Y)$; the image of w by the function corresponding to H_i will be denoted by w^{H_i}. With 1 being the empty word, and noting that $H_i w$ and $H_i x^{-1}$ are themselves cosets, the inductive definition (simultaneously for all i) is the following:

$$1^{H_i} = 1,\ x^{H_i} = y_{ix},\ (x^{-1})^{H_i} = (x^{H_i x^{-1}})^{-1}, \text{ and}$$

$$(wx^{\varepsilon})^{H_i} = w^{H_i}(x^{\varepsilon})^{H_i w} \text{ for all } w \text{ and } \varepsilon = \pm 1.$$

It is easy to see by induction on the length of v that, for any words u and v, $(uv)^{H_i} = u^{H_i}\, v^{H_i u}$, and that $(v^{-1})^{H_i} = (v^{H_i v^{-1}})^{-1}$. It then follows that $(ux^{\varepsilon}x^{-\varepsilon}v)^{H_i} = (uv)^{H_i}$. Hence H_i defines a mapping, which we still denote by H_i, from $F(X)$ to $F(Y)$.

Let ψ be the restriction to H of the map which sends u to u^H. Plainly ψ is a homomorphism.

By induction on length we see that, for any word w, $(w^{H_i})\vartheta = \overline{H_i}\, w(\overline{H_i w})^{-1}$. In particular, for any $h \in H$, $h\psi\vartheta = h$, so that ψ is one-one and ϑ is onto. It follows that $F(Y)\vartheta\psi = H\psi$, and this is isomorphic to H.

Consider $\vartheta\psi$ on $F(Y)$. Since $(\vartheta\psi)(\vartheta\psi) = \vartheta(\psi\vartheta)\psi = \vartheta\psi$, Lemma 6 tells us that $H = \langle y_{ix};\ y_{ix} = y_{ix}\vartheta\psi\rangle$.

For any $u \in F(X)$, $u^H \in \ker\vartheta\psi$ iff $u^H \in \ker\vartheta$ iff $\overline{H}\, u\, \overline{Hu}^{-1} = 1$; that is, since $\overline{H} = 1$, iff $u \in T$.

Further, letting $u = \overline{H_i}$ and $v = \overline{H_i x}$ we see that

$$y_{ix}\vartheta\psi = (uxv^{-1})\psi = (uxv^{-1})^H = u^H x^{Hu} (v^{-1})^{Hv} = u^H y_{ix} (v^H)^{-1}.$$

Hence $y_{ix}(y_{ix}\vartheta\psi)^{-1}$ is a consequence of $\{u^H; u \in T\}$. It follows that H has a presentation $\langle y_{ix}\, ; \, u^H$ for all $u \in T\rangle$.

This presentation is more general than the one in Schreier's Theorem, since T is an arbitrary transversal. It is not clear from this presentation that H is free. However, we shall not bother to obtain the previous presentation from this one when T is a Schreier transversal, since that can be obtained from the graph-theoretic interpretation below of the current approach; also a related simplification is obtained in the algebraic proof of Kurosh's Theorem.

Let Γ and $\tilde{\Gamma}$ be the graphs previously considered. Let $\tilde{v}_i$ be the vertex corresponding to the coset H_i, and let $\tilde{e}_{ix}$ be the edge of $\tilde{\Gamma}$ starting at $\tilde{v}_i$ which lies above the edge of Γ with label x. The element $\overline{H_i}$ of T, regarded as an element of $\pi_1(\Gamma)$, lifts to a path p_i in $\tilde{\Gamma}$ from $\tilde{v}$ to $\tilde{v}_i$.

We know, by Theorem 5.16, that $\pi_1(\tilde{\Gamma})$ has a presentation with generators Y, where y_{ix} maps to the class of the path $p_i\tilde{e}_{ix}p_j^{-1}$, where $\tilde{v}_j$ is $\tilde{e}_{ix}\tau$; also the relators correspond to the edge-paths p_i. If we identify $\pi_1(\tilde{\Gamma})$ with H using the map induced by the projection from $\tilde{\Gamma}$ to Γ then the map from $F(Y)$ to $\pi_1(\tilde{\Gamma})$ becomes the map ϑ considered above.

Paths in Γ and elements of W can be regarded as the same. To each word w and each coset H_i we can find a unique path in $\tilde{\Gamma}$ starting at $\tilde{v}_i$ and lying above w. This path defines, by replacing each $\tilde{e}_{ix}$ by the corresponding y_{ix}, an element of $F(Y)$ which is just w^{H_i}. The property $(uv)^{H_i} = u^{H_i}v^{H_i u}$ then corresponds to an obvious formula about the lift of a product of two paths. Also, when $w \in T$, the element w^H of $F(Y)$ corresponds to the path p_i, where H_i is Hw.

Thus the algebraic construction of a presentation for H is just obtained by interpreting the graph-theoretic construction of the presentation obtained in Theorem 5.16. We obtained in Theorem 5.17 a slightly simpler presentation than the one in Theorem 5.16 by taking a maximal tree and using paths in it rather than arbitrary paths. Since Schreier transversals correspond to maximal trees, this presentation is the one in Schreier's Theorem, and it can be obtained from the presentation just constructed algebraically in exactly the same way that Theorem 5.17 is obtained from Theorem 5.16..

A Schreier transversal can be obtained algebraically as follows. Define the length of a coset to be the shortest length of any element of that coset. Let $\overline{H} = 1$, and suppose that we have defined representatives for all cosets of length $< r$, so as to satisfy the Schreier condition. Let Hw be a coset of length r, and choose some w of length r in the coset. Write w as ux^{ε}, where u has length r-1. Then Hu has a representative v already chosen, since Hu has length at most r-1. Now $vx^{\varepsilon} \in Hw$, and vx^{ε} has length at most r. Hence its length must be exactly r, and it must be reduced as written. We can then take vx^{ε} as the representative of Hw, and the Schreier condition still holds. Thus we obtain a Schreier transversal inductively.

We can extend Schreier's Theorem to obtain a presentation for a subgroup of an arbitrary group. Let G be a group with presentation $\langle X;R\rangle^{\varphi}$. Let H be a subgroup of G, and let M be $H\varphi^{-1}$. Let T be a transversal for M in $F(X)$. We use the previous notation and results for M. We obtain the following theorem, which can also be obtained by applying Theorem 5.16 to the covering $\tilde{L}$ corresponding to H of the complex L which is obtained from Γ by adding cell-pairs corresponding to the relators.

Theorem 7 (Reidemeister-Schreier Theorem) *H has the presentation* $\langle y_{ix};\ u^{M}\ (u \in T),\ r^{M_i}\ (r \in R,\ all\ i)\rangle^{\vartheta\varphi}$.

Proof Let $N = \ker\varphi$. Then N is the normal closure in $F(X)$ of R. This is easily seen to be the normal closure in M of $\{\overline{M_i}\ r\overline{M_i}^{\,-1}\}$. Now $r \in N \subseteq M$, and $N \triangleleft F(X)$. Hence, for any w, $Mwr = M(wrw^{-1})w = Mw$, and so $\overline{M_i r} = \overline{M_i}$. It follows that $\overline{M_i}\ r(\overline{M_i})^{-1} = (r^{M_i})\vartheta$. We get the stated presentation for H, since M has the presentation $\langle y_{ix};\ u^{M}\ (u \in T)\rangle^{\vartheta}$ .//

Corollary *If T is chosen to be a Schreier transversal then H has the presentation* $\langle Y;\ Y_1,\ r^{M_i}\rangle^{\vartheta\varphi}$, *where* $Y_1 = \{y_{ix}$ *such that* $\overline{M_i x} = \overline{M_i}\ x\}$.//

In order to apply the theorem and its corollary, we need to compute the elements r^{M_i}. This element may also be written as $(\overline{M}_i r\overline{M}_i^{\,-1})^{M}$. Thus, what we need to do is compute u^{H}, where H is a subgroup of the free group F and $u \in H$. This can be done inductively from the definition. An alternative approach, which is essentially the same but with a different notation, is to use the following way of expressing u in terms of the generators of H given from our chosen Schreier transversal T.

Let $u = x_{i_1}^{\varepsilon_1} \dots x_{i_n}^{\varepsilon_n}$. Let $t_r \in T$ be such that $x_{i_1}^{\varepsilon_1} \dots x_{i_{r-1}}^{\varepsilon_{r-1}} \in Ht_r$, and let $u \in Ht_{n+1}$; in particular, $t_{n+1} = 1$ if $u \in H$, and $t_0 = 1$ always. Then $u = [\,\Pi(t_r x_{i_r}^{\varepsilon_r} t_{r+1}^{-1})]t_{n+1}$. If $\varepsilon_r = 1$ then $t_r x_{i_r} t_{r+1}^{-1}$ is one of the generators of H given by Schreier's Theorem (but it may be 1), while if $\varepsilon_r = -1$ then $t_r x_{i_r}^{-1} t_{r-1}^{-1}$ is the inverse of a generator. Thus we have written any $u \in H$ in terms of the generators of H.

Examples Let F be free on $\{a,b\}$. Let A be $\langle a\rangle^F$. Then A has $\{b^i$, all $i\}$ as a Schreier transversal. Hence A has basis $\{b^i a b^{-i}\}$.

The commutator subgroup F' has $\{a^i b^j$, all i and $j\}$ as a Schreier transversal. The representatives of the cosets containing $a^i b^j b$ and $a^i b^j a$ are $a^i b^{j+1}$ and $a^{i+1} b^j$, respectively. Hence F' has basis $\{a^i b^j a b^{-j} a^{-(i+1)}\}$.

Let N be the kernel of the homomorphism from F onto the symmetric group of degree 3 which sends a and b to the permutations (1 2) and (1 2 3). Then N has the set $\{1,b,b^2,a,ba,b^2a\}$ as a Schreier transversal. Now a^2, b^3, and $(ab)^2$ are in N, and so $bab \in Na$ and $b^2ab \in Nba$. It follows that N has basis $\{b^3,\ a^2,\ ba^2b^{-1},\ b^2a^2b^{-2},\ aba^{-1}b^{-2},\ baba^{-1},\ b^2aba^{-1}b^{-1}\}$. Another Schreier transversal is $\{1,b,b^2,a,ab,ab^2\}$, and the corresponding basis is $\{b^3,\ ab^3a^{-1},\ bab^{-2}a^{-1},\ b^2ab^{-1}a^{-1},\ a^2,\ abab^{-2},\ ab^2ab^{-1}\}$.

We now apply similar, but more complicated, techniques to determine the structure of subgroups of a free product. A rather simpler (but fundamentally similar) proof of the result is given in the next chapter.

Theorem 8 (Kurosh Subgroup Theorem) *Let H be a subgroup of the free product $*G_\alpha$. Then $H = F * *(uG_\alpha u^{-1} \cap H)$, where F is free and there is one factor $uG_\alpha u^{-1} \cap H$ for each α and each u in a suitably chosen set of representatives of the (H,G_α) double cosets.*

Proof (using covering spaces) Take disjoint connected complexes L_α with a vertex v_α in L_α such that $\pi_1(L_\alpha, v_\alpha) = G_\alpha$. Take a new vertex v and new edge-pairs $\{e_\alpha, \bar{e}_\alpha\}$ such that $e_\alpha\sigma = v$ and $e_\alpha\tau = v_\alpha$ for all α, and let K_α be $L_\alpha \cup \{v, e_\alpha, \bar{e}_\alpha\}$. Let K be $\cup K_\alpha$, so that $\pi_1(K) = *G_\alpha$.

Let $p:\tilde{K} \to K$ be the covering corresponding to H. Let the components of $L_\alpha p^{-1}$ be $\tilde{L}_{\alpha i}$ (where i runs over some index set depending on α), and let $\tilde{v}_{\alpha i}$ be a vertex of vp^{-1} in $\tilde{L}_{\alpha i}$. It is easy to check that the components of $K_\alpha p^{-1}$ will be complexes $\tilde{K}_{\alpha i}$, where $\tilde{K}_{\alpha i}$ is the union of $\tilde{L}_{\alpha i}$ with certain edge-pairs in $\{e_\alpha, \bar{e}_\alpha\}p^{-1}$ and the vertices of these; further, each such edge-pair has exactly one vertex in $\tilde{L}_{\alpha i}$ and no two have a vertex in common.

Let $T_{\alpha i}$ be a maximal tree in $\tilde{L}_{\alpha i}$. Then $\cup T_{\alpha i}$ is a forest. Let Y be the subgraph of $\tilde{K}$ whose edges are the edges of $\cup T_{\alpha i}$ together with the edges of $\{e_\alpha, \bar{e}_\alpha\}p^{-1}$ for all α. Then $V(Y) = V(\tilde{K})$, since $V(T_{\alpha i}) = V(\tilde{L}_{\alpha i})$. Now any two vertices of Y may be joined by a path in $\tilde{K}$, each edge of which is either in $\{e_\alpha, \bar{e}_\alpha\}p^{-1}$ for some α or is in $\tilde{L}_{\alpha i}$ for some α and i. In the latter case there is a path in $T_{\alpha i}$ beginning and ending at the ends of this edge. Replacing each such edge by the corresponding path, we obtain a path in Y. It follows that Y is connected, and so $\bigcup T_{\alpha i}$ lies in a maximal tree T of Y.

The intersection of any two of the connected complexes $\tilde{L}_{\alpha i} \cup T$ is T. Also $Y \cap (\tilde{L}_{\alpha i} \cup T) = T$, and $\tilde{K} = Y \cup \bigcup \tilde{L}_{\alpha i}$. Take a vertex $\tilde{v} \in vp^{-1}$. Using the presentations of the various fundamental groups given by Theorem 5.17 and the maximal tree T, we see that $\pi_1(\tilde{K},\tilde{v}) = \pi_1(Y,\tilde{v}) * *\pi_1(\tilde{L}_{\alpha i} \cup T,\tilde{v})$.

Since $p*$ maps $\pi_1(\tilde{K},\tilde{v})$ isomorphically to H, and $\pi_1(Y)$ is free, the theorem follows once we have shown that $\pi_1(\tilde{L}_{\alpha i} \cup T,\tilde{v})p*$ is of the required form.

Any edge with one vertex in L_α either lies in L_α or is e_α or $\bar{e}_\alpha$. It follows that any edge with one vertex in $\tilde{L}_{\alpha i}$ either lies in $\tilde{L}_{\alpha i}$ or has its other vertex in vp^{-1}. We then see that a path in T from $\tilde{v}$ which ends in $\tilde{L}_{\alpha i}$ and has minimal length must be of the form $\tilde{\gamma}.\tilde{e}_\alpha$, where $\tilde{\gamma}$ ends in a vertex of vp^{-1} and $\tilde{e}_\alpha p = e_\alpha$. Let $\lambda \in \pi_1(K,v) = G$ be represented by $\tilde{\gamma}p$. (Strictly, we should refer to paths $\tilde{\gamma}_{\alpha i}$ and edges $\tilde{e}_{\alpha i}$, but this complicates the notation too much.)

Since $\tilde{L}_{\alpha i} \cap T = T_{\alpha i}$, we see by van Kampen's Theorem (since $\pi_1(T)$ is trivial) that $\pi_1(\tilde{L}_{\alpha i} \cup T,\tilde{v}_{\alpha i}) = \pi_1(\tilde{L}_{\alpha i},\tilde{v}_{\alpha i})$. By Proposition 6.12 this is mapped isomorphically by $p*$ to the group $\pi_1(L_\alpha,v_\alpha) \cap \pi_1(\tilde{K},\tilde{v}_{\alpha i})p*$ This image equals

$$\pi_1(K_\alpha,v_\alpha) \cap \pi_1(\tilde{K},\tilde{v}_{\alpha i})p* = \pi_1(K_\alpha,v)e_\alpha^{\#} \cap [\pi_1(\tilde{K},\tilde{v})(\tilde{\gamma}.\tilde{e}_\alpha)^{\#}]p*$$

which in turn equals $(G_\alpha \cap \pi_1(\tilde{K},\tilde{v})p*\lambda^{\#})e_\alpha^{\#}$, which, since $\lambda^{\#}$ is just conjugation by λ, can be written as $(\lambda G_\alpha \lambda^{-1} \cap H)\lambda^{\#} e_\alpha^{\#}$.

But we may also write

$$\pi_1(\tilde{L}_{\alpha i} \cup T,\tilde{v}_{\alpha i})p* = \pi_1(\tilde{L}_{\alpha i} \cup T,\tilde{v})(\tilde{\gamma}.\tilde{e}_\alpha)^{\#} p* = \pi_1(\tilde{L}_{\alpha i} \cup T,\tilde{v})p*\lambda^{\#} e_\alpha^{\#}.$$

We deduce that $\pi_1(\tilde{L}_{\alpha i} \cup T,\tilde{v})p* = \lambda G_\alpha \lambda^{-1} \cap H$, and so $\pi_1(\tilde{L}_{\alpha i} \cup T,\tilde{v})$ is isomorphic to $\lambda G_\alpha \lambda^{-1} \cap H$.

This is almost what we want. We still have to show that the elements λ for a given α form a set of double coset representatives. This is not difficult, as the subcomplexes $\tilde{K}_{\alpha i}$ can easily be shown to correspond to double cosets. Thuis will be left as an exercise, as the algebraic proof below gives more specific information, and a simplification of the above proof is given in the next chapter.//

Proof (using purely algebraic methods) Let $G = *G_\alpha$, where α runs over some index set A. To begin with we shall assume that each G_α is free and has basis X_α. We shall also choose one element of A, which we denote by 1. We shall obtain a presentation of H (which is, of course, a free group in this case) from which the result for arbitrary G_α can be deduced.

Let $H \subseteq G$, and let H_i (for i in some index set I) be the cosets of H, with H_1 being H. For each α take an element ${}^\alpha\overline{H_i}$ in the coset H_i; this element will *be* called an α-representative of H_i. We require that ${}^\alpha\overline{H} = 1$ for all α; no other conditions are required at present, though they will be imposed later.

Let Y be a set with elements y_{ix} for all i and all $x \in \cup X_\alpha$. Let Z be a set with elements $z_{i\alpha}$ for all $i \neq 1$ and all $\alpha \neq 1$. We define a homomorphism $\vartheta: F(Y) * F(Z) \to H$ by

$$y_{ix}\vartheta = {}^\alpha\overline{H_i}\, x\,({}^\alpha\overline{H_i x}\,)^{-1} \text{ where } x \in X_\alpha, \text{ and } z_{i\alpha}\vartheta = {}^\alpha\overline{H_i}({}^1\overline{H_i}\,)^{-1}.$$

For convenience, let $z_{i\alpha}$ be 1 if $i = 1$ or $\alpha = 1$; the formula for $z_{i\alpha}\vartheta$ will hold in this case also.

We define, for each H_i, a function, whose value on w will be denoted by w^{H_i}, from the set of all (not necessarily reduced) words on $\cup X_\alpha \cup (\cup X_\alpha)^{-1}$ to $F(Y) * F(Z)$. These functions, are defined inductively by

$$1^{H_i} = 1,\ x^{H_i} = y_{ix},\ (x^{-1})^{H_i} = (x^{H_j})^{-1} \text{ where } H_i = H_j x, \text{ and}$$

$$(ux^\varepsilon)^{H_i} = u^{H_i} z_{k\beta} z_{k\gamma}^{-1}(x^\varepsilon)^{H_k}, \text{ where } \varepsilon = \pm 1,\ H_k = H_i u,\ x \in X_\gamma,$$

and the last letter of u is in G_β (when u has normal form $g_1 \ldots g_n$ the elements g_1 and g_n are called the first and last letters of u, and when $u = 1$ we say that the first and last letters of u are in G_α for all α).

As in the algebraic proof of Schreier's Theorem, we can easily show inductively that $(uv)^{H_i} = u^{H_i} z_{j\beta} z_{j\gamma}^{-1} v^{H_j}$ for all u and v, where $H_j = H_i u$, and the last letter of u is in G_β and the first letter of v is in G_γ. In particular, $(ux^\varepsilon x^{-\varepsilon} v)^{H_i} = (uv)^{H_i}$, and so these functions induce functions from G to $F(Y) * F(Z)$ with the same properties.

If the first letter of u is in G_α and the last is in G_β we find (inductively) that $(u^{H_i})\vartheta = {}^\alpha\overline{H_i}\, u\,({}^\beta\overline{H_i u}\,)^{-1}$. It follows that the restriction ψ to H of the map sending u to u^H is a homomorphism such that $\psi\vartheta$ is the identity. As before, this shows that H has the presentation

$$\langle y_{ix} \text{ (all } i \text{ and } x),\ z_{i\alpha} \text{ (all } i \neq 1 \text{ and } \alpha \neq 1);\ y_{ix} = y_{ix}\vartheta\psi,\ z_{i\alpha} = z_{i\alpha}\vartheta\psi\rangle.$$

We now proceed to simplify this by a good choice of representatives (called a *Kurosh system* of representatives) satisfying the following conditions:

(i) One coset in the double coset HgG_α has an α-representative which is either 1 or has its last syllable not in G_α; this element will be called an *α-representative in the large.* When $H_i \subseteq HgG_\alpha$ and g is the α-representative in the large, then ${}^\alpha\overline{H_i} \in gG_\alpha$.

(ii) If g is an α-representative in the large whose last syllable is in G_β then ${}^\beta\overline{Hg}$ is g.

Our construction will provide a Kurosh system for a subgroup of an arbitrary free product, not just for a free product of free groups. This will be of importance later.

We construct these representatives by induction on the shortest length of elements of HgG_α (length in the expression of G as the free product $*G_\alpha$, not as a free group). Once we have chosen an α-representative in the large for such a double coset it is easy to choose α-representatives of each coset in this double coset so as to satisfy the condition (i).

Take g of shortest length, r say, in the double coset HgG_α. If $r=0$ then $g=1$, and we take 1 to be the α-representative in the large. If $r\neq 0$ we may write g as ug_β where $1\neq g_\beta \in G_\beta$ and u has length r-1. We cannot have $\beta=\alpha$, as then u would be a shorter element in the same double coset. Now HgG_β contains u of length r-1, so, by our inductive assumption, we already have a β-representative in the large of HgG_β. Let this element be v. Then v has length at most r-1. As remarked, we may also assume that a β-representative of form vw with $w \in G_\beta$ has been chosen for the coset Hg. We then choose vw to be the α-representative in the large of HgG_α. Plainly condition (ii) is satisfied for this element, and we have completed the inductive definition.

If $u^H \in \ker\vartheta\psi$ then $u^H \in \ker\vartheta$, and so $u({}^\beta\overline{Hu})^{-1}=1$, where the last letter of u is in G_β (since all representatives of H are 1, it does not matter where the first letter of u lies). Thus u is a representative. Conversely, if u is a β-representative whose last syllable is in G_γ then either $\beta=\gamma$, so that $u={}^\gamma\overline{Hu}$, or, because of (i) and (ii), u is the β-representative in the large, in which case we again have $u={}^\gamma\overline{Hu}$. Hence $u^H \in \ker\vartheta\psi$ for any representative u.

Let u be an α-representative in the large whose last syllable is in G_β, and let H_i denote Hu. Since ${}^\alpha\overline{H_i}=u={}^\beta\overline{H_i}$, we see that $z_{i\alpha}z_{i\beta}^{-1} \in \ker\vartheta\psi$. We call the elements $z_{i\alpha}z_{i\beta}^{-1}$ relators of the first kind. Note that if u is any α-representative, and the last syllable of u is in G_β, then either $\alpha=\beta$ and so $z_{i\alpha}z_{i\beta}^{-1}=1$ or else $z_{i\alpha}z_{i\beta}^{-1}$ is a relator of the first kind.

Take any i and α, and any $x \in X_\alpha$, and write $u = {}^\alpha\overline{H_i}$ and $v = {}^\alpha\overline{H_i x}$ and $H_j = Hv = Hux$. Let the last syllable of u be in G_β, the last syllable of ux in G_γ, and the last syllable of v in G_δ. Then $\gamma = \alpha$ unless the last syllable of u is x^{-1}, and in this case ux is the α-representative in the large, by the properties (i) and (ii). Thus $z_{j\alpha} z_{j\gamma}^{-1}$ is either 1 or a relator of the first kind.

Observe that $1 = (vv^{-1})^H = v^H (v^{-1})^{Hv}$, so that $(v^{-1})^{Hux} = (v^H)^{-1}$. Hence

$$y_{ix}\vartheta\psi = (uxv^{-1})\psi = u^H z_{i\beta} z_{i\alpha}^{-1} y_{ix} z_{j\gamma} z_{j\delta}^{-1} (v^H)^{-1},$$

where $z_{j\gamma} z_{j\delta}^{-1}$, being $(z_{j\alpha} z_{j\gamma}^{-1})^{-1}(z_{j\alpha} z_{j\delta}^{-1})$, is a consequence of relators of the first kind.

Similarly, we find that $z_{i\alpha}\vartheta\psi = u^H z_{i\beta} z_{i\gamma}^{-1} (v^H)^{-1}$ for some representatives u and v, where $z_{i\alpha} z_{i\beta}^{-1}$ is either 1 or a relator of the first kind, as is $z_{i\gamma} z_{i1}^{-1}$. Recall that $z_{i1} = 1$.

Thus the defining relators of the presentation are consequences of the relators of the first kind and the relators $\{u^H$; all representatives $u\}$. We proceed to simplify the second type of relator further.

Take any α and any α-representative in the large u. Let $Y_{u\alpha}$ be $\{y_{ix}\colon x \in X_\alpha$ and $H_i \subseteq HuG_\alpha\}$. It is easy to check (inductively on its length as a word in $X_\alpha \cup X_\alpha^{-1}$) that for any $g \in G_\alpha$ the element g^{Hu} is in $F(Y_{u\alpha})$.

Let v be any α-representative, and write v as ug, where u is the α-representative in the large and $g \in G_\alpha$. Then $v^H = u^H z_{j\beta} z_{j\alpha}^{-1} g^{Hu}$, where H_j is Hu and the last letter of u is in G_β. It follows that this element g^{Hu} is a relator. It also follows, by induction on the length of a relator, that every relator is a consequence of relators of the first kind and those relators which are of form g^{Hu}.

We have therefore shown that H has a presentation on the generators $z_{i\alpha}$ and y_{ix} in which each defining relator is either of the first kind or is in some $F(Y_{u\alpha})$. Hence $H = (F(Z)\vartheta) * *(F(Y_{u\alpha})\vartheta)$. Also $F(Z)\vartheta$ is free, since the relators of the first kind simply identify certain elements of Z.

Let u be an α-representative in the large, and let $H_i \subseteq HuG_\alpha$ and let $x \in X_\alpha$. Then the representatives ${}^\alpha\overline{H_i}$ and ${}^\alpha\overline{H_i x}$ are ug and ug' for some g and $g' \in G_\alpha$. Thus $y_{ix}\vartheta = (ug)x(ug')^{-1} \in uG_\alpha u^{-1}$, and, of course, $y_{ix}\vartheta \in H$. Conversely, let $h \in H \cap uG_\alpha u^{-1}$. Write h as ugu^{-1}. Then $(ug)^H = u^H \xi g^{Hu}$, where ξ is a relator of the first kind, and $(hu)^H = (h\psi)u^H$, since $h \in H$. As remarked, g^{Hu} is in $F(Y_{u\alpha})$. Since $(hu)^H = (ug)^H$, we see that $h = h\psi\vartheta = g^{Hu}\vartheta \in F(Y_{u\alpha})\vartheta$. It follows that $F(Y_{u\alpha})\vartheta = H \cap uG_a u^{-1}$. This proves Kurosh's Theorem when each G_α is free.

Now let $G = *G_\alpha$, where the groups G_α are arbitrary. Take a presentation $\langle X_\alpha ; R_\alpha \rangle^{\varphi_\alpha}$ of G_α, and let $\varphi: *F(X_\alpha) \to *G_\alpha$ be induced by the φ_α. For each $g \in G_\alpha$ take an element $u \in F(X_\alpha)$ with $u\varphi_\alpha = g$, where $u = 1$ if $g = 1$. Then, for each element of G. say $g_1 \ldots g_n$ where g_i and g_{i+1} are in different factors, we have a corresponding element of $*F(X_\alpha)$ mapping to it under φ, namely the element $u_1 \ldots u_n$ where u_i is the element chosen to map to g_i.

Let H be a subgroup of G, and let K be $H\varphi^{-1}$. Take a Kurosh system for H in G. It is easy to check that the corresponding elements of $*F(X_\alpha)$ form a Kurosh system for K in $*F(X_\alpha)$.

The presentation already obtained for K gives rise to a presentation for H. Now H is obtained from K by factoring out by the normal subgroup generated by the elements $vr_\alpha v^{-1}$ for all α, all $r_\alpha \in R_\alpha$, and all α-representatives v of cosets of K. These elements may be written as $(r_\alpha{}^{Kv})\vartheta$. Since the presentation of K is obtained using ϑ, we see that H is obtained from K by adding the extra relators $r_\alpha{}^{Kv}$. Now let w be the α-representative in the large such that $v \in KwF(X_\alpha)$. It is not difficult to check (by induction on the length in the generators X_α) that for any $u_\alpha \in F(X_\alpha)$ we have $u_\alpha{}^{Kv} \in F(Y_{w\alpha})$. Thus the expression of K as a free product gives rise to an expression of H as a free product. This is what we require, noting also that $(K \cap wF(X_\alpha)w^{-1})\varphi = H \cap gG_\alpha g^{-1}$, where $g = w\varphi$.//

We note also that any subgroup $H \cap gG_\alpha g^{-1}$ is conjugate in H to one of the free factors. To see this, we need only write g as hcg_α, where $h \in H$ and c is an α-representative in the large of H.

The above algebraic construction has a significance similar to that in Schreier's Theorem. Take complexes L_α such that $\pi_1(L_\alpha) = G_\alpha$, and such that L_α has only one vertex and has one edge-pair corresponding to each member of X_α. Let $L = \cup L_\alpha$, the vertices being identified, and let $p: \tilde{L} \to L$ be the covering corresponding to H. Choose a vertex $\tilde{v}$ of $\tilde{L}$, and the vertices then correspond to cosets of H. For each component of $L_\alpha p^{-1}$ choose some path from $\tilde{v}$ to a vertex in $L_\alpha p^{-1}$, and then for each vertex $\tilde{v}_i$ of $L_\alpha p^{-1}$ take a path $\tilde{f}_{i\alpha}$ (trivial if $\tilde{v}_i = \tilde{v}$) from $\tilde{v}$ to $\tilde{v}_i$ which is the product of the chosen path and a path in $L_\alpha p^{-1}$. Then by Theorem 5.16 $H = \pi_1(\tilde{L})$ is generated by the classes of the loops $\tilde{f}_{i1} \cdot \tilde{e}_\alpha \cdot \tilde{f}_{j1}{}^{-1}$, where $\tilde{e}_\alpha$ is an edge in $L_\alpha p^{-1}$ from $\tilde{v}_i$ to $\tilde{v}_j$. Thus H is generated by the classes of the loops $\tilde{f}_{i\alpha} \cdot \tilde{e}_\alpha \cdot \tilde{f}_{j\alpha}{}^{-1}$ together with the classes of the loops $\tilde{f}_{i\alpha} \cdot \tilde{f}_{i1}{}^{-1}$. These correspond to y_{ix_α} and $z_{i\alpha}$ respectively. Given any irreducible edge-loop u in L whose first edge is in L_α and whose last is in L_β

(equivalently, any element of G whose first syllable is in G_α and whose last is in G_β) and given any vertex $\tilde{v}_i$ of $\tilde{L}$, we may lift u to a path $\tilde{u}$ starting at $\tilde{v}_i$, and we denote its last vertex by $\tilde{v}_j$. Then u^{H_i} is the expression for the loop $\tilde{f}_{i\alpha}.\tilde{u}.\tilde{f}_{j\beta}^{-1}$ in terms of the generators. Note that this construction is not quite the same as the one used in the covering space argument. The complexes used here can be obtained from the ones previously used by contracting subtrees.

Exercise 1 Let G be $\langle a_1, \ldots, a_n ; a_1^2 \ldots a_n^2 \rangle$. Let $\varphi, \psi: G \rightarrow \mathbb{Z}_2$ be the homomorphisms given by $a_1\varphi = 1$, $a_i\varphi = 0$ for $i \neq 1$, and $a_i\psi = 1$ for all i. Use the Reidemeister-Schreier Theorem to find presentations of kerφ and kerψ with $2(n-1)$ generators.

Exercise 2 Let G be $\langle a, b;\ a^8, b^8, (ab)^2 \rangle$, and let N be $\langle ab^{-3} \rangle^G$. Show that $\{a^i, 0 \leq i \leq 7\}$ is a Schreier transversal for N. Show that N has a presentation $\langle x, y, z, w;\ xyzw = wzyx \rangle$, where x, y, z, and w are among the generators of N coming from the stated Schreier transversal. Find a presentation for N in terms of the generators x, y, u, and v, where $u = yxz$ and $v = wz$.

Exercise 3 Let G be $\langle a, b;\ a^2, b^3 \rangle$. Find a normal subgroup of index 6, for which the Reidemeister-Schreier method (and some Tietze transformations) shows that the subgroup is free. Find another proof that the subgroup is free.

8 BASS-SERRE THEORY

In 1968-69 Serre, in a course of lectures at the College de France, introduced a new technique into combinatorial group theory. Some of the more general results of the theory were due to Bass, so that the material is known as Bass-Serre theory. The most convenient source for the original material is in the book by Serre (1980).

We shall see that the theory is to a great extent just a disguised form of the theory of covering complexes. However, it permits us to work with graphs only, rather than arbitrary complexes, and with an analogue of the universal covering rather than an arbitrary covering. These simplifications permit us to avoid the notational complications of the earlier theory, and so make possible powerful techniques for obtaining a range of results.

8.1 *TREES AND FREE GROUPS*

An *action* of a group G on a set X is a homomorphism φ from G to Sym$X$, the group of permutations of X. We let Sym$X$ act on the left (so as to be consistent with other writers on the subject), and we write gx instead of $(g\varphi)x$, where $g \in G$ and $x \in X$. When G acts on X we say that X is a *G-set.*

A graph Γ is called a *G-graph* if both $V(\Gamma)$ and $E(\Gamma)$ are G-sets, and $\overline{ge} = g\overline{e}$ and $(ge)\tau = g(e\tau)$ for all $g \in G$ and $e \in E(\Gamma)$.

Example Let G be any group, and let S be any subset of G. Let $\Gamma(G,S)$ be the graph with vertex set G and with edge set $G \times S \times \{-1,1\}$, where $(g,s,1)\sigma = g$ and $(g,s,1)\tau = gs$, and with $\overline{(g,s,\varepsilon)} = (g,s,-\varepsilon)$ for $\varepsilon = \pm 1$. Plainly $\Gamma(G,S)$ is a graph on which G acts. We call $\Gamma(G,S)$ the *Cayley graph* of G with respect to S.

Lemma 1 (*i*) *$\Gamma(G,S)$ is connected iff S generates G.* (*ii*)*$\Gamma(G,S)$ is a forest iff S freely generates $\langle S \rangle$.* (*iii*)*$\Gamma(G,S)$ is a tree iff G is free with basis S.*

Proof These results follow immediately if we show that there is a bijection between ways of writing $g^{-1}h$ (for arbitrary g and h in G) as $s_1^{\varepsilon_1} \dots s_n^{\varepsilon_n}$ with n arbitrary, $s_i \in S$, and $\varepsilon = \pm 1$, and paths in $\Gamma(G,S)$ from g to h, with the bijection making reduced words correspond to irreducible paths. The required bijection maps $s_1^{\varepsilon_1} \dots s_n^{\varepsilon_n}$ to the path $e_1, \dots, e_n$, where e_r is $(g_r, s_r, \varepsilon_r)$, with g_r being $gs_1^{\varepsilon_1} \dots s_{r-1}^{\varepsilon_{r-1}}$ if $\varepsilon_r = 1$ and g_r being $gs_1^{\varepsilon_1} \dots s_r^{\varepsilon_r}$ if $\varepsilon_r = -1$.//

Note that when G is free with basis S, the tree $\Gamma(G,S)$ is the graph obtained in Chapter 5 as the universal cover of the graph with only one vertex and with one edge-pair corresponding to each element of S.

Let the group G act on the graph X. If there is some $g \in G$ and some edge e such that $ge = \bar{e}$ we say that G acts *with inversions*; otherwise we say that G acts *without inversions*. Plainly G acts on $\Gamma(G,S)$ without inversions.

We discussed in section 5.5 how to construct from any complex K another complex K', which we called its barycentric subdivision. When K is a graph, which is the only case that will concern us, K' is just obtained by subdivision of edges (for a general complex we have to subdivide edges, and then subdivide cells). It is easy to see that if G acts on a graph X then it also acts on its subdivision X'. Also, G always acts on X' without inversions.

Because of this, in future when we are given an action of a group on a graph we shall always assume that the group acts without inversions.

Let G act on X without inversions. We form a graph called the *quotient graph of X by G*, and written $G\backslash X$ this notation is preferable to writing X/G, which suggests that G acts on the right), as follows. On the vertex set $V(X)$ we have an equivalence relation given by $w \equiv v$ if $w = gv$ for some g, and we have the similar equivalence relation on the edge set $E(X)$. We denote the equivalence class of v or e by $[v]$ or $[e]$. Then the vertex set of $G\backslash X$ is defined to be the set of equivalence classes $[v]$ for $v \in V(X)$, and the edge set of $G\backslash X$ is defined to be the set of equivalence classes $[e]$ for $e \in E(X)$. We define $\overline{[e]}$ to be $[\bar{e}]$, and $[e]\tau$ and $[e]\sigma$ to be $[e\tau]$ and $[e\sigma]$. It is clear that these only depend on $[e]$, and not on the choice of e. We need G to act without inversions to ensure that we have $\overline{[e]} \neq [e]$.

The next three lemmas could be extracted from material in Chapters 5 and 6, but I prefer to give an explicit proof. Only the first lemma is needed immediately. We begin with two definitions.

Let $p:X \to Y$ be a map of graphs. We call p *locally surjective* if for every vertex v of X and every edge y of Y such that $y\sigma = vp$ there is an edge x of X with $x\sigma = v$ and $xp = y$. We call p *locally injective* if for any two distinct edges x_1 and x_2 of X such that $x_1\sigma = x_2\sigma$ the edges x_1p and x_2p of Y are distinct. We call p *locally bijective* if it is both locally surjective and locally injective.

Lemma 2 *Let $p:X \to Y$ be locally surjective. Let T be a tree and let $f:T \to Y$ be a map. Let a be a vertex of T, and let v be a vertex of X such that $vp = af$. Then there is a map $\varphi:T \to X$ such that $\varphi p = f$ and $a\varphi = v$.*

Proof Consider pairs (T_1, φ_1) such that T_1 is a subtree of T with $a \in T_1$ and φ_1 is a map from T_1 to X such that $a\varphi_1 = v$ and $\varphi_1 p$ equals f on T_1. Such pairs exist; for instance, T_1 could consists solely of the vertex a. We may give a partial ordering to these pairs, writing $(T_1, \varphi_1) < (T_2, \varphi_2)$ if $T_1 \subseteq T_2$ and φ_1 is the restriction of φ_2 to T_1. By Zorn's Lemma, there is a maximal pair (T_0, φ_0), and we need only show that $T_0 = T$.

Since T is a tree, either $T_0 = T$ or some vertex of T is not in T_0. In the latter case there is an edge e with $e\sigma \in T_0$ and $e\tau \notin T_0$ (this must hold for some edge in a path starting in T_0 and ending outside T_0). By hypothesis, there is an edge x of X starting at $e\sigma\varphi_0$ such that $xp = ef$. We can then enlarge T_0 to the tree $T_1 = T_0 \cup \{e, \bar{e}, e\tau\}$, and we can extend φ_0 to a map φ_1 on T_1 such that $\varphi_1 p = f$ on T_1 by defining $e\varphi_1$ to be x, $\bar{e}\varphi_1$ to be $\bar{x}$, and $e\tau\varphi_1$ to be $x\tau$. This contradicts the maximality of T_0.//

Lemma 3 *Let $p:X \to Y$ be locally injective. Let φ and ψ be maps from a connected graph Z to X such that $\varphi p = \psi p$. If there is some vertex z such that $z\varphi = z\psi$ then $\varphi = \psi$.*

Proof Let S be $\{v \in V(Z);\ v\varphi = v\psi\}$. Take an edge e with $e\sigma \in S$. Then $e\varphi$ and $e\psi$ both begin at $e\sigma\varphi$ ($= e\sigma\psi$, by hypothesis) and have the same image under p. As p is locally injective, this implies that $e\varphi = e\psi$. In particular, $e\tau\varphi = e\varphi\tau = e\psi\tau = e\tau\psi$, and so $e\tau \in S$. Since Z is connected, and $S \neq \emptyset$ by hypothesis, Lemma 5.1 tells us that $S = V(Z)$. This, combined with what we have just proved about edges, tells us that $\varphi = \psi$.//

Lemma 4 *Let $p:S \to Y$ and $q:T \to Y$ be local bijections, and let S and T be trees. Let v and w be vertices of S and T respectively such that $vp = wq$. Then there is a unique isomorphism $\varphi:S \to T$ such that $v\varphi = w$ and $\varphi q = p$.*

Proof Since S is a tree and q is locally surjective, by Lemma 2 there is a map $\varphi:S \to T$ such that $v\varphi = w$ and $\varphi q = p$. Since q is locally injective, φ is unique, by Lemma 3. Reversing the roles of S and T, there is also a map $\psi:T \to S$ such that $w\psi = v$ and $\psi p = q$.

Then $v\varphi\psi = v = vI_S$ and $\varphi\psi p = p = I_S p$, where I_S is the identity on S. Since p is locally injective, Lemma 3 tells us that $\varphi\psi = I_S$. Similarly, $\psi\varphi = I_T$, and so φ is an isomorphism.//

Lemma 5 *Let $f:X \to Y$ be a map of graphs.* (*i*) *If f is locally surjective and Y is connected then f is surjective.* (*ii*) *If f is locally injective, X is connected, and Y is a tree then f is injective.* (*iii*) *If f is locally bijective and X and Y are trees then f is bijective.*

Proof (i) Let e be an edge of Y such that $e\sigma \in V(X)f$. Take a vertex v of X such that $vf = e\sigma$. Since f is ocally surjective, there is an edge x of X such that $x\sigma = v$ and $xf = e$. It follows that $e\tau \in V(X)f$, and the result follows by Lemma 5.1.

(ii) Since f is locally injective, it is injective on edges if it is injective on vertices.

Suppose, if possible, that there are distinct vertices v and w such that $vf = wf$. Since X is connected, there is an irreducible path $e_1,\dots,e_n$ from v to w. Then $e_1f,\dots,e_nf$ will be a loop in Y. Since Y is a tree, there must be some i such that $e_{i+1}f = \bar{e}_i f$. Because f is locally injective, this gives $e_{i+1} = \bar{e}_i$, contradicting the irreducibility of the path $e_1,\dots,e_n$.//

Let X be a G-graph, and let p be the natural map from X to $G\backslash X$. Let T be a maximal tree in $G\backslash X$. We see from Lemma 2 that there is a map $j:T \to X$ such that jp is the inclusion of T in $G\backslash X$. In particular, p defines an isomorphism from Tj to T. Hence Tj is a tree, which we call a *representative tree* for the action of G on X.

We say that G acts *freely* on X if for any $g \in G - \{1\}$ and any vertex v of X we have $gv \neq v$. For instance, for any group G and any subset S of G, we see that G acts freely on $\Gamma(G,S)$.

An *orientation* of a graph is a set of edges containing exactly one edge from each pair. Let G act on a graph X. Let B be an orientation of $G\backslash X$, and let A be Bp^{-1}. Then A is an orientation of X and $GA = A$. This property is needed below.

Theorem 6 *Let G act freely on a tree X. Let T be a representative tree for the action of G, and let A be an orientation of V such that $GA = A$. Let S be $\{g \in G - \{1\};\ \exists e \in A$ with $e\sigma \in T$ and $e\tau \in gT\}$. Then G is free with basis S.*

Proof Observe first that $gT \cap hT = \emptyset$ for $g \neq h$. Otherwise we can choose vertices v and w in T such that $gv = hw$. Since T is a representative tree it contains only one vertex from each orbit, and so $v = w$. As G acts freely, this shows that $g = h$, as required. Further, every vertex is in gT for some g, since T is a representative tree.

Let Y be obtained from X by contracting each of the trees gT. By Lema 5.12, Y is a tree. By Lemma 2, it is enough to show that $\Gamma(G,S)$ is isomorphic to Y.

Map $\Gamma(G,S)$ to Y as follows. The vertices of Y can be identified with the trees gT, and we map the vertex g of $\Gamma(G,S)$ to the vertex gT of Y.

To each $s \in S$ there is an edge e of X in A with $e\sigma \in T$ and $e\tau \in sT$. There can only be one such edge, as two such edges would give two distinct edges in Y with the same start and finish. The edge ge of X gives an edge of Y from gT to gsT. We now map the edge $(g,s,1)$ of $\Gamma(G,S)$ to this edge of Y, and the inverse $(g,s,-1)$ of $(g,s,1)$ to the inverse of this edge.

This function is clearly a map of graphs which is bijective on the vertex set. The edges $(g,s,1)$ and $(h,s',-1)$ have different images, since the image of the first is in $GA = A$, while the image of the second is in $G\bar{A} = \bar{A}$. The images of two distinct edges $(g,s,1)$ and $(h,s',1)$ have different starting points gT and hT if $g \neq h$, while if $g = h$ they have different finishing points gsT and $gs'T$. Thus the map is injective on edges.

Now take any edge of Y. By definition of Y, it is the image of an edge e with $e\sigma \in gT$ and $e\tau \in hT$ for some g and h in G with $g \neq h$. Replacing the edge by its inverse if necessary, we may assume that $e \in A$. Then $g^{-1}e \in GA = A$, and $g^{-1}h \in S$, since $g^{-1}e$ is an edge starting in T and finishing in $g^{-1}hT$. Then $(g,g^{-1}h,1)$ is an edge of $\Gamma(G,S)$ whose image in Y is the given edge. Hence our map is also surjective on edges, as required.//

Combining Theorem 6 and Lemma 1, we see that *a group is free iff it acts freely on a tree.* When a group acts freely on a tree, any subgroup also acts freely on the same tree, and we find once more that *any subgroup of a free group is free.*

By going into a little more detail, we may obtain a Schreier basis for a subgroup of a free group. Let F be free with basis U, and let G be a subgroup of F. Let X be $\Gamma(F,U)$, with the orientation consisting of the edges $(f,u,1)$. Let T be a representative tree for the action of G on X such that $1 \in T$.

Since $V(X) = F$, we see that $V(T)$ must be a transversal of G. Let $f \in T$, and let $f = hu^{\varepsilon}$, reduced as written. Then the unique irreducible path in X from 1 to f goes through h. This path must lie in the tree T, and so $h \in T$, showing that $V(T)$ is a Schreier transversal.

Conversely, take any Schreier transversal. As a subset of F, we can regard it as a set of vertices of X, and we let T be the full subgraph spanned by these vertices. Because we have a Schreier transversal, the irreducible path in X from 1 to any vertex of T lies entirely in T. Hence T is connected, and so T is a tree because X is a tree. Plainly T is a representative tree, since $V(T)$ contains exactly one vertex in each orbit of G. Thus Schreier transversals and representative trees containing 1 are essentially the same things.

The basis S for G consists of those $g \neq 1$ such that there is some $t \in T$ and $u \in U$ such that $(t,u,1)$ ends in gT. This means that $tu \in gT$, and so there must be some $t_1 \in T$ with $tu = gt_1$. Thus $g = tut_1^{-1}$, where t_1 is the element of T in the coset Gtu. This is exactly the basis we have previously obtained.

This proof is essentially just a translation of the previous proofs. The tree X is the universal covering of the graph with one vertex only and with one edge-pair for each element of U. The graph $G\backslash X$ is just the covering of this graph corresponding to the subgroup G. The projection of X onto $G\backslash X$ maps T bijectively to its image, by definition of T, so this image is just a maximal tree of $G\backslash X$, as used before.

The current proof is slightly simpler than the previous ones, mainly because we work only with the universal covering rather than with arbitrary coverings, and we do not have to talk about fundamental groups at all. Also, we gain some notational simplifications by obtaining a basis for G by considering an arbitrary tree on which it acts freely, and only later specialising to the tree $\Gamma(F,U)$.

We shall develop in later sections the theory of groups which act non-freely on trees. We shall obtain subgroup theorems which generalise the Kurosh subgroup theorem, and which include subgroup theorems for amalgamated free products and HNN extensions.

8.2 *NIELSEN'S METHOD*

We have seen that every subgroup of a free group is free. In order to find a basis for a subgroup G, the approach we have considered requires us to find a transversal for G. We are quite likely in explicit examples to be given a subset S which generates G, and in such contexts reference to a transversal is not very natural. We would prefer a method which gives a sufficient condition for S to be a basis of G, and which gives a procedure for replacing any S by another generating set which satisfies such a condition. Nielsen's approach and its generalisations provide a method for doing this.

We call the subset S of the free group $F(X)$ *Nielsen reduced* if it satisfies the following three conditions: (N0) $1 \notin S$, (N1) $|vw| \geq |v|$ for all v and w in $S \cup S^{-1}$ such that $vw \neq 1$, (N2) $|uvw| > |u| + |w| - |v|$ for all u, v, and w in $S \cup S^{-1}$ such that $uv \neq 1 \neq vw$. Note that condition (N1) includes the condition $|w^{-1}v^{-1}| > |w^{-1}|$, which may be rewritten as $|vw| > |w|$.

Condition (N1) holds iff in reducing vw at most half of w (and v) is cancelled. In the presence of (N1), condition (N2) holds iff either the left half of v does not cancel in reducing uv or the right half of v does not cancel in reducing vw.

For $v \in S \cup S^{-1}$, let $\lambda(v)$ be the longest initial segment of v which cancels in reducing some uv with $u \in S \cup S^{-1}$ and $uv \neq 1$, and let $\rho(v)$ be the longest final segment of v which cancels in reducing some vw with $w \in S \cup S^{-1}$ and $vw \neq 1$. Plainly $\rho(v) = \lambda(v^{-1})^{-1}$. Note that $\lambda(v)$ and $\rho(v)$ may be trivial.

From the remarks already made, we see that if S is Nielsen reduced then for each $v \in S \cup S^{-1}$ there is a non-trivial segment $\mu(v)$ of v such that $v = \lambda(v)\mu(v)\rho(v)$, reduced as written.

Lemma 7 *Let S be Nielsen reduced, and take $u_1, \ldots, u_n \in S \cup S^{-1}$ such that $u_{i+1} \neq u_i^{-1}$ for $i < n$. When the product $u_1 \ldots u_n$ is reduced, the cancellations do not affect any $\mu(u_i)$. Also $|u_1 \ldots u_n| \geq |u_2 \ldots u_n|$ and $|u_1 \ldots u_n| \geq |u_1 \ldots u_{n-1}|$.*

Corollary 1 *With the above notation, $|u_1 \ldots u_n| \geq n$, and $|u_1 \ldots u_n| \geq |u_i|$ for all i.*

Corollary 2 *When S is Nielsen reduced it freely generates $\langle S \rangle$.*

Proof Corollary 2 and the first part of Corollary 1 are obvious consequences of the lemma. The second part of Corollary 1 comes by an obvious induction from the lemma.

We prove the lemma by induction. Suppose that the cancellations required to reduce the product $u_2 \ldots u_n$ do not affect any $\mu(u_i)$ for $i = 2,\ldots,n$. Since no part of $\mu(u_2)$ gets cancelled, the product $u_2 \ldots u_n$ must begin with $\lambda(u_2)\mu(u_2)$. By definition of $\rho(u_1)$ and $\lambda(u_2)$, in the product u_1u_2 cancellation cannot affect $\mu(u_1)$ or $\mu(u_2)$. (Note that we require S to be Nielsen reduced in order that the segments $\mu(u_i)$ are defined.) Since both $u_2 \ldots u_n$ and u_2 begin with $\lambda(u_2)\mu(u_2)$, it follows that in the product $u_1u_2 \ldots u_n$ cancellation cannot affect $\mu(u_1)$ or $\mu(u_2)$. Since it does not affect $\mu(u_2)$, it cannot affect any of the segments $\mu(u_i)$ for $i > 2$, as these occur to the right of $\mu(u_2)$.

Since $2|\rho(u_1)| \le |u_1|$, we see that we also have $|u_1 \ldots u_n| \ge |u_2 \ldots u_n|$. Applying this to $(u_1 \ldots u_n)^{-1}$, we also obtain $|u_1 \ldots u_n| \ge |u_1 \ldots u_{n-1}|$.//

We will show that every subgroup of $F(X)$ has a Nielsen reduced basis. We will need to define a fairly complicated ordering of the reduced words.

We begin with an arbitrary well-ordering $<$ of X. Let $w = x_{i_1}^{\varepsilon_1} \ldots x_{i_n}^{\varepsilon_n}$ and $v = x_{j_1}^{\delta_1} \ldots x_{j_m}^{\delta_m}$ be reduced words. Write $w << v$ if either $n < m$ or $n = m$ and there is some k such that $i_r = j_r$ and $\varepsilon_r = \delta_r$ for $r < k$ and either $x_{i_k} < x_{j_k}$ or $i_k = j_k$ and $\varepsilon_k = 1$ and $\delta_k = -1$. It is easy to see that if $w << v$ then $uw << uv$ provided that uw and uv are reduced as written and $wu << vu_1$ provided that $|u| \le |u_1|$ and wu and vu_1 are reduced as written. The relation $<<$ is a well-ordering, since given any set S of words we can look at those members of S of smallest length and among them look at those with smallest first letter, then the subset of these with smallest second letter and so on.

We define the *left half* $h(w)$ of a word w of length $2n$ to be its initial segment of length n, and the left half of a word of length $2n+1$ to be its initial segment of length $n+1$. Finally, we define $w < v$ if $|w| < |v|$ or $|w| = |v|$ and either $h(w) << h(v)$ or $h(w) = h(v)$ and $h(w^{-1}) << h(v^{-1})$. Plainly $<$ is a well-ordering.

Proposition 8 *Let G be a subgroup of $F(X)$. For $a \in G$, let G_a be $\langle g \in G;\ g < a\rangle$. Let A be $\{a \in G;\ a \notin G_a\}$. Then A is a Nielsen-reduced set of generators for G.*

Theorem 9 (Nielsen's Theorem) *Any subgroup of a free group is free.*

Proof The theorem is immediate from the proposition and Lemma 7.

We first show, by transfinite induction, that $G = \langle A\rangle$. Take any $a \in G$, and suppose that $g \in \langle A\rangle$ for all $g < a$. Then $G_a \subseteq \langle A\rangle$, so $a \in \langle A\rangle$ if $a \in G_a$, while $a \in A$ by definition if $a \notin G_a$.

We now show that A is Nielsen reduced. Since $1 \in \langle\emptyset\rangle$, we see that

$1 \notin A$. Suppose that we had distinct x and y in A such that $|x^\varepsilon y^\delta| < |y|$ for some choice of ε and δ as ± 1. Since $x \in \langle x^\varepsilon y^\delta, y\rangle$ and $y \in \langle x^\varepsilon y^\delta, x\rangle$, this contradicts the assumption that $x \in A$ if $y < x$ and the assumption that $y \in A$ if $x < y$. Thus condition (N1) holds for A.

Since (N1) holds, (*N*2) will also hold unless we have elements x, y, and z in A with $x \neq y^{-1} \neq z$ such that $x^\varepsilon = up^{-1}$, $y^\delta = pq^{-1}$, and $z^\eta = qv$, all reduced as written, for some choice of ε, δ, and η as ± 1, where there is no cancellation between p and v or between u and q^{-1}, and $|p| = |q|$ and $|p| \leq |u|$ and $|p| \leq |v|$. Taking inverses if necessary, we may assume that $\delta = 1$. Since $y \in A$, we have $y < y^{-1}$, and so $p << q$.

If $\eta = 1$ we have $y < z$ and $yz = pv < z$, contradicting $z \in A$. If $\eta = -1$ and $|y| < |z|$ we have $y < z$ and $zy^{-1} = v^{-1}p^{-1} < v^{-1}q^{-1} = z$, with the same contradiction. Suppose that $\eta = -1$ and $|y| = |z|$, and so $|p| = |v|$. If $p << v^{-1}$ we again have $y < z$ and $zy^{-1} < z$, while if $v^{-1} << p$ we have $z < y$ and $zy^{-1} < y$, this time contradicting $y \in A$.//

We shall later simplify this argument, obtaining a basis for G which satisfies a condition weaker than being Nielsen reduced. While we have now proved that every subgroup of a free group has a Nielsen reduced basis, the current argument is not entirely satisfactory, since it does not give a straightforward method of finding such a basis when G is $\langle S\rangle$ for some given finite set S. We now show how to do this.

It is often convenient to regard a finite set S as having its elements given in some specified order, and we shall assume this done without further mention.

Let S be the ordered set $\{s_1, \ldots, s_n\}$. The ***elementary Nielsen transformations*** on S are the following: (T1) replace some s_i by s_i^{-1}, (T2) replace some s_i by $s_i s_k$ for some $k \neq i$, (T3) delete s_i provided that $s_i = 1$; in all three cases s_j remains unchanged if $j \neq i$. A product of elementary Nielsen transformations (on different sets) is called a ***Nielsen transformation***, and a Nielsen transformation is called ***regular*** if it does not involve any transformation of type (T3).

It is easy to check that each transformation of type (T1) or (T2) has an inverse of the same type (applied to a different set!). It is also not difficult to show that every permutation of S is a Nielsen transformation, as are the transformations fixing s_j for $j \neq i$ and replacing s_i by any of $s_i^\varepsilon s_k^\delta$ or $s_k^\delta s_i^\varepsilon$, where ε and δ are ± 1. We shall call any product of the latter transformations for fixed i (but varying k) a ***regular Nielsen transformation on S fixing S-$\{s_i\}$***.

If T is obtained from S by a Nielsen transformation then $\langle T\rangle = \langle S\rangle$, and if T is obtained from S by a regular Nielsen transformation then T is a basis for $\langle S\rangle$ iff S is a basis for $\langle S\rangle$. These results are obvious for an elementary Nielsen transformation, and follow by induction for the general case.

Theorem 10 *For any finite subset S of $F(X)$ there is a Nielsen reduced subset T which comes from S by a Nielsen transformation.*

Remark Our proof will show how to find the elementary Nielsen transformations which obtain T from S.

Proof If $1 \in S$, delete it. If there are x and y in S such that $|x^{\varepsilon}y^{\delta}| < |y|$ (or $< |x|$, similarly), then there is a Nielsen transformation which replaces y by $x^{\varepsilon}y^{\delta}$.

Suppose that neither of the first two possibilities occurs. Replacing elements of S by their inverses if necessary, we may assume that $s < s^{-1}$ for all $s \in S$. Suppose that this has been done, and that there are x, y and z in S such that $x \neq y^{-1} \neq z$ with $x^{\varepsilon} = up^{-1}$, $y^{\delta} = pq^{-1}$, and $z^{\eta} = qv$, all reduced as written, for some choice of ε, δ, and η as ± 1, where there is no cancellation between p and v or between u and q^{-1}, and $|p| = |q|$ and $|p| \leq |u|$ and $|p| \leq |v|$. Taking inverses if necessary, we may assume that $\delta = 1$. Since $y < y^{-1}$, this gives $p < < q$. If $\eta = 1$ we replace z by yz, noting that $yz < z$. If $\eta = -1$ we replace $z = v^{-1}q^{-1}$ by $zy^{-1} = v^{-1}p^{-1}$, noting that $zy^{-1} < z$.

By construction, if none of these operations can be performed then S is Nielsen reduced. Hence it is enough to show that by repeatedly performing these operations we ultimately reach a set to which none of the operations can be applied.

Deleting an occurrence of 1 reduces the number of elements of the set. The second kind of operation reduces the sum of the lengths of the members of the set. The third and fourth operations do not alter either of these two numbers. Hence the first two operations can only be performed a finite number of times in all. So we need only show that between any two operations of the first and second kinds there can only be a finite number of operations of the third and fourth kinds.

We may define an ordering $<$ on n-tuples of elements by $(s_1, \ldots, s_n) < (t_1, \ldots, t_n)$ iff there is some k with $s_i = t_i$ for $i < k$ and $s_k < t_k$. Since $<$ is a well-ordering of elements, the ordering $<$ on n-tuples is also a well-ordering. Now operations of the third and fourth kinds by construction replace an n-tuple by another n-tuple which is smaller in this ordering. Hence only a finite number of them can be applied before no further such operation can be used.//

Lemma 11 *Let Y be a Nielsen reduced basis of $F(X)$. Then $Y = A \cup B$, where $A \subseteq X$ and $B^{-1} = X - A$.*

Proof Any element x of X can be written as the product of elements of $Y \cup Y^{-1}$. By Lemma 7, such a product (assumed not involving any $y^{\varepsilon}y^{-\varepsilon}$) must have length 1. That is, $X \subseteq Y \cup Y^{-1}$, from which the result is immediate.//

Any regular elementary Nielsen transformation on X extends to an automorphism of $F(X)$, which we call an *elementary Nielsen X-automorphism.*

Proposition 12 *Let X be finite. Then the elementary Nielsen X-automorphisms generate the group of automorphisms of $F(X)$.*

Proof Let α be an automorphism of $F(X)$, and let Y be $X\alpha$. By Theorem 10 and Lemma 11, using the length function given by the basis Y rather than that given by X, X can be reduced to a permutation of Y by applying elementary Nielsen transformations. These must all be regular, since otherwise $F(X) = F(Y)$ could be generated by fewer than $|X|$ elements.

At first glance this seems to prove the result. However, the Nielsen transformations applied are not Nielsen transformations on X, so we have to be more careful.

There are ordered subsets $X = X_1, \ldots, X_n, Y$ of $F(X)$ such that, for $1 \le i < n$, X_{i+1} is obtained from X_i by applying a regular elementary Nielsen transformation and Y is a permutation of X_n.

Now any permutation is the product of transpositions. We show that any transposition on a set U can be obtained as the product of regular elementary Nielsen transformations.

Consider the sequence of pairs

$$u_i,u_j;\ (u_iu_j)^{-1},u_j;\ u_j^{-1}u_i^{-1},u_ju_j^{-1}u_i^{-1};\ u_j^{-1}u_i^{-1},(u_i^{-1})^{-1};\ u_j^{-1}u_i^{-1}u_i,u_i;\ u_j,u_i.$$

There are elementary Nielsen transformations which send each pair onto the next one and which fix u_k for $i \ne k \ne j$. The product of these will transpose u_i and u_j.

It follows that we can find $X_{n+1}, \ldots, X_m = Y$ such that X_{i+1} comes from X_i by a regular elementary Nielsen transformation for all $i < m$.

We show, by induction on r, that $X_r = X\alpha_r$ for some automorphism α_r which is the product of elementary Nielsen X-automorphisms. In particular, $X\alpha_m = Y = X\alpha$, and so $\alpha = \alpha_m$, since $\alpha_m\alpha^{-1}$ is an automorphism of $F(X)$ which fixes X.

Let X_r be $\{u_1, \ldots, u_n\}$ and let X_{r+1} be $\{v_1, \ldots, v_n\}$, the order being taken into account. Suppose that $v_i = u_iu_j$ and $v_k = u_k$ for $k \neq i$ (the other case is similar). Let β_{ij} be the Nielsen X-automorphism which maps x_i to x_ix_j and fixes the remaining elements of X. By assumption, $u_k = x_k\alpha_r$ for all k. Hence, for $k \neq i$, $v_k = x_k\alpha_r = x_k\beta_{ij}\alpha_r$, while $v_i = (x_i\alpha_r)(x_j\alpha_r) = (x_ix_j)\alpha_r = x_i\beta_{ij}\alpha_r$. It follows that we may define α_{r+1} to be $\beta_{ij}\alpha_r$.//

We should investigate the relationship between Nielsen's method and Schreier's method. It turns out (Lyndon 1963) that Nielsen's method can be applied to any group G which has an abstract length function satisfying certain conditions (in particular, when G is a subgroup of a free group F with the length function on G induced from that on F), and that given such a length function on G we can always embed G in a free group so that the given length function is the induced one. Chiswell (1976a) also showed how, given such a length function on G, we can obtain a tree on which G acts freely. This would again show that a group with such a length function is free, and the basis obtained from Nielsen's approach is that obtained from a suitable representative tree; equivalently, that it is obtained as in Schreier's method. The details of this interesting material would take us too far afield.

Reducing Schreier's method to Nielsen's is easier, as is shown by the following result.

Proposition 13 *Let G be a subgroup of $F(X)$, and let T be a Schreier transversal for G such that each representative is of shortest length in its coset. Then the set of non-trivial generators of G obtained from T is Nielsen reduced.*

Remark We know that such a Schreier transversal can be obtained.

Proof Let $t \in T$ and $x \in X$. Let u represent the coset Gtx. We first show that txu^{-1} is reduced as written if it is not the identity.

If x cancels into t then t is vx^{-1}, reduced as written. Since T is a Schreier transversal, $v \in T$, and so we have $u = v$ since $v \in Gtx$. Similarly, if x

cancels into u^{-1} then u is wx, reduced as written, and then $w \in T$, which gives $w = t$ since $t \in Gux^{-1}$. In both these cases $txu^{-1} = 1$.

Now we have $|u| \le |t| + 1$, since tx and u are in the same coset, and u is the shortest element of its coset, and, for a similar reason, we also have $|t| \le |u| + 1$. It follows that in txu^{-1}, the element x is the middle element when txu^{-1} has odd length and is one of the two middle elements when txu^{-1} has even length.

Notice that the inverse of a generator is of the form $px^{-1}q^{-1}$, where p and q are in T and $q \in Gpx^{-1}$.

Now suppose that we have two non-identity elements $tx^{\varepsilon}u^{-1}$ and $py^{\delta}q^{-1}$ which are generators or inverses of generators. It is enough to show that in the product $tx^{\varepsilon}u^{-1}py^{\delta}q^{-1}$ cancellation cannot reach the factors x^{ε} and y^{δ} unless the two elements are inverses of each other. For this property, together with the fact that x^{ε} and y^{δ} are (one of the two) middle elements is equivalent to Nielsen's conditions.

Suppose, if possible, that this does not hold, and assume that cancellation reaches x^{ε} first. We cannot have $u = p$, as y^{δ} cannot then cancel against x^{ε} unless our elements are inverses of each other. So, by assumption, $ux^{-\varepsilon}$ must be an initial segment of p. Since T is a Schreier transversal, this means that $ux^{-\varepsilon} \in T$. Since $t \in Gux^{-\varepsilon}$ and $t \in T$, we find $t = ux^{-\varepsilon}$, contary to our assumption that $tx^{\varepsilon}u^{-1} \ne 1$.//

We now look at a condition on subsets of $F(X)$ which is weaker than being Nielsen reduced, but which enables us to prove similar results.

Definition Let U be a subset of $F(X)$ with $1 \notin U$. We call U *weakly reduced* if for every n and every $u_0, \ldots u_n$ in $U \cup U^{-1}$ with $u_{i+1} \ne u_i^{-1}$ for all i we have $|u_0 u_1 \ldots u_n| \ge |u_1 \ldots u_n|$. Notice that, looking at inverses, we must also have $|u_0 \ldots u_n| \ge |u_0 \ldots u_{n-1}|$. It follows inductively that $|u_0 \ldots u_n| \ge |u_i|$ for all i.

By Lemma 7, any Nielsen reduced set is weakly reduced. In the free group $F(a,b,c,d)$ the subset $\{ab^{-1}, bc^{-1}, cd^{-1}\}$ is not Nielsen reduced, but later results will show that it is weakly reduced. Plainly, any weakly reduced set is a basis of the subgroup it generates.

For any v and w in $F(X)$ we define the *distance* $d(v,w)$ between v and w to be the length of the maximum common initial segment of v and w. This equals the amount of cancellation between v and w^{-1}, and so equals $(|v| + |w| - |vw^{-1}|)/2$. Plainly, if $d(u,v) \geq m$ and $d(v,w) \geq m$ then $d(u,w) \geq m$. (1)

Since $d(u,v^{-1})$ is the length of that portion of v which cancels in reducing uv and $d(v,w^{-1})$ is the length of that portion of v which cancels in reducing vw, we see that

if $d(u,v^{-1}) + d(v.w^{-1}) \geq |v|$ then $|uvw| \leq |u| + |w| - |v|$, (2)

since the whole of v must cancel. Also,

if $d(u,v^{-1}) + d(v,w^{-1}) < |v|$ then $d(uv,w^{-1}) = d(v,w^{-1})$, (3)

since in this case the whole of v does not cancel in reducing uvw. Further,

if $|u| = |v|$ and $d(u,v) + d(u^{-1},v^{-1}) \geq |u|$ then $u = v$ (4)

Lemma 14 *Let* $b_0, \ldots, b_n$ *be in* $F(X)$*, and write* $a_i = b_0 \ldots b_{i-1}$ *for* $0 < i \leq n$*, and* $c_i = b_{i+1} \ldots b_n$ *for* $0 \leq i < n$*. Suppose that, for* $0 < i < n$*, we have* $|a_{i+1}| \geq |a_i|$*,* $|a_{i+1}| \geq |b_i|$*, and* $|c_{i-1}| \geq |c_i|$*. Suppose further that* $|b_0 \ldots b_n| < |b_1 \ldots b_n|$*. Then, for* $0 < i < n$*,* $|a_{i+1}| = |a_i|$ *and* $|c_{i-1}| = |c_i|$*.*

Proof There is nothing to prove if $n = 1$, so take $n > 1$. By hypothesis, for $0 < i < n$, we have $|a_{i+1}| + |c_{i-1}| \geq |a_i| + |c_i|$ with equality iff $|a_{i+1}| = |a_i|$ and $|c_{i-1}| = |c_i|$. We shall show inductively that we have equality for all i.

So we suppose that we have equality for $i < r$. We will assume that we have strict inequality for $i = r$ and derive a contradiction. Since $a_{r+1} = a_r b_r$ and $c_{r-1} = b_r c_r$, we have

$$(|a_r| + |b_r| - |a_r b_r|) + (|b_r| + |c_r| - |b_r c_r|) < 2|b_r|,$$

by the assumed inequality. That is, $d(a_r, b_r^{-1}) + d(b_r, c_r^{-1}) < |b_r|$. Consequently, $d(a_r b_r, c_r^{-1}) = d(b_r, c_r^{-1})$; that is, $|a_r b_r| + |c_r| - |a_r b_r c_r| = |b_r| + |c_r| - |b_r c_r|$. Since we are given that $|a_r b_r| \geq |b_r|$, we see that $|a_r b_r c_r| \geq |b_r c_r|$. Now $a_r b_r c_r = b_0 \ldots b_n$, while $b_r c_r = c_{r-1}$. Since our choice of r tells us that $|c_{r-1}| = |c_0|$ and $c_0 = b_1 \ldots b_n$ this contradicts the hypotheses.//

Lemma 15 *Let U be weakly reduced, let v satisfy $|v| \geq |u|$ for all $u \in U$, and suppose that $U \cup \{v\}$ is not weakly reduced. Then there is a regular Nielsen transformation on $U \cup \{v\}$ which fixes U and replaces v by an element of shorter length.*

Proof Since $U \cup \{v\}$ is not weakly reduced, we can find n and $b_0, \ldots, b_n$ in $U \cup U^{-1} \cup \{v, v^{-1}\}$, with $b_{i+1} \neq b_i^{-1}$ for all i, such that $|b_0 \ldots b_n| < |b_1 \ldots b_n|$. Take n minimal subject to this condition. Then no subword or its inverse satisfies the corresponding inequality. In particular, the hypotheses of Lemma 14 hold. Hence $|c_0| = |c_1| = \ldots = |c_{n-1}|$, where $c_i = b_{i+1} \ldots b_n$.

Now $b_i \in \{v, v^{-1}\}$ for some i, since U is weakly reduced. If there is only one such i, then we have a regular Nielsen transformation on $U \cup \{v\}$ which fixes U and sends b_i to $b_0 \ldots b_n$ (this works whether b_i is v or v^{-1}). By hypothesis, $|v| \geq |b_n|$. Now $b_n = c_{n-1}$, and $c_0 = b_1 \ldots b_n$. Since $|c_{n-1}| = |c_0|$, our Nielsen transformation replaces v by a shorter element.

So it is enough to show that there is at most one i with $b_i \in \{v, v^{-1}\}$. Suppose that there are at least two such values, which we denote by r and s where $r < s$.

Since $|b_r| \geq |b_i|$ for all i, we see that, for $0 < i < n$, $d(c_{i-1}, c_i) = (|c_{i-1}| + |c_i| - |b_i|)/2 \geq (2|c_0| - |b_r|)/2$. It then follows that for any i and j we also have $d(c_i, c_j) \geq (2|c_0| - |b_r|)/2$, by (1). Also $d(c_i^{-1}, b_i) = (|c_i| + |b_i| - |c_{i-1}|)/2 = |b_i|/2$.

There are a number of cases to consider. First suppose that $b_r = b_s$ and $s < n$. Applying the final result of the last paragraph both for $i = r$ and for $i = s$, and then using (1), we see that $d(c_r^{-1}, c_s^{-1}) \geq |b_r|/2$. Thus $d(c_r, c_s) + d(c_r^{-1}, c_s^{-1}) \geq |c_0| = |c_r| = |c_s|$. By (4), this gives $c_r = c_s$, and so $b_{r+1} \ldots b_s = 1$. This contradicts the minimality of n.

Next suppose that $b_r = b_n$. Since $c_{n-1} = b_n$, we have $d(c_r^{-1}, c_{n-1}) = d(c_r^{-1}, b_r) = |b_r|/2$, while $d(c_r, c_{n-1}) \geq (2|c_0| - |b_r|)/2 = (2|c_{n-1}| - |b_r|)/2 = |b_r|/2$. By (1), this gives $d(c_r, c_r^{-1}) \geq |b_r|/2$. Since $b_r = b_n$, by assumption, and $b_n = c_{n-1}$, by definition, and $|c_{n-1}| = |c_r|$, this simplifies to $|c_r^2| \leq |c_r|$. This can only hold if $c_r = 1$, which again contradicts the minimality of n.

Finally, suppose that $b_r = b_s^{-1}$. We must then have $r < s-1$. Now $d(c_{s-1}^{-1}, b_s^{-1}) = (|c_{s-1}| + |b_s| - |c_s|)/2 \geq |b_s|/2 = |b_r|/2$ (here, if $s = n$ we define c_n to be 1). Also $|b_0 c_0| < |c_0|$ by hypothesis, while $|b_i c_i| = |c_{i-1}| = |c_i|$ for $0 < i < n$. Hence $d(c_r^{-1}, b_r) = (|c_r| + |b_r| - |b_r c_r|)/2 \geq |b_r|/2$. By (1), this gives $d(c_r^{-1}, c_{s-1}^{-1}) \geq |b_r|/2$. Also, $d(c_r, c_{s-1}) \geq (2|c_0| - |b_r|)/2$. Hence $d(c_r, c_{s-1}) + d(c_r^{-1}, c_{s-1}^{-1}) \geq |c| = |c_r| = |c_{s-1}|$. By (4) this gives $c_r = c_{s-1}$, leading to the same contradiction as before.//

We now have a simpler and stronger version of Proposition 8.

Theorem 16 *Let G be a subgroup of $F(X)$. Take any well-ordering of G such that $g < h$ whenever $|g| < |h|$. For any $a \in G$, let G_a be $\langle b \in G;\ b < a\rangle$ and let S be $\{a \in G;\ a \notin G_a\}$. Then S is a basis for G.*

Proof As in Proposition 8, S generates G. It is enough to show that S is weakly reduced.

Suppose not. Then there is a least element $v \in A$ such that the set $\{a \in A; a \leq v\}$ is not weakly reduced. In particular, letting U be $\{a \in A; a < v\}$, we see that any finite subset of U will be weakly reduced, and so U itself is weakly reduced. Also, by our choice of well-ordering, $|v| \geq |u|$ for all $u \in A$. Then Lemma 15 tells us that there are u and w in $\langle U\rangle$ such that $|uvw| < |v|$. Since $U \subseteq G_v$, this gives $v \in G_v$, contradicting $v \in A$.//

Lemma 17 *Let U be a finite subset of $F(X)$ with m elements. Then U is weakly reduced iff we cannot find $b_0, \dots, b_n$ in $U \cup U^{-1}$ with $b_{i+1} \neq b_i^{-1}$ and $|b_0 b_1 \dots b_n| < |b_1 \dots b_n|$ and $n < 2^m$.*

Remark By definition, U is weakly reduced iff there is no n for which there are $b_0, \dots, b_n$ with these properties. The point of the lemma is that we have an explicit bound for n, and so we can write down all sequences $b_0, \dots, b_n$ of elements of $U \cup U^{-1}$ with $n < 2^m$ and check whether or not any of them satisfy the conditions of the lemma. Consequently we have a finite process for checking whether or not U is weakly reduced. In particular, this could be used to show that the subset $\{ab^{-1}, bc^{-1}, cd^{-1}\}$ of $F(a,b,c,d)$ is weakly reduced.

Proof If U is weakly reduced then there is no n for which there are $b_0, \dots, b_n$ in $U \cup U^{-1}$ with $b_{i+1} \neq b_i^{-1}$ and $|b_0 \dots b_n| < |b_1 \dots b_n|$, so there is certainly no $n < 2^m$ for which such elements exist.

Suppose that U is not weakly reduced. As before, we take n minimal subject to $|b_0 b_1 \dots b_n| < |b_1 \dots b_n|$. We need to show that $n < 2^m$.

The proof of Lemma 15 shows that we cannot have r and s with $r < s$ and $b_s \in \{b_r, b_r^{-1}\}$ such that $|b_i| \leq |b_r|$ for $r < i < s$.

Let u_i be that member of $\{b_i, b_i^{-1}\}$ which is in U. It will be enough to show that if $n \geq 2^m$ then there are r and s such that $0 < r < s$, and $u_r = u_s$ and for $r < i < s$ either $u_i = u_r$ or $|u_i| \leq |u_r|$. This is in fact a property of arbitrary partial ordered sets (the partial ordering being given by $u < v$ if $|u| < |v|$).

Let a be an element of U of maximal length. If a occurs twice in the sequence $u_1,\ldots,u_n$ then we have the required r and s. If a occurs only once (or not at all) take k such that $u_k = a$. If $k > 2^{m-1}$ then $u_1,\ldots,u_{k-1}$ is a sequence of at least 2^{m-1} elements of $U - \{a\}$, while if $k \le 2^{m-1}$ then $u_{k+1},\ldots,u_n$ is again a sequence of at least 2^{m-1} elements of $U - \{a\}$. Inductively, such a sequence will contain a suitable pair of elements u_r and u_s.//

8.3 GRAPHS OF GROUPS

Definition A *graph of groups* $\mathfrak{G}$ consists of

(i) a connected graph X,

(ii) for each vertex v of X a group G_v, and for each edge e of X a group G_e such that $G_{\bar{e}} = G_e$ for all e, and

(iii) for each edge e of X a monomorphism from G_e to $G_{e\tau}$. If we have (i) and (ii), but (iii) is replaced by

(iii') for each edge e a homomorphism from G_e to $G_{e\tau}$

we refer to a *generalised graph of groups*. The groups G_v and G_e are called *vertex groups* and *edge groups* of $\mathfrak{G}$, respectively. Note that, because $G_{\bar{e}} = G_e$ and $\bar{e}\tau = e\sigma$, we also have a monomorphism (or homomorphism, in the generalised case) from G_e to $G_{e\sigma}$. We shall use τ and σ to denote the homomorphisms from G_e to $G_{e\tau}$ and $G_{e\sigma}$. If we wish to draw attention to the graph X we will say that $\mathfrak{G}$ is a *graph of groups on* X, or that the pair $(\mathfrak{G},X)$ is a graph of groups.

Let $\mathfrak{G}$ be a generalised graph of groups. Let E be the free group with basis $E(X)$, the set of edges of X. Let $F(\mathfrak{G})$ be the group $(E * {*}_{v \in V(X)} G_v)/N$, where N is the normal closure of the subset $\{e^{-1}(a\sigma)e(a\tau)^{-1};\ e \in E(X),\ a \in G_e\} \cup \{e\bar{e}; e \in E(X)\}$. Note that when $\mathfrak{G}$ is a graph of groups (rather than a generalised one) $F(\mathfrak{G})$ is an HNN extension of ${*}G_v$ with one stable letter for each edge pair. Notice that we can regard any path in X as an element of E, and hence as an element of $F(\mathfrak{G})$.

Let T be a maximal tree in X. We define $\pi(\mathfrak{G},X,T)$ to be $F(\mathfrak{G})/M$, where M is the normal closure of $\{e; e \in T\}$; that is, $\pi(\mathfrak{G},X,T)$ is obtained from $F(\mathfrak{G})$ by adding relations $e = 1$ for all $e \in T$.

Let $v_0 \in V(X)$. We define $\pi(\mathfrak{G},X,v_0)$ to be the subset of $F(\mathfrak{G})$ consisting of all elements which can be written in the form $g_0e_1g_1e_2\ldots e_ng_n$ such that $e_1,\ldots,e_n$ is a loop based at v_0 and $g_0 \in G_{v_0}$ and $g_i \in G_{e_i\tau}$ for $1 \le i \le n$. We allow $n = 0$, in which case we just have an element of G_{v_0}. This subset is plainly a subgroup.

The next result is a generalisation of Theorem 5.17.

Proposition 18 *Let* $\mathfrak{G}$ *be a generalised graph of groups, let* v_0 *be in* $V(X)$, *and let* T *be a maximal tree of* X. *Then the natural map from* $F(\mathfrak{G})$ *to* $\pi(\mathfrak{G},X,T)$ *induces an isomorphism from* $\pi(\mathfrak{G},X,v_0)$ *to* $\pi(\mathfrak{G},X,T)$

Proof For each $v \in V(X)$ let p_v be the irreducible path in T from v_0 to v. In particular, p_{v_0} is the trivial path, and so, when it is regarded as an element of E, is the identity. We may define a homomorphism from $E * *G_v$ to itself which sends $g \in G_v$ to $p_v g p_v^{-1}$ and which sends e to $p_{e\sigma} e p_{e\tau}^{-1}$. By von Dyck's Theorem, this induces a homomorphism from $F(\mathfrak{G})$ to itself, which in fact maps into $\pi(\mathfrak{G},X,v_0)$. If $e \in T$ then either $p_{e\tau} = p_{e\sigma}e$ or $p_{e\sigma} = p_{e\tau}\bar{e}$ (as paths, or as elements of E). It follows that our homomorphism sends any e in T to 1, and so it induces a homomorphism ϑ from $\pi(\mathfrak{G},X,T)$ to $\pi(\mathfrak{G},X,v_0)$. It is easy to check, as in Theorem 5.16, that ϑ is the inverse of the natural map from $\pi(\mathfrak{G},X,v_0)$ to $\pi(\mathfrak{G},X,T)$.//

It follows that, up to isomorphism, $\pi(\mathfrak{G},X,T)$ is independent of T, and $\pi(\mathfrak{G},X,v_0)$ is independent of v_0. If we are only interested in these groups up to isomorphism we denote them by $\pi(\mathfrak{G},X)$ or by $\pi(\mathfrak{G})$, calling them the *fundamental group of* $\mathfrak{G}$. Notice that $\pi_1(X)$ is a quotient of $\pi(\mathfrak{G})$.

Examples Let all the vertex groups and edge groups be trivial. Then $\pi(\mathfrak{G})$ is just $\pi_1(X)$.

Let all the edge groups be trivial. Then $\pi(\mathfrak{G})$ is the free product of $\pi_1(X)$ and all the vertex groups. In particular, suppose that we have groups G_i for i in some index set I. Let 0 be a symbol not in I. Let X be the graph whose vertex set is $I \cup \{0\}$, with one edge pair $\{e_i,\bar{e}_i\}$ for each i, where $e_i\sigma = 0$ and $e_i\tau = i$. Let $\mathfrak{G}$ be the graph of groups on X in which all the edge groups are trivial, G_i is the given group for $i \in I$, and G_0 is trivial. Then $\pi(\mathfrak{G})$ is the free product of the G_i.

Alternatively, choose some $i_0 \in I$. Let Y be the graph with vertex set I, and one edge pair $\{e_i,\bar{e}_i\}$ for all $i \neq i_0$, where $e_i\sigma = i_0$ and $e_i\tau = i$. Let $\mathfrak{H}$ be the graph of groups on Y with all edge groups trivial and with G_i being the given group for each vertex i. Then $\pi(\mathfrak{H})$ is also the free product of the G_i. The advantage of the first formulation is that all vertices play the same role, whereas in the second formulation we have to treat one vertex in a special way (but we need no additional vertices).

Now let us take a graph of groups $\mathfrak{G}$ on a graph which has only one edge pair. Let the edge e have two distinct vertices v and w. Then $\pi(\mathfrak{G})$ is the amalgamated free product of G_v and G_w with the subgroups $G_e\sigma$ and $G_e\tau$ being amalgamated. If the edge e has its two vertices the same, both being v, then $\pi(\mathfrak{G})$ is the HNN extension of G_v with associated subgroups $G_e\sigma$ and $G_e\tau$. When we have a generalised graph of groups whose graph has only one edge pair, the corresponding fundamental groups are the pushout and the pseudo-HNN extension.

Let $\mathfrak{G}$ be a generalised graph of groups on X, and let Y be a connected subgraph of X. Then $\mathfrak{G}|Y$ denotes the generalised graph of groups whose underlying graph is Y, and whose vertex groups, edge groups, and homomorphisms are the same as those for $\mathfrak{G}$.

Let $\mathfrak{G}$ be a generalised graph of groups on X, and let Y be a connected subgraph of X. Let $v_0 \in V(Y)$, let S be a maximal tree of Y, and let T be a maximal tree of X which contains S (such a tree exists, by the corollary to Proposition 5.8). It is easy to check that there are natural homomorphisms from $\pi(\mathfrak{G}|Y,Y,v_0)$ to $\pi(\mathfrak{G},X,v_0)$ and from $\pi(\mathfrak{G}|Y,Y,S)$ to $\pi(\mathfrak{G},X,T)$. It is also easy to check that these homomorphisms transform to each other when the isomorphism of Proposition 18 is applied. So we may regard these homomorphisms as going from $\pi(\mathfrak{G}|Y)$ to $\pi(\mathfrak{G})$.

Lemma 19 *Let $\mathfrak{G}$ be a graph of groups on X, and let Y be a connected subgraph of X. With the above notations, the homomorphism from $\pi(\mathfrak{G}|Y)$ to $\pi(\mathfrak{G})$ is a monomorphism. In particular, for any vertex v of X, the natural homomorphism from G_v to $\pi(\mathfrak{G})$ is a monomorphism.*

Proof Suppose that we have shown that for any graph of groups $\mathfrak{H}$ on a tree T and any subtree S of T the map from $\pi(\mathfrak{H}|S,S,S)$ to $\pi(\mathfrak{H},T,T)$ is a monomorphism. In particular it will follow that the map from H_v to $\pi(\mathfrak{H},T,T)$ is a monomorphism.

Now $\pi(\mathfrak{G},X,T)$ is defined to be a pseudo-HNN extension of $\pi(\mathfrak{G}|T,T,T)$, and this will be an HNN extension if each G_v embeds in $\pi(\mathfrak{G}|T,T,T)$. For the same reason, $\pi(\mathfrak{G}|Y,Y,S)$ will be an HNN extension of $\pi(\mathfrak{G}|S,S,S)$. If we also know that $\pi(\mathfrak{G}|S,S,S)$ embeds in $\pi(\mathfrak{G}|T,T,T)$ it will follow from Proposition 1.34 that $\pi(\mathfrak{G}|Y,Y,S)$ embeds in $\pi(\mathfrak{G},X,T)$, as required.

So we start with a graph of groups $\mathfrak{H}$ on some tree T, and a subtree S of T. We first consider the case when T is finite, and use induction on the number of vertices of T.

If $S = T$ there is nothing to prove, so assume $S \neq T$. There must be an edge e with $e\sigma \in S$ and $e\tau \notin S$ (some edge on a path starting in S and ending outside S has this property). By Exercise 5.1, $T - \{e, \bar{e}\}$ is the disjoint union of two trees S_1 and S_2 one of which, say S_1, contains S. Now the definition tells us that $\pi(\mathfrak{H}, T, T)$ is the pushout of $H_e \to H_{e\sigma} \to \pi(\mathfrak{H}|S_1, S_1, S_1)$ and $H_e \to H_{e\tau} \to \pi(\mathfrak{H}|S_2, S_2, S_2)$. Inductively, $H_{e\sigma}$ and $H_{e\tau}$ embed in $\pi(\mathfrak{H}|S_1, S_1, S_1)$ and $\pi(\mathfrak{H}|S_2, S_2, S_2)$, respectively, and also $\pi(\mathfrak{H}|S, S, S)$ embeds in $\pi(\mathfrak{H}|S_1, S_1, S_1)$. Thus the pushout is an amlagamated free product, and so the factors embed in $\pi(\mathfrak{H}, T, T)$, and $\pi(\mathfrak{H}|S, S, S)$ also embeds in $\pi(\mathfrak{H}, T, T)$, as required.

Next let T be arbitrary and let S be finite. Then $\pi(\mathfrak{H}, T, T)$ is the direct limit of the groups $\pi(\mathfrak{H}|T_1, T_1, T_1)$ for all finite subtrees T_1 of T which contain S. Since we have just seen that $\pi(\mathfrak{H}|S, S, S)$ embeds in $\pi(H|T_1, T_1, T_1)$ for any such T_1, Proposition 3.3 tells us that $\pi(\mathfrak{H}|S, S, S)$ also embeds in $\pi(\mathfrak{H}, T, T)$.

Now let T and S be arbitrary. Choose a vertex $v \in S$. Then $\pi(\mathfrak{H}|S, S, S)$ is the direct limit of the groups $\pi(\mathfrak{H}|S_1, S_1, S_1)$ over all finite subtrees S_1 of S which contain v. We saw in the last paragraph that $\pi(\mathfrak{H}|S_1, S_1, S_1)$ embeds in $\pi(\mathfrak{H}, T, T)$, and so $\pi(\mathfrak{H}|S, S, S)$ itself must embed in $\pi(\mathfrak{H}, T, T)$, by Proposition 3.3.//

In the course of proving the lemma, we have seen that when X is a finite tree the group $\pi(\mathfrak{G})$ is obtained (inductively) by a process of repeating amalgamated free products. This can be done in various ways, since the removal of any edge-pair gives two disjoint trees smaller than X, and $\pi(\mathfrak{G})$ is the amalgamated free product of the fundamental groups of the restrictions of $\mathfrak{G}$ to these trees. In particular, by Lemma 5.10, we can remove an edge-pair in such a way that one of the trees consists of a single vertex.

We have also seen that when X is an infinite tree then $\pi(\mathfrak{G})$ is the direct limit of groups $\pi(\mathfrak{G}|Y)$, where Y is a finite subtree of X. When X is a tree, we often refer to $\pi(\mathfrak{G})$ as a *tree product* of the vertex groups G_v.

For arbitrary X, we know that $\pi(\mathfrak{G})$ is an HNN extension of $\pi(\mathfrak{G}|T)$, where T is a maximal tree of X. Hence $\pi(\mathfrak{G})$ is an HNN extension of a tree product, and we often refer to it as a *treed HNN group*.

Lemma 20 *Let $\mathfrak{G}$ be a graph of groups., and let H be a group. Suppose that there are homomorphisms $\varphi_v : G_v \to H$ for each vertex v, and elements α_e of H for each edge e such that $\alpha_{\bar{e}} = \alpha_e^{-1}$ and, for all $g \in G_e$, $\alpha_e^{-1}(g\sigma\varphi_{e\sigma})\alpha_e = (g\tau)\varphi_{e\tau}$. Then there is a homomorphism from $\pi(\mathfrak{G})$ to H whose restriction to G_v is a conjugate of φ_v.*

Proof By von Dyck's Theorem, if $\alpha_e = 1$ for $e \in T$ then there is, as required, a homomorphism from $\pi(\mathfrak{G},X,T)$ to H which is φ_v on G_v (and which sends the generator e to α_e).

In the general case, we will show how to find elements β_e of H and homomorphisms $\psi_v: G_v \to H$ such that ψ_v is a conjugate of φ_v and $\beta_e^{-1}(g\sigma\psi_{e\sigma})\beta_e = g\tau\psi_{e\tau}$ for $g \in G_e$, and with $\beta_e = 1$ for $e \in T$. By the previous paragraph, this will prove the result,

Let p be the path $e_1, \ldots, e_n$. Let α_p be $\alpha_{e_1} \ldots \alpha_{e_n}$. Choose some vertex v_0, and, for each vertex v, let α_v be α_p where p is the irreducible path in T from v_0 to v. Let β_e be $\alpha_v \alpha_e \alpha_w^{-1}$, where $e\sigma = v$ and $e\tau = w$. Plainly, $\beta_e = 1$ for $e \in T$. We then define ψ_v by $g\psi_v = \alpha_v(g\varphi_v)\alpha_v^{-1}$ for $g \in G_v$.//

An *isomorphism* from the generalised graph of groups $\mathfrak{G}$ on X to the generalised graph of groups $\mathfrak{H}$ on Y consists of an isomorphism φ from X to Y, for each vertex v of X an isomorphism (still called φ) from G_v to $H_{v\varphi}$, and for each edge e of X an isomorphism from G_e to $H_{e\varphi}$ such that the isomorphisms from G_e to $H_{e\varphi}$ and from $G_{\bar{e}}$ to $H_{\bar{e}\varphi}$ are the same and the homomorphisms $G_e \to G_{e\tau} \to H_{e\tau\varphi}$ and $G_e \to H_{e\varphi} \to H_{e\varphi\tau}$ are the same. Plainly an isomorphism from $\mathfrak{G}$ to $\mathfrak{H}$ induces an isomorphism from $\pi(\mathfrak{G})$ to $\pi(\mathfrak{H})$.

The generalised graph of groups $\mathfrak{H}$ is *conjugate* to the generalised graph of groups $\mathfrak{G}$ if they have the same underlying graph, the same vertex groups and the same edge groups, and the homomorphism from H_e to $H_{e\tau}$ is the homomorphism from G_e to $G_{e\tau}$ followed by conjugation by an element of $G_{e\tau}$. It is easy to check that conjugacy is an equivalence relation between generalised graphs of groups. When $\mathfrak{H}$ is isomorphic to a conjugate of $\mathfrak{G}$ we say that $\mathfrak{H}$ is *conjugate isomorphic* to $\mathfrak{G}$.

Lemma 21 *Let $\mathfrak{G}$ and $\mathfrak{H}$ be conjugate generalised graphs of groups. Let T be a maximal tree of the underlying graph X. Then there is an isomorphism $\vartheta: \pi(\mathfrak{G},X,T) \to \pi(\mathfrak{H},X,T)$ such that for each vertex v of X there is an element c_v of $\pi(\mathfrak{H},X,T)$ with $x\vartheta = c_v^{-1}xc_v$ for all $x \in G_v$. Further, given a vertex v_0 of X, we may take c_{v_0} to be 1.*

Remark Strictly speaking, we should write $xi\vartheta = c_v^{-1}xjc_v$, where i and j are the natural maps from G_v into $\pi(\mathfrak{G},X,T)$ and $\pi(\mathfrak{H},X,T0$. However, i and j are inclusions when we have graphs of groups rather than generalised graphs of groups. Even in the general situation, it is more convenient not to mention i and j explicitly, though the reader should be aware that strictly speaking they are necessary.

Proof The previous lemma applies, where each φ_v is the inclusion of G_v into $\pi(\mathfrak{H},X,T)$, to give a homomorphism $\vartheta:\pi(\mathfrak{G},X,T)\to\pi(\mathfrak{H},X,T)$. Similarly, we may construct a homomorphism from $\pi(\mathfrak{H},X,T)$ to $\pi(\mathfrak{G},X,T)$. These two homomorphisms are plainly inverses of each other.//

Let $\mathfrak{G}$ be a (generalised) graph of groups on X, and let X' be the barycentric subdivision of X. Thus X' has as vertices the vertices of X together with new vertices $v_e = v_{\bar{e}}$ for each edge-pair $\{e,\bar{e}\}$, and has two edges e_- and e_+ for each edge e of X, where e_- starts at $e\sigma$ and ends at v_e and e_+ starts at v_e and ends at $e\tau$. We define a (generalised) graph of groups $\mathfrak{G}'$ on X' as follows. The vertex group of $\mathfrak{G}'$ at the vertex v is G_v and its vertex group at v_e is G_e. The edge groups of $\mathfrak{G}'$ at the edges e_- and e_+ are both G_e. The homomorphisms from an edge group of $\mathfrak{G}'$ to the vertex group at v_e is the identity, while the homomorphisms from an edge group to the vertex group at a vertex of X are the corresponding homomorphisms for $\mathfrak{G}$. It is easy to check that $\pi(\mathfrak{G}')$ is isomorphic to $\pi(\mathfrak{G})$. The purpose of using $\mathfrak{G}'$ is that X' contains no circles, so the proposition below can be used.

We now show how generalised graphs of groups can be associated with certain complexes. Let K be a connected complex which is the union of connected subcomplexes K_v for v in some index set V, such that $K_u \cap K_v \cap K_w = \emptyset$ for distinct u,v, and w. Let X be a graph with $V(X) = V$ and with one edge-pair joining v and w for each component of $K_v \cap K_w$. Thus X has no circles, and X is connected because K is connected. We denote by K_e the component of $K_v \cap K_w$ which corresponds to the edge e. Let $\mathfrak{G}$ be the generalised graph of groups on X such that $G_v = \pi_1(K_v)$ and $G_e = \pi_1(K_e)$ for each vertex v and edge e of X, with the homomorphism from G_e to $G_{e\tau}$ being induced by the inclusion of K_e in $K_{e\tau}$, followed if necessary by a map induced by change of base-point. Thus $\mathfrak{G}$ is determined up to conjugate isomorphism, since the map induced by change of base-point is determined up to conjugacy.

Proposition 22 *Any generalised graph of groups whose graph has no circles is conjugate isomorphic to a graph obtained from a complex by the procedure above.*

Proof For each $e \in E(X)$ take a complex K_e with $\pi_1(K_e)$ isomorphic to G_e, and such that $K_{\bar{e}}$ is isomorphic to K_e for all e. For each vertex v there is, by the corollary to Theorem 5.19, a complex K_v with $\pi_1(K_v)$ isomorphic to G_v such that K_v contains disjoint copies of those complexes K_e for which $e\tau = v$ and also such that the homomorphism from G_e to G_v is obtained by inclusion of K_e

in K_v followed by a map induced by change of base-point.

Let K be obtained from the disjoint union of the K_v by identifying $K_{\bar{e}}$ with K_e for all e. Then $K_v \cap K_w$ is the disjoint union of those K_e for which e joins v and w, and so $K_u \cap K_v \cap K_w = \emptyset$ for distinct u, v, and w. It is easy to check that the generalised graph of groups constructed from K is conjugate isomorphic to $\mathfrak{G}$.//

Proposition 23 *Let $\mathfrak{G}$ be the generalised graph of groups associated with a complex K by the procedure above. Then $\pi(\mathfrak{G})$ is isomorphic to $\pi_1(K)$. In particular, X is a tree if K is simply connected.*

Proof Since $\pi_1(X)$ is a quotient of $\pi(\mathfrak{G})$, the second part follows at once from the first part.

Let T be a maximal tree of X. Let L be the complex obtained from the disjoint union of the complexes K_v by identifying K_e and $K_{\bar{e}}$ for all $e \in T$. Then K is obtained from L by identifying K_e with $K_{\bar{e}}$ for all $e \notin T$.

Notice that if e and f are distinct edges and $f \neq \bar{e}$ then $K_f \cap K_e = \emptyset$. For either f and e have three or four distinct end-points, in which case the result holds because $K_u \cap K_v \cap K_w = \emptyset$, or else f and e both have v and w as end-points, in which case K_f and K_e are distinct components of $K_v \cap K_w$.

By van Kampen's Theorem, and the usual direct limit argument, $\pi_1(L)$ is isomorphic to $\pi(\mathfrak{G}|T,T,T)$. It then follows by Proposition 5.21 and a direct limit argument that $\pi_1(K)$ is isomorphic to $\pi(\mathfrak{G})$.//

8.4 *THE STRUCTURE THEOREMS*

Let the group G act on the connected graph X (we will primarily be interested in the case when X is a tree, but the construction applies generally). Let Y be $G\backslash X$, p the projection, T a maximal tree of Y, and A be orientation of Y. Let $j:T \to X$ be a map such that jp is the identity on T (we have called Tj a representative tree for the action).

We extend j to a function from Y to X. This function, which we still denote by j, will not be a map of graphs, and Yj will not be a subgraph of X. Since T is a maximal tree, j is defined already on all the vertices. Let $e \in A - T$. Since p is locally surjective, there is at least one edge x of X such that $xp = e$ and $x\sigma = e\sigma j$. We define ej to be such an edge, and we define $\bar{e}j$ to be $\overline{ej}$. Notice that j is still an injection.

For $e \in A$ we have $ej\tau p = ejp\tau = e\tau = e\tau jp$. It follows that there is some

$\gamma_e \in G$ such that $ej\tau = \gamma_e(e\tau j)$. For $e \in A \cap T$ we have $ej\tau = e\tau j$, and we choose γ_e to be 1 in this case. We also define $\gamma_{\bar{e}}$ to be γ_e^{-1}.

The *stabiliser* stab$x$ of a vertex or edge x of X is defined to be the subgroup $\{g;\ gx = x\}$.

We can now define a graph of groups $\mathfrak{G}$ on Y as follows. For any vertex or edge y of Y, we define G_y to be stab(yj). We need to define monomorphisms from G_e to $G_{e\sigma}$ and to $G_{e\tau}$ for each edge $e \in A$.

For any $e \in A$ we have $ej\sigma = e\sigma j$. Hence stab(ej) $\subseteq$ stab($e\sigma j$); that is, $G_e \subseteq G_{e\sigma}$, and the required monomorphism is this inclusion. Also stab(ej) $\subseteq$ stab($ej\tau$) $= \gamma_e(\text{stab}\,e\tau j)\gamma_e^{-1}$. We define the monomorphism from G_e to $G_{e\tau}$ to send g to $\gamma_e^{-1}g\gamma_e$.

One of our main theorems, to be proved later, is that G is isomorphic to $\pi(\mathfrak{G})$ when X is a tree.

Conversely, let us start with a graph of groups $\mathfrak{G}$ on the graph Y, and let T be a maximal tree of Y, and let A be an orientation of Y. Let G be $\pi(\mathfrak{G},Y,T)$.

We proceed to define a graph $\tilde{Y}$, which we call *the universal cover of the graph of groups* $\mathfrak{G}$. We consider pairs (g,v) where $g \in G$ and v is a vertex of Y. We say that (h,w) is equivalent to (g,v) if $w = v$ and $h \in gG_v$. This is plainly an equivalence relation, and the vertices of $\tilde{Y}$ are defined to be the equivalence classes $[g,v]$. Similarly, we consider pairs (g,y) and $(g,\bar{y})$, where $g \in G$ and $y \in A$. We say that (h,z) is equivalent to (g,y) and $(h,\bar{z})$ is equivalent to $(g,\bar{y})$ if $z = y$ and $h \in g(G_y\sigma)$. This is an equivalence, and the edges of $\tilde{Y}$ are defined to be the equivalence classes $[g,y]$ and $[g,\bar{y}]$. We leave the definitions of σ, τ, and $^-$ on $\tilde{Y}$ for the moment.

We define $p:\tilde{Y} \to Y$ and $j:Y \to \tilde{Y}$ by $[g,y]p = y$ and $yj = [1,y]$ for any vertex or edge y of Y. We make G act on $\tilde{Y}$ in the obvious manner; that is, $g[h,y] = [gh,y]$ for any vertex or edge y of Y. This is easily seen to be well-defined, and to be an action without inversions. Also we have $[g,y] = g(yj)$.

We define $\overline{[g,y]}$ to be $[g,\bar{y}]$. For $y \in A$ we define $[g,y]\sigma$ to be $[g,y\sigma]$ and $[g,y]\tau$ to be $[gy,y\tau]$ (here y is being used both for an edge of Y and for the corresponding element of G). It is easy to check that, because $y^{-1}(G_y\sigma)y = G_y\tau$, these definitions depend only on $[g,y]$ and not on the particular choice of g. Thus they make $\tilde{Y}$ into a graph, and p into a map of graphs. Also, because the group element y is 1 for $y \in T$, $j:T \to \tilde{Y}$ is a map of graphs, though $j:Y \to \tilde{Y}$ is not a map of graphs.

It is convenient to have an expression for $[g,y]\sigma$ and $[g,y]\tau$ that can be used whether or not $y \in A$. Recalling that the group element $\bar{y}$ equals y^{-1}, it is easy to check that we always have $[g,y]\sigma = h[1,y\sigma]$ and $[g,y]\tau = hy[1,y\tau]$, where $h = g$ if $y \in A$ and $h = gy^{-1}$ if $y \notin A$.

Evidently, for any vertex v, $g[1,v] = [1,v]$ iff $g \in G_v$, while for $y \in A$ we have $g[1,y] = [1,y]$ iff $g \in G_y\sigma$. Since $\sigma: G_y \to G_{y\sigma}$ is a monomorphism, it is easy to check that the graph of groups associated with the action of G on $\tilde{Y}$ is isomorphic to $\mathfrak{G}$ itself.

We now proceed to show that $\tilde{Y}$ is a tree. The proof will require a considerable amount of discussion. We begin by showing that $\tilde{Y}$ is connected.

Let Z be the subgraph of $\tilde{Y}$ whose edges are yj and $\bar{y}j$ for all $y \in A$ and whose vertices are the endpoints of all such edges. Then $Tj \subseteq Z$, and every edge of Z has at least one vertex in Tj. Thus Z is connected. Plainly $\tilde{Y} = GZ$.

Let $g \in G_v$ for some vertex v. Then $g[1,v] = [1,v]$, and so $Z \cap gZ \neq \emptyset$, since $[1,v] \in Tj \subseteq Z$. Also, for any $y \in A$ we have $[1,y]\tau = [y,y\tau] = y[1,y\tau]$. This shows that $Z \cap yZ \neq \emptyset$, and so also $Z \cap y^{-1}Z \neq \emptyset$.

It follows by induction on n, using the remark before Lemma 5.1, that for any n and any $s_1,\ldots,s_n$ in $\cup\{G_v$; all vertices $v\} \cup \{y, y^{-1}$; all $y \in A\}$, the subgraph $Z \cup s_1Z \cup s_1s_2Z \cup \ldots \cup s_1\ldots s_nZ$ is connected. Hence the union of all these subgraphs is itself connected. Since G is generated by the subgroups G_v and the elements y, this subgraph is just GZ, as required.

The fact that $\tilde{Y}$ is a tree, and not an arbitrary connected graph, is equivalent to a reduced form theorem for G. There is no theoretical difficulty in this, since we already have reduced form theorems for amalgamated free products and HNN extensions, and G is built up out of these. However, the reader will easily see that such a reduced form theorem is likely to be notationally complicated. Hence it is better to find a proof not using such a result (which can then be used to obtain the reduced form theorem if one wishes).

The easiest method is to use the results of Propositions 22 and 23. First note that we may assume that Y has no circles. For it is not difficult to see that the graph Z associated with the graph of groups $\mathfrak{G}'$ is just the subdivision of $\tilde{Y}$, and hence $\tilde{Y}$ is a tree if Z is a tree.

Let Y have no circles. Let K be the complex associated with $\mathfrak{G}$ by Proposition 22. Let $p: \tilde{K} \to K$ be the universal cover of K. Then $\tilde{K}$ is the union of the components of the complexes K_vp^{-1}, and we may obtain the graph of groups $(\mathfrak{H}, W)$ associated with this expression for $\tilde{K}$. By Proposition 23, we see that W is a tree. It is easy to check, using Exercise 6.5, that W is isomorphic to $\check{Y}$, showing that $\check{Y}$ is a tree.

While this is probably the shortest proof, it is not in the spirit of this chapter, where we want to talk about graphs and not about arbitrary complexes. Consequently I give another proof, which may be longer than proofs using the reduced form theorem but which avoids to some extent the notational complexities of such proofs, though it is still quite complex.

To begin with, we look at the case where $\mathfrak{G}$ is the graph of trivial groups on Y. In this case, we just refer to $\check{Y}$ as the universal cover of Y. (It is the same as the universal cover discussed in Chapter 7, but we do not need the results of that chapter here.)

Let $[g_1,e_1],\ldots,[g_n,e_n]$ be a loop in $\check{Y}$ based at $[1,v]$. It is enough to show that there is some i such that $e_{i+1}=\bar{e}_i$. For then, because the vertex groups are trivial, it is easy to see that, because $[g_{i+1},e_{i+1}]\sigma=[g_i,e_i]\tau$, we must also have $g_{i+1}=g_i$, and so the path will be reducible.

Let $h_i=g_i$ if $e_i\in A$ and $h_i=g_ie_i^{-1}$ if $e_i\notin A$. Then $[g_i,e_i]\tau=h_ie_i[1,e_i\tau]$ and $[g_i,e_i]\sigma=h_i[1,e_i\sigma]$. Thus we have, since the vertex groups are trivial, $h_{i+1}=h_ie_i$ for $i<n$, and $h_ne_n=h_1$. That is, $e_1\ldots e_n=1$ in G. Since G is free with basis the set of edges of A not in T, we have three possibilities. Either there is some r with $e_{r+1}=\bar{e}_r$ or there are r and s with $r+1<s$ such that $e_s=\bar{e}_r$ and $e_i\in T$ for $r<i<s$ or $e_i\in T$ for all i. In the latter two cases we have a non-trivial loop in T, and so, as this loop is reducible, there must be some i with $e_{i+1}=\bar{e}_i$, as required.

Thus the universal cover of any connected graph is a tree. Let Z be the universal cover of Y, and let $p:Z\to Y$ be the projection. Because the vertex groups are trivial, p is locally injective. Let α be an automorphism of Y, and let z and z' be vertices of Z such that $z'p=zp\alpha$. Then Lemma 4 may be applied to p and $p\alpha$ to give a unique automorphism β of Z such that $z\beta=z'$ and $\beta p=p\alpha$. We call β an *extension* of α.

We now return to the general situation. Let W be the universal cover of $\check{Y}$, and let q be the projection. There is a map $m:Tj\to W$ such that mq is the identity on Tj. For $y\in A$, we also define yjm to be the unique edge of W starting at $y\sigma jm$ and mapping onto yj by q. We also define $\bar{y}jm$ to be $\overline{yjm}$. Denote jm by k. Although k is not a map of graphs, we have $yk\sigma=y\sigma k$ for all $y\in A$, by construction.

Let v be a vertex of Y, and let $g\in G_v$. Then g gives an automorphism of $\check{Y}$ fixing vj, and this extends to an automorphism of W fixing

vk. We denote this automorphism by φ_g and we will write it on the left. If h is also in G_v we see that both $\varphi_g \varphi_h$ and φ_{gh} fix vk and extend gh. By uniqueness, $\varphi_g \varphi_h = \varphi_{gh}$.

Now take any $y \in A$. The element y of G gives an automorphism of $\tilde{Y}$ which sends $y\tau j$ to $yj\tau$. This extends uniquely to an automorphism φ_y of W such that φ_y sends $y\tau k$ to $yk\tau$.

Let y be in A and g in $G_y\sigma$. Then $(\varphi_g(yk))\sigma = \varphi_g(yk\sigma) = \varphi_g(y\sigma k) = y\sigma k$ (by the definition of φ_g) which equals $yk\sigma$. Also $(\varphi_g(yk))q = g(yj) = yj$, because $g \in G_y\sigma$, and this equals $(yk)q$. Since q is locally injective, φ_g fixes yk.

Let y be in A and h in G_y. Then $\varphi_{h\tau}$ extends $h\tau$ and fixes $y\tau k$, since $h\tau \in G_{y\tau}$. We have seen that $\varphi_{h\sigma}$ fixes yk, and consequently fixes $yk\tau$. Thus $\varphi_y^{-1}\varphi_{h\sigma}\varphi_y$ fixes $y\tau k$ (remember that these automorphisms act on the left!) and it also extends $y^{-1}(h\sigma)y$, which equals $h\tau$ in G. Then uniqueness of extensions shows that $\varphi_y^{-1}\varphi_{h\sigma}\varphi_y = \varphi_{h\tau}$. Also, if $y \in T$ we have $y\tau j = yj\tau$ and $y\tau k = yk\tau$, and so φ_y is the identity.

Thus we have homomorphisms from each G_v into Aut$W$, and elements φ_e of Aut$W$. From the facts above and Lemma 20, these give a homomorphism from G to Aut$W$; that is, they give an action of G on W. Any element of $\tilde{Y}$ (whether vertex or edge) can be written as $g(yj)$ for some unique vertex or edge y of Y. The construction of $\tilde{Y}$ and the remarks made about elements of W fixed by certain elements of G then show that the element $g(yk)$ of W depends only on the chosen element of $\tilde{Y}$ and not on the choice of g. Thus we have a function $s:\tilde{Y} \to W$. It is easy to check that s is a map of graphs such that sq is the identity on $\tilde{Y}$. Thus $\tilde{Y}$ maps injectively into the tree W. Since we already know that $\tilde{Y}$ is connected, it follows that $\tilde{Y}$ is a tree. That is, we have proved the following theorem.

Theorem 24 (First Structure Theorem) *The universal cover of a graph of groups is a tree.//*

We have now made two constructions. The first starts with a group acting on a connected graph and constructs a graph of groups, while the second starts with a graph of groups and constructs a tree on which a group acts. We need to elucidate the relationship between these two constructions.

As already remarked, if we start with a graph of groups, construct from it a group acting on a tree, and then construct from that a new graph of groups, then the two graphs of groups are isomorphic.

Conversely, let us start with a group G acting on a connected graph X. Let Y be $G\backslash X$, and let $p:X\to Y$ be the projection. Let T be a maximal tree in Y, and let A be an orientation of Y. Let $j:Y\to X$ be as before. Let $\mathfrak{G}$ be the graph of groups on Y constructed using T, A, and j. Let π denote $\pi(\mathfrak{G},Y,T)$, and let $\tilde{Y}$ be the universal cover of $\mathfrak{G}$. Thus $\tilde{Y}$ is a tree on which π acts.

Plainly, X cannot be isomorphic to $\tilde{Y}$ unless X is a tree. We shall show that if X is a tree then X is isomorphic to $\tilde{Y}$, that G is isomorphic to π, and that the isomorphisms respect the actions.

Each vertex group G_v of $\mathfrak{G}$ embeds in G. To each edge $e\in A$ we have associated an element γ_e of G, with $\gamma_e=1$ for $e\in T$. Also, for $e\in A$ and $g\in G_e$, we have homomorphisms $\sigma:G_e\to G_{e\sigma}$ and $\tau:G_e\to\mathfrak{G}_{e\tau}$, such that $g\sigma=g$ and $g\tau=\gamma_e^{-1}g\gamma_e$. It follows that there is a homomorphism $\varphi:\pi\to G$ which is inclusion on each G_v and sends e to γ_e for $e\in A$.

The vertices and edges of $\tilde{Y}$ are classes $[h,y]$, where $h\in\pi$ and y is a vertex or edge of Y. When y is a vertex of Y or an edge in A we have $[h,y]=[k,z]$ iff $y=z$ and $k\in hG_y$, while if y is an edge not in A we have $[h,y]=[k,z]$ iff $[h,\bar{y}]=[k,\bar{z}]$. Also, for y an edge in A or a vertex we have $G_y=\mathrm{stab}(yj)$, and the restriction of φ to G_y is inclusion in G. It follows that we may define a function ψ from $\tilde{Y}$ to X by $[h,y]\psi=(h\varphi)(yj)$.

For $e\in A$ we have $[h,e]\sigma=[h,e\sigma]$ and $[h,e]\tau=[he,e\tau]$ Also $ej\sigma=e\sigma j$, while $ej\tau=\gamma_e e\tau j$, and $e\varphi=\gamma_e$. Using these, we see easily that ψ is a map of graphs.

Since $X=G(Yj)$, it is immediate that ψ maps onto X provided that φ maps onto G. We shall show later that φ always maps onto G.

We next show that ψ is locally injective. Take any y and z in A, and any h and k in π. We cannot have $[h,y]\psi=[k,\bar{z}]\psi$, since $y\neq\bar{z}$. Suppose that $[h,y]\psi=[k,z]\psi$ and $[h,y]\sigma=[k,z]\sigma$. Then $y=[h,y]\psi p=[k,z]\psi p=z$ and $h^{-1}k\in G_{y\sigma}$ and also $(h^{-1}k)\varphi\in\mathrm{stab}(yj)$. Since φ maps $G_{y\sigma}$ isomorphically onto $\mathrm{stab}(y\sigma j)$, and the subgroup G_y of $G_{y\sigma}$ is mapped identically to the subgroup $\mathrm{stab}(yj)$ of $\mathrm{stab}(y\sigma j)$, we see that $h^{-1}k\in G_y$, and so $[h,y]=[k,z]$.

Similarly, suppose that $[h,\bar{y}]\psi=[k,\bar{z}]\psi$ and $[h,\bar{y}]\sigma=[k,\bar{z}]\sigma$. Then $[h,y]\psi=[k,z]\psi$ and $[h,y]\tau=[k,z]\tau$. As before, $y=z$ and $(h^{-1}k)\varphi\in\mathrm{stab}(yj)$.Further, $(hy)^{-1}(ky)\in G_{y\tau}$. Now, by construction, G_y is just $\mathrm{stab}(yj)$, and G_y is given as a subgroup of $G_{y\sigma}$, on which φ is just inclusion. Thus $(h^{-1}k)\varphi\in G_y\varphi$. Now φ is one-one on $G_{y\tau}$ and so is also one-one on $yG_{y\tau}y^{-1}$. This subgroup contains $h^{-1}k$, and also contains the subgroup $yG_y\tau y^{-1}=G_y\sigma$ (as a subgroup of π, using the definition of π), which is just G_y, by the definition of σ. Thus we have $h^{-1}k\in G_y$, which ensures that $[h,\bar{y}]=[k,\bar{z}]$, as required.

Lemma 25 *Let H be a subgroup of G. Let X_1 and X_2 be subgraphs of X such that $X_2 \subseteq X_1$, $GX_1 = X$, $V(X_1) \subseteq HV(X_2)$, and $gV(X_2) \cap V(X_2) = \emptyset$ for $g \notin H$. Then $H = G$.*

Proof We show first that $HX_1 = X$. Since X is connected, it is enough, by Lemma 5.1, to show that if e is an edge with $e\sigma \in HX_1$ then $e \in HX_1$. Plainly, we may assume that $e\sigma \in X_1$. There exist $g \in G$ and $x \in E(X_1)$ with $e = gx$. There are h and k in H such that $h(e\sigma)$ and $k(x\sigma)$ are in X_2. Then $hgk^{-1}(k(x\sigma)) = hg(x\sigma) = h(e\sigma) \in hgk^{-1}V(X_2) \cap V(X_2)$, so $hgk^{-1} \in H$. Thus $g \in H$, and so $e = gx \in HX_1$.

In particular, $HV(X_1) = V(X)$. Since $V(X_1) \subseteq HV(X_2)$, we see that $V(X) = HV(X_2)$.

Take $g \in G$. Choose any vertex v of X_2. Then there will be some $h \in H$ and $w \in V(X_2)$ such that $gv = hw$. Thus $w = h^{-1}gv \in V(X_2) \cap h^{-1}gV(X_2)$. Hence $h^{-1}g \in H$, and so $g \in H$, as required.//

Corollary *φ maps π onto G.*

Proof Let X_2 be Tj, let X_1 be the subgraph consisting of all edges in Yj and all endpoints of these edges, and let H be $\pi\varphi$. Since $Y = G\backslash X$, we have $X = GX_1$. Every vertex of Y is a vertex of T. Thus $V(X_1) = V(X_2) \cup \{ej\tau;\ e \in A\}$. For $e \in A$ we have $ej\tau = \gamma_e(e\tau j) \in HV(X_2)$, since $\gamma_e \in H$. Also, if $gV(X_2) \cap V(X_2) \neq \emptyset$ there will be vertices y and z of Y such that $g(yj) = zj$. By definition of j, this requires that $y = z$. We will then have $g \in \mathrm{stab}(yj) = G_y$, and so $g \in H$. Thus the conditions of the lemma are satisfied.//

We can now prove the main theorem, using the notations above.

Theorem 26 (Second Structure Theorem, or Main Theorem of Bass-Serre Theory) *The following are equivalent: (i) X is a tree, (ii) ψ is an isomorphism of graphs, (iii) φ is an isomorphism of groups.*

Proof We have seen that φ maps onto G and consequently that ψ maps onto X. We have also seen that ψ is locally injective. Thus $\tilde{Y}\psi \supseteq G(Yj) = X$. That is, ψ maps onto X.

It then follows from Lemma 5 that ψ is an isomorphism if X is a tree. Conversely, by Theorem 24, X is a tree if ψ is an isomorphism.

Suppose that φ is an isomorphism. Now ψ, being locally injective, will be an isomorphism if it is injective on $V(\tilde{Y})$. So let $[h,v]\psi = [k,w]\psi$, where h and k are in π and v and w are vertices of Y. Then $(h\varphi)(vj) = (k\varphi)(wj)$. Applying p, we see that $v = w$. Also $(h^{-1}k)\varphi \in \text{stab}\, vj = G_v\varphi$. Since φ is an isomorphism, this gives $h^{-1}k \in G_v$, and so $[h,v] = [k,w]$, as required.

Now suppose that ψ is an isomorphism. Let h be in $\ker\varphi$. Take any vertex v of Y. Then $[h,v]\psi = (h\varphi)(vj) = vj = [1,v]\psi$. Hence $[h,v] = [1,v]$, and so $h \in G_v = \text{stab}(vj)$. Since we know that φ restricted to G_v is just inclusion, this gives $h = 1$, as needed.//

Exercise 1 Let $\mathfrak{G}$ be a graph of groups on Y, where Y has only one edge (it may have one or two vertices). Show explicitly, using the Reduced Form Theorems for HNN extensions and amalgamated free products, that $\tilde{Y}$ is a tree.

Exercise 2 A sketch argument using covering complexes was given to show that $\tilde{Y}$ is a tree for any graph of groups on any graph Y. Fill in the details of this proof.

8.5 *APPLICATIONS OF THE STRUCTURE THEOREMS*

Evidently, if a group acts on a tree then every subgroup also acts on a tree. By the Structure Theorems, we see that *any subgroup of the fundamental group of a graph of groups is itself the fundamental group of some graph of groups.* However, in this generality the result is vacuous, since every group is the fundamental group of a graph of groups; we need only take the graph to consist of one vertex and no edges, with the vertex group being the given group. To obtain a significant result, we need to be more precise about the relationship between the original graph of groups and that corresponding to the subgroup.

Theorem 27 *Let $(\mathfrak{G},X)$ be a graph of groups, and let H be a subgroup of $\pi(\mathfrak{G})$. Then $H = \pi(\mathfrak{H})$, where the vertex groups of $\mathfrak{H}$ are $H \cap gG_vg^{-1}$ for all vertices v of X, where g runs over a suitable set of (H,G_v) double coset representatives, and the edge groups are $H \cap gG_eg^{-1}$ for all edges e of X. where g runs over a suitable set of (H,G_e) double coset representatives.*

Proof We know that $\pi(\mathfrak{G})$ acts on the tree $\tilde{X}$, and that the vertices of $\tilde{X}$ are $[g,v]$ with stabilisers gG_vg^{-1}, where g is given up to its coset gG_v. The H-stabiliser of $[g,v]$ is $H \cap gG_vg^{-1}$, and $[g,v]$ and $[g_1,v]$ are in the same H-orbit iff g and g_1 are in the same (H,G_v) double coset. We know that $H = \pi(\mathfrak{H})$, where the underlying graph of $\mathfrak{H}$ is $H\backslash\tilde{X}$, and that the vertex group at a vertex of $H\backslash\tilde{X}$ is the stabiliser of a corresponding vertex of $\tilde{X}$. A similar result holds for edges. This proves the theorem.//

The following corollaries are obvious.

Corollary 1 *Let H be a subgroup of the fundamental group of a graph of groups such that H meets every conjugate of a vertex group in the trivial group. Then H is free.//*

Corollary 2 *Let $\mathfrak{G}$ be a graph of groups. Suppose that there is a group H and a homomorphism $\varphi:\pi(\mathfrak{G})\to H$ such that φ is one-one on each vertex group. Then $\ker\varphi$ is free.//*

This theorem includes the Schreier Subgroup Theorem, taking $\mathfrak{G}$ to have its vertex and edge groups trivial, and it includes the Kurosh Subgroup Theorem, taking the edge groups of $\mathfrak{G}$ to be trivial.

Exercise 3 Let H be a subgroup of finite index m in the free product $G_1*\ldots*G_n$, and let m_i be the number of (H,G_i) double cosets for $1\le i\le n$. Show that the free part of H in the expression of H as a free product according to the Kurosh Subgroup Theorem has rank $1+m(n-1)-\Sigma m_i$.

What groups can act on trees? Plainly, any group can act trivially on any tree. More generally, the exercises show that for any group there are many trees on which the group acts non-trivially but stabilising some vertex. Thus the interesting question is what groups can act on trees with the action stabilising no vertex.

Any HNN extension and any proper amalgamated free product can be written as the fundamental group of a graph of groups with only one edge-pair. The universal cover of this graph of groups is then a tree on which the group acts without stabilising a vertex.

Let G be the union of a strictly increasing sequence of subgroups G_n. Let X be the graph with one edge-pair $\{e_n,\bar{e}_n\}$ for each n, and one vertex v_n for each n, where $e_n\sigma=v_n$ and $e_n\tau=v_{n+1}$. Then G is the fundamental group of the graph of groups on X with the groups at the vertex v_n and at the edge e_n both being G_n. Again, the universal cover of this graph of groups is a tree on which G acts without stabilising a vertex.

We shall see in Theorem 29 below that these are the only possible groups which can act on a tree without stabilising a vertex. However, the trees involved may correspond to more complicated expressions of the group than the ones above.

Let T be a G-tree. We say that g *translates the edge* e if there is an irreducible path $e_1,\ldots,e_n$ with $e_1 \in \{e,\bar{e}\}$ such that $g(e_1\sigma) = e_n\tau$ and $ge_1 \neq \bar{e}_n$.

Proposition 28 *Let T be a G-tree, and let g be in G. Then the following are equivalent: (i) g stabilises no vertex, (ii) g translates some edge, (iii) there are edges e_i (all $i \in \mathbb{Z}$) such that $e_{i+1}\sigma = e_i\tau$ and $e_{i+1} \neq \bar{e}_i$ and a number n such that $ge_i = e_{i+n}$ for all i.*

Proof If (iii) holds, the path $e_1,\ldots,e_n$ shows that g translates e_1. Conversely, if g translates an edge, then from the irreducible path $e_1,\ldots,e_n$ whose existence is assumed we get a seqeunce of edges satisfying (iii) by defining e_{i+nk} to be $g^k e_i$ for all $k \in \mathbb{Z}$ and $1 \leq i \leq n$.

Suppose that g stabilises no vertex. Choose the vertex v so that the irreducible path from v to gv is as short as possible. Let this path be $e_1,\ldots,e_n$. Then $g(e_1\sigma) = gv = e_n\tau$, and (ii) holds provided that $ge_1 \neq \bar{e}_n$. Since G acts without inversions, we cannot have $ge_1 = \bar{e}_n$ if $n = 1$. If $n > 1$ then $ge_1 = \bar{e}_n$ would tell us that the path $e_2,\ldots,e_{n-1}$ started at $e_1\tau$ and ended at $g(e_1\tau)$, contradicting the choice of v.

Finally, suppose that (iii) holds. Take any vertex v. Take i so that the irreducible path from v to $e_i\sigma$ is as short as possible. Let this path be $x_1,\ldots,x_m$. Evidently, we have $e_{i-1} \neq x_m \neq \bar{e}_i$. It follows that the path $x_1,\ldots,x_m, e_i,\ldots,e_{i+n-1}, g\bar{x}_m,\ldots,g\bar{x}_1$ is an irreducible path from v to gv, and so $v \neq gv$.//

Theorem 29 *Let T be a G-tree. Then exactly one of the following occurs: (i) G stabilises a vertex, (ii) there is an infinite sequence of edges $e_1, e_2,\ldots$ with $e_{n+1}\sigma = e_n\tau$ and $e_{n+1} \neq \bar{e}_n$ for all n and such that $stab(e_n\sigma) \subseteq stab(e_{n+1}\sigma)$ for all n (from which we see that $stab(e_n\sigma) = stab\, e_n$) and $G = \bigcup stab(e_n)$ but $G \neq stab(e_n)$, (iii) some element of G translates some edge e. In case (iii) G is either a proper amalgamated free product with G_e as amalgamated subgroup or is an HNN extension with G_e as an associated subgroup.*

Proof In case (ii), every element of G stabilises some vertex. Hence, by Proposition 28, if (iii) holds neither (i) nor (ii) can hold.

Suppose that (ii) holds, and let v be any vertex. Take m so that the irreducible path from v to $e_m\sigma$ is as short as possible. It is then easy to check that, for all $n>m$, the irreducible path from v to $e_n\sigma$ contains e_m. In particular, any g which stabilises both v and $e_n\sigma$ will (since T is a tree) also stabilise e_m. It is now easy to see that (i) cannot hold. Thus at most one of the three possibilites can hold.

Suppose that (i) and (iii) do not hold. Take any vertex v. Since (i) does not hold, $G \neq \text{stab}v$. Take any $g \notin \text{stab}v$. Since (iii) does not hold, it follows from Proposition 28 that g stabilises some vertex. Take w so that $g \in \text{stab}w$ and such that the irreducible path from v to w is as short as possible subject to this.

Let this path begin with the edge e and end with the edge x. Because w is chosen so that the path is as short as possible. we cannot have $gx = x$. It follows that we have an irreducible path $e, \ldots, x, g\bar{x}, \ldots, g\bar{e}$ whose first edge is e and whose last edge is $g\bar{e}$.

We next show that the edge e does not depend on the choice of g or w. Let us choose another $g_1 \notin \text{stab}v$ (we permit $g_1 = g$ to allow for a different choice of w) and the corresponding edge e_1. Then there is an irreducible path starting with e_1 and ending with $g_1\bar{e}_1$. If $e_1 \neq e$ then the path $g_1\bar{e}, g_1e_1, \ldots, \bar{e}_1, e, \ldots, g\bar{e}$ is irreduicble. Thus $g^{-1}g_1$ translates $\bar{e}$, contradicting our assumption that no element of G translates any edge.

Thus e depends only on v, and we have a G-map $\varphi: V \to V$, where $v\varphi = e\tau$. Since φ is a G-map, $\text{stab}v \subseteq \text{stab}(v\varphi)$ for all v. By construction, for any $g \notin \text{stab}v$, any w which is as close as possible to v subject to the condition $gw = w$ has $v\varphi$ on the irreducible path from v to w, and so does not have v on the irreducible path from $v\varphi$ to w. Now take $g \notin \text{stab}(v\varphi)$, and we see from this that $(v\varphi)\varphi \neq v$. It then follows that, letting e_n be the edge corresponding to $v\varphi^{n-1}$, we have $e_n\tau = e_{n+1}\sigma$ and $e_{n+1} \neq \bar{e}_n$, which is part of what we need.

It is also clear from the construction and the remarks of the previous paragraph, that when w is chosen as close as possible to v subject to being stabilised by g then w is also a vertex as close as possible to $v\varphi$ subject to being stabilised by g. It then follows, inductively, that $w = v\varphi^n$ for some n. In particular, g, which was any element not in $\text{stab}v$, lies in $\text{stab}(v\varphi^n)$ for some n. Thus $G = \bigcup \text{stab}(v\varphi^n)$, as needed. Finally, G cannot stabilise any edge, since it would then have to stabilise some vertex.

We still have to obtain the decomposition in case (iii). Now $T - (Ge \cup G\bar{e})$ is a G-forest. Contracting each of the trees in this forest, we obtain, by Lemma 5.12, a G-tree S with exactly one orbit of edge-pairs. Since e is translated in T by some element of G, e will also be translated in S by this element. Thus, by Proposition 28, G stabilises no vertex of S. The Second Structure Theorem now gives the result.//

Exercise 4 Let G be any group, and let T be any tree and v any vertex of T. Take disjoint copies T_g of T for all $g \in G$. Take the union of these with the vertices v_g for all g being identified to a single vertex. Show that this is a G-tree such that G stabilises a vertex, and such that no $g \neq 1$ stabilises any other vertex.

Exercise 5 Let G be any group, and let T be any tree and v any vertex of T. Show that there is a graph of groups $\mathfrak{G}$ on T whose fundamental group is G. What other graphs of groups can you find whose fundamental group is G? (There are some easy modifications of the graph of groups already found which apply for all G. There are also other examples, depending on the subgroup structure of G.)

We know that a finite subgroup of an amalgamated free product lies in a conjugate of one of the factors, with a similar result for HNN extensions. We now prove a stronger result.

Let T be a tree. A subset S of V is *bounded* if there is a number k such that, for any two vertices v and w in S, the irreducible path from v to w has length at most k; the minimum such k will be called the *diameter of S*. In particular, if T is a G-tree, it is easy to check that if one orbit of vertices is bounded then every orbit of vertices is bounded. We then say that *G has bounded action on T*. More generally, we can refer to a subgroup of G as having bounded action on T. Plainly any finite subgroup has bounded action.

Proposition 30 *Let G have bounded action on a tree T. Then G stabilises some vertex of T.*

Corollary *Let H be a finite subgroup of a fundamental group of a graph of groups. Then H is contained in a conjugate of some vertex group.*

Proof The corollary is immediate from the proposition, using the First Structure Theorem and the remarks already made.

Let G have bounded action on T, and let v be a vertex of T. Let X be the subtree of T consisting of all vertices and edges lying on some irreducible path joining two vertices of Gv. Plainly, X is a G-tree.

By hypothesis, there is some k such that the irreducible path joining two vertices of Gv has length at most k. By definition of X, it follows that to any vertex x of X there is a vertex of Gv such that the irreducible path from x to it has length at most $k/2$. We then see that any two vertices of X have an irreducible path joining them of length at most $k/2 + k + k/2$; that is, X is bounded.

We shall show that to any bounded G-subtree of T with diameter >1 there is a G-subtree of smaller diameter. Since an action of G on a tree of diameter 0 or 1 plainly stabilises the vertices, the result follows by induction.

We showed in Lemma 5.10 that in any non-trivial finite tree there is a vertex of degree 1 (that is, a vertex which is the endpoint of only one edge-pair). It is easy to see that the proof of Lemma 5.10 also shows that the same result holds for a bounded tree. In particular, let Y be the set of vertices of X of degree 1. Then Y is a non-empty G-set. Suppose that X has diameter >1, so that X does not consist of only one vertex, nor of two vertices joined by an edge. Let X_1 be the graph obtained from X by deleting the vertices of Y and deleing those edges that have a member of Y as an endpoint. Then $X_1 \neq \emptyset$. Plainly X_1 is a G-subset. As in Lemma 5.11, X_1 is also a tree, and it will have smaller diameter than X, because two vertices of X joined by an irreducible path of length the diameter of X must lie in Y.//

We now look at finitely generated groups acting on trees. We need some general definitions about trees.

Let v be a vertex of a tree T. We say that the edge e *points towards* v if the irreducible path from $e\sigma$ to v begins with e, and *points away from* v otherwise. It is easy to check that the set of edges pointing towards v forms an orientation of T, and that if we replace v by another vertex then we change the orientation of only finitely many edges (namely, those on the irreducible path between the two vertices).

Evidently there is no edge starting at v which points towards v, but for every other vertex w there is exactly one edge starting at w and pointing towards v; we denote this edge by e_w. This orientation has also been considered in section 5.3.

For any irreducible path $e_1, \ldots, e_n$ in T there will be some r such that e_i points towards v for $i \leq r$ and away from v for $i > r$ (possibly $r = 0$). For otherwise there will be some k such that e_k points away from v and e_{k+1} points towards v. Then both $\bar{e}_k$ and e_{k+1} point towards v and both start at $e_k\tau$. As already remarked, this would make $\bar{e}_k = e_{k+1}$, contradicting the irreducibility of the path.

Let S be a subtree of T with $v \in S$. Then for any vertex w of S we have $e_w \in S$, since the irreducible path in S from w to v must begin with e_w. It follows that if we have an irreducible path in T all of whose edges point towards v and the starting vertex of the path is in S then all vertices and edges of the path are in S.

Now let X be a G-tree, and let Y, T, and j have the same meaning as in the previous section. Orient X towards some vertex a of Tj. We say that g is *negative for* e if ge and e are differently oriented. We say that a G-orbit of edges is *reversing* if for some (and then for every) edge e in the orbit there is some g such that g is negative for e.

Lemma 31 *There are finitely many reversing orbits iff there are finitely many edges of Y not in T and also there are only finitely many vertices v of Tj such that $stab(v) \neq stab(e_v)$.*

Proof In an orbit above an edge of Y not in T there will be an edge e such that $e\sigma \in Tj$ but $e\tau \notin Tj$. There will be some g such that $g(e\tau) \in Tj$, and then $g(e\sigma) \notin Tj$. As already remarked, this tells us that both e and $g\bar{e}$ point away from v, and so the orbit is reversing. In particular, if there are only finitely many reversing orbits then there can be only finitely many edges of $Y-T$.

Let v be a vertex of Tj such that $\mathrm{stab}v \neq \mathrm{stab}e_v$. Since e_v is the only edge starting at v and pointing towards a, it is immediate that any element of $\mathrm{stab}v - \mathrm{stab}e_v$ is negative for e_v. Hence e_v is in a reversing orbit. Also, if w is another vertex of Tj then e_v and e_w are in different orbits, since all vertices of e_v and e_w are in Tj and distinct vertices of Tj are in distinct orbits. Thus there are only finitely many vertices of Tj such that $\mathrm{stab}v \neq \mathrm{stab}e_v$ if there are only finitely many reversing orbits.

Suppose that there are only finitely many edges of $Y-T$ and only finitely many vertices v of Tj such that $\mathrm{stab}v \neq \mathrm{stab}e_v$. Let Z be a finite subtree of X containing a, all vertices v of Tj such that $\mathrm{stab}v \neq \mathrm{stab}e_v$, and all vertices yj with y a vertex of an edge of $Y-T$. We will prove the result by showing that any reversing orbit either lies above an edge of $Y-T$ or else has an edge in Z.

Take a reversing orbit lying above an edge of T. Take an edge e of Tj in this orbit, and let g be negative for e. Take an irreducible path from some vertex of e to some vertex of ge. If this path starts at $e\sigma$ but does not begin with e we may add the edge $\bar{e}$ at the beginning of the path. Similarly, if it starts at $e\tau$ but does not begin with $\bar{e}$ we may add the edge e at the beginning. We may make a similar addition at the end. Finally, replacing e with $\bar{e}$ if necessary, we may assume that we have an irreducible path $e_1,\dots,e_n$ such that $e_1 = e$ and $e_n \in \{ge, g\bar{e}\}$. Write $v_0 = e_1\sigma$, and $v_i = e_i\tau$ for $1 \leq i \leq n$.

As we have seen, if e_n points towards a then every edge e_i points towards a. Since e and ge have different orientations, we would then have $e_n = g\bar{e}$, and so $v_n = gv_0$; of course, $v_n \neq v_0$, as $n > 0$ and X has no irreducible loops. Since $v_0 \in Tj$, we know that every vertex of the path, and so v_n in particular, lies in Tj. This is impossible, as Tj contains only one vertex in each orbit.

Suppose that there is some r with $0<r<n$ such that e_i points towards a for $i \leq r$ and away from a for $i>r$. In particular, e_1 and e_n have different orientations. As g is negative for e, we see that $e_n = ge$. If $v_0 \in Z$, we know that $e_i \in Z$ for all $i \leq r$, since e_i points towards a; in particular, $e \in Z$. If $v_0 \notin Z$, then any edge with vertex v_0 lies above an edge in T. In particular, the edge $g^{-1}e_{n-1}$ which ends at $g^{-1}(e_n\sigma) = v_0$ is hx for some $h \in G$ and $x \in Tj$. Since $x\tau \in Tj$ and $h(x\tau) = v_0$, and Tj contains only one element in each orbit, we must have $x\tau = v_0$, and so $hv_0 = v_0$. Let v_{-1} be $hg^{-1}v_{n-2}$, and let e_0 be $hg^{-1}e_{n-1}$. Consider the path $e_0, e_1, \ldots, e_{n-1}$. Since e points towards a and $v_0 \notin Z$, we know that $\text{stab}v_0 = \text{stab}e$. Thus $he_1 = e_1$, and since $ge_1 \neq \bar{e}_{n-1}$, we have $e_1 \neq \bar{e}_0$, so that the path is irreducible and $\bar{e}_0$ points away from a. By induction on $n-r$, we see that there must be an irreducible path whose first $r+1$ edges point towards a, with e among these $r+1$ edges, and whose first vertex is in Z. As already remarked, it then follows that each of these $r+1$ edges, and so e in particular, is in Z.

We are left with the case when e_i points away from a for all i. We must have $e_n = g\bar{e}$, and so $v_n = gv_0$. Since $v_n \neq v_0$ and $v_0 \in Tj$, we have $v_n \notin Tj$. Take r so that $v_i \in Tj$ for $i \leq r$ but $v_{r+1} \notin Tj$; $r>0$, by our assumption on e. If e_r lies above an edge not in T then $v_r \in Z$, by definition of Z. We then find that $e \in Z$, by our remarks about the orientation. If e_r lies above an edge in T, $\exists h$ with $he_r \in Tj$. Then $hv_r \in Tj$, and so $hv_r = v_r$. Now take the path $e_1, \ldots, e_r, he_{r+1}, \ldots, he_n$. If this path is irreducible, then, since e_1 points away from a, all edges point away from a, and e is in Z, by induction on $n-r$. If this path is reducible we must have $he_{r+1} = \bar{e}_r$, which points towards a, and so e_{v_r} is he_{r+1}. As $e_{r+1} \neq \bar{e}_r$, we have $he_{r+1} \neq e_{r+1}$. Thus, by definition, $v_r \in Z$, and, as before, $e \in Z$.//

Lemma 32 *Let G be finitely generated. Then G has finitely many reversing orbits. If, in addition, each edge has finitely generated stabiliser then each vertex has finitely generated stabiliser. Conversely, if G has finitely many reversing orbits and each vertex has finitely generated stabiliser then G is finitely generated.*

Proof G has as homomorphic image the free group $\pi_1(Y)$, which has one basis element γ_y for each edge-pair $\{y, \bar{y}\}$ not in T. Hence there are only finitely many edges not in T if G is finitely generated.

The finitely many generators of G will involve the γ_y and elements from the stabilisers of finitely many vertices. Thus we may take a finite subtree Z of Tj containing a, all these vertices, and all vertices of Tj above a vertex of an edge not in T. It follows that G, which is the HNN extension with stable letters $\{\gamma_y\}$ of the tree product over Tj of the groups stab$v$, is equal to its subgroup in which the base group is the tree product over Z. This requires, by Proposition 1.34, that the tree products over Tj and over Z are the same. Obviously, then, the tree products over Z and over any finite tree Z_1 with $Z \subseteq Z_1 \subseteq Tj$ will be the same. By an easy induction over the length of the irreducible path from v to a, we see that for $v \notin Z$ we must have stab$v$ = stabe_v, since otherwise adding the vertex v and the edge-pair $\{e_v, \bar{e}_v\}$ will enlarge the group. The result now follows by Lemma 31.

Now suppose that every edge has finitely generated stabiliser. Any vertex is in the orbit of some vertex v of Tj, and so has stabiliser conjugate to stab$v$. If $v \notin Z$ then stab$v$ = stabe_v is finitely generated. Now G is an HNN extension with finitely many stable letters of the tree product of the groups stab$v$ for $v \in Z$, and so, by Proposition 1.35 and Proposition 1.29 and induction, each stab$v$ for $v \in Z$ is finitely generated.

Conversely, suppose that G has finitely many reversing orbits. By Lemma 31, the expression of G given by the Structure Theorems is an HNN extension with finitely many stable letters of a tree product of vertex stabilisers, and this tree product can be regarded as a product over a finite tree. The result follows.//

A group is said to have the *Howson property* (or *finitely generated intersection property*) if the intersection of any two finitely generated subgroups is finitely generated. Plainly, any finite group has the Howson property, as does the infinite cyclic group. Howson (1954) proved that free groups have the Howson property. This is immediate from the stronger result below.

Theorem 33 *Let $\mathfrak{G}$ be a graph of groups in which each edge group is finite and each vertex group has the Howson property. Then $\pi(\mathfrak{G})$ has the Howson property.*

Proof Let H and K be finitely generated subgroups of $\pi(\mathfrak{G})$. Let X be the universal cover of $\mathfrak{G}$. Then both H and K act on X.

By Lemma 32, there are only finitely many reversing H-orbits, and

$H \cap \mathrm{stab}v$ is finitely generated for each vertex v, and the same for K. Since $\mathrm{stab}v$ has the Howson property for all v, by hypothesis, we know that $H \cap K \cap \mathrm{stab}v$ is finitely generated for all v. Using Lemma 32 again, the theorem follows if we can show that there are only finitely many reversing $(H \cap K)$-orbits.

Now any reversing $(H \cap K)$-orbit is plainly in the intersection of a reversing H-orbit and a reversing K-orbit. Since there are only finitely many of each of these, it is enough to prove that the intersection of an H-orbit and a K-orbit contains only finitely many $(H \cap K)$-orbits.

Let e be an edge with stabiliser A. Then $ge \in He \cap Ke$ iff $g \in HA \cap KA$. Now $HA \cap KA$ is the union of the sets $Ha \cap Kb$ for a and b in A. Each of these sets is either empty or a coset of $H \cap K$. Since A is finite, $HA \cap KA$ consists of the union of finitely many cosets of $H \cap K$. In particular, it is the union of finitely many double cosets $(H \cap K)g_iA$, and so ge is in the $(H \cap K)$-orbit of one of finitely many edges g_ie.//

A more complicated analysis on generally similar lines would enable us to prove the following. *Let H be a finitely generated subgroup of the fundamental group G of a graph of groups with finite edge groups. If H contains an infinite subnormal subgroup of G then H has finite index in G.*

Howson's argument for free groups was very different from the above proof, and gives a bound on the rank of $H \cap K$ in terms of the ranks of H and K. We give a graphical version of Howson's proof.

Let Γ be any connected graph, and let T be a maximal tree in Γ. Let B be the set of edges not in T. Let Γ^* be the subgraph of Γ consisting of B together with all vertices and edges on the irreducible paths in T joining two vertices of edges in B. We say that an irreducible loop $e_1,\ldots,e_n$ is *cyclically irreducible* if $e_1 \neq \bar{e}_n$.

Lemma 34 (*i*) *Any edge of Γ^* lies in a cyclically irreducible loop.* (*ii*) *Any edge in a cyclically irreducible loop lies in Γ^*.* (*iii*) *$\Gamma^* \cap T$ is connected* (*and hence is a maximal tree of Γ^**).

Proof Any edge of B lies in a circuit, and so lies in a cyclically irreducible loop. If e is an edge of Γ^* not in B then there is an irreducible path $e_1,\ldots,e_n$ in T containing e and beginning and ending at vertices of edges in B. Thus there are edges e_0 and e_{n+1} of B with $e_0\tau = e_1\sigma$ and $e_{n+1}\sigma = e_n\tau$. If $e_{n+1} = e_0$ then $e_0,e_1,\ldots,e_n$ is a cyclically irreducible loop. Otherwise there is also a path $x_1,\ldots,x_m$ in T from $e_{n+1}\tau$ to $e_0\sigma$, and $e_0, e_1,\ldots,e_n, e_{n+1}, x_1,\ldots,x_m$ is a cyclically irreducible loop.

Conversely, let e be in a cyclically irreducible loop $e_1,\ldots,e_n$. We may assume $e \notin B$. Thus we may assume, taking a cyclic permutation, that $e_1 \in B$, and that $e = e_r$ for some $r > 1$. Then we may find $p < r < q \leq n+1$ so that e_p and e_q are in B (where e_{n+1} is e_1) and $e_i \notin B$ for $p < i < q$. Then the irreducible path $e_{p+1},\ldots,e_{q-1}$ shows that $e \in \Gamma^*$.

Now take any two vertices of Γ^*. There is a path in Γ joining them, and this path will, by definition of Γ^*, consist of edges either in B or in $\Gamma^* \cap T$. Each of the edges in B can be replaced by an irreducible path in T with the same endpoints, and this path lies in Γ^*, by definition. Thus we obtain a path in $\Gamma^* \cap T$, as required.//

From (i) and (ii), Γ^* can be described as the set of edges lying in some cyclically irreducible loop, together with their vertices. In particular, Γ^* does not depend on the choice of T. We call Γ^* the *core* of Γ.

The inclusion of Γ^* in Γ induces an isomorphism of $\pi_1(\Gamma^*)$ to $\pi_1(\Gamma)$, since by (iii) and Corollary 1 to Theorem 5.17 both have the same basis.

If $\pi_1(\Gamma)$ is finitely generated then B is finite, and so Γ^* is finite.

Let Γ be a finite connected graph with m vertices and n edges. Let d_i (or $d_i(\Gamma)$, if we need to consider several graphs) be the number of vertices with i edges starting at the vertex. Then $m = \Sigma d_i$, while $2n = \Sigma i d_i$, since there are $2n$ edges in all, and for each i there are d_i vertices at which exactly i edges start. By Corollary 1 to Theorem 5.17, $\pi_1(\Gamma)$ has rank $n - m + 1$. Thus $2(\text{rank}\,\pi_1(\Gamma) - 1) = \Sigma(i-2)d_i$. In particular, if Γ has vertices of degrees 2, 3, and 4 only then $2(\text{rank}\,\pi_1(\Gamma) - 1) = d_3 + 2d_4$.

Proposition 35 *Let H and K be finitely generated subgroups of a free group. Then $H \cap K$ is finitely generated and $\text{rank}(H \cap K) - 1 \leq 2(\text{rank}H - 1)(\text{rank}K - 1)$.*

Proof Since H and K are finitely generated, there is a finitely generated free group containing them both (for instance, the group on those basis elements of the given free group which are involved in the generators of H and K). Since any finitely generated free group is contained in a free group of rank 2, we may assume that H and K are subgroups of a group F with basis $\{x, y\}$.

Then F acts freely on a tree T. We may label the edges of T with the symbols x, y, x^{-1}, and y^{-1}, and starting at each vertex there is exactly one edge with a given label.

We know that $H\backslash T$ and $K\backslash T$ have H and K respectively as their fundamental groups. Let Γ and Δ be the cores of $H\backslash T$ and $K\backslash T$, Since H and K are finitely generated, both Γ and Δ are finite. By the definition of a core, no vertex of either Γ or Δ can have degree 1, while by construction of T no vertex can have degree greater than 4.

Let $\Gamma\times\Delta$ be the graph such that $V(\Gamma\times\Delta) = V(\Gamma)\times V(\Delta)$, there being an edge of $\Gamma\times\Delta$ from (a,b) to (p,q) iff there is an edge in Γ from a to p and an edge in Δ from b to q, with both edges having the same label. Plainly $d_4(\Gamma\times\Delta)\le d_4(\Gamma)d_4(\Delta)$, while $d_3(\Gamma\times\Delta)\le d_3(\Gamma)d_3(\Delta)+d_3(\Gamma)d_4(\Delta)+d_4(\Gamma)d_3(\Delta)$. From this it follows that $d_3(\Gamma\times\Delta)+2d_4(\Gamma\times\Delta)\le[d_3(\Gamma)+2d_4(\Gamma)][d_3(\Delta)+2d_4(\Delta)]$. From the previous discussion, it follows that for any connected graph Z which has a one-one map into $\Gamma\times\Delta$ we have $\operatorname{rank}\pi_1(Z)-1\le 2(\operatorname{rank}H-1)(\operatorname{rank}K-1)$.

Thus it is enough to show that there is a one-one map from the core of $(H\cap K)\backslash T$ into $\Gamma\times\Delta$. We plainly have a map from $(H\cap K)\backslash T$ into each of $H\backslash T$ and $K\backslash T$. Further, it is easy to see that a loop in any of these three graphs is cyclically reduced iff the element of F corresponding to it (when we replace each edge by its label) is cyclically reduced. Thus the core Z of $(H\cap K)\backslash T$ maps into the cores Γ and Δ of $H\backslash T$ and $K\backslash T$. This plainly gives a map of Z into $\Gamma\times\Delta$. Now the vertices of T can be regarded as the elements of F, and the vertices of $H\backslash T$ will be the cosets of H, and similarly for K and $H\cap K$. Since the intersection of a coset of H and a coset of K is either empty or consists of exactly one coset of $H\cap K$, the map of Z into $\Gamma\times\Delta$ will be one-one on vertices. Because the edges are labelled, it will then also be one-one on edges.//

Exercise 6 Let H and K be finitely generated subgroups of a free group F, and suppose that H has finite index in F. Show that $\operatorname{rank}(H\cap K)-1\le(\operatorname{rank}H-1)(\operatorname{rank}K-1)$.

Exercise 7 Find an example of a map from a graph $\Gamma$to a graph Δ which does not map Γ^* into Δ^*.

We now prove an important result about free products. We will need to begin with some results about graphs.

Let X be a G-graph (since G can be trivial, the results below hold for arbitrary graphs). Let $\equiv$ be an equivalence relation on E such that $ge_1\equiv ge$ and $\bar{e}_1\equiv\bar{e}$ if $e_1\equiv e$ and such that we never have $\bar{e}\equiv ge$. We may define an equivalence relation on V, which we still denote by $\equiv$, to be the smallest equivalence relation such that $e_1\tau\equiv e\tau$ whenever $e_1\equiv e$ (we will also have $e_1\sigma\equiv e\sigma$). Thus $gv_1\equiv gv$ if $v_1\equiv v$. It is clear that we have a G-graph, which we denote by $X/\equiv$, whose vertex set is the set of equivalence classes of V and whose edge set is the set of equivalence classes of E, and that we have a G-map p from X to $X/\equiv$. We need $\bar{e}$ not to be equivalent to ge in order to ensure that G acts without inversions on $X/\equiv$ (and to ensure that $^{-}$ does not fix any edge of $X/\equiv$).

Let X be a G-graph. A *fold* consists of a pair of edges e_1 and e_2 such that $e_1\sigma = e_2\sigma$ and $e_2 \notin Ge_1 \cup G\bar{e}_1$. When $f:X \to Z$ is a map of G-graphs and we also have $e_1 f = e_2 f$, we refer to a *fold along f*.

Let X have a fold. Let $\equiv$ be the smallest equivalence relation on E such that $e_2 \equiv e_1$ and such that if $x \equiv y$ then $\bar{x} \equiv \bar{y}$ and $gx \equiv gy$. We construct the G-graph $X/\equiv$ as above, and we denote $X/\equiv$ by Y. The condition that $e_2 \notin G\bar{e}_1$ is needed to ensure that G acts without inversions on Y (and to ensure that $^{-}$ fixes no edge of Y), while the condition $e_2 \notin Ge_1$ is used to ensure that if no edge of X is fixed by a non-trivial element of G then the same holds for Y. Plainly, if we have a fold along the map f then there is a map $f':Y \to Z$ such that $f = pf'$.

We will call X *edge-free* if every edge has trivial stabiliser.

Lemma 36 *The graph obtained by folding a G-tree is a G-tree.*

Proof We use Lemma 5.13. Thus we define $C_0(X)$ to be the free abelian group on $E(X)$, $C_1(X)$ to be the quotient of the free abelian group on $E(X)$ by the subgroup generated by all $e + \bar{e}$, $\varepsilon_X:C_0(X) \to \mathbf{Z}$ to be given by $v\varepsilon_X = 1$ for all vertexes v, and $\partial_X:C_1(X) \to C_0(X)$ to be given by $e\partial_X = e\tau - e\sigma$ for all edges e of X, and we make similar definitions for Y.

Since X is a G-graph, both $C_0(X)$ and $C_1(X)$ are G-modules, and ∂_X and ε_X are homomorphisms of G-modules (where $\mathbf{Z}$ has the trivial module structure), and similarly for Y. It is easy to check that $C_1(Y)$ is the quotient of $C_1(X)$ by the submodule generated by $e_2 - e_1$, and that $C_0(Y)$ is the quotient of $C_0(X)$ by the submodule generated by $e_2\tau - e_1\tau$. Since $e_2\sigma = e_1\sigma$, we have $(e_2 - e_1)\partial = e_2\tau - e_1\tau$. We see that ∂_Y and ε_Y are induced from ∂_X and ε_X.

By Lemma 5.13, since X is a tree, we have $\ker\partial_X = 0$ and $\ker\varepsilon_X = \operatorname{im}\partial_X$. It follows easily that $\ker\partial_Y = 0$ and $\ker\varepsilon_Y = \operatorname{im}\partial_Y$, which, using Lemma 5.13 again, shows that Y is a tree.//

Lemma 37 *Let $f:X \to Z$ be a map of G-trees, and let X be edge-free. Then there is an edge-free G-tree X', such that X' is $X/\equiv$ for some equivalence relation for which $xf = yf$ whenever $x \equiv y$, and such that X' has no fold along f', where $f':X' \to Z$ is the map such that $f = pf'$.*

Proof We will need to look at various equivalences on E. We shall therefore regard an equivalence as a subset R of $E \times E$, and we denote the quotient graph by X/R. All our equivalences will be such that if $x \equiv y$ then $xf = yf$, $gx \equiv gy$, and $\bar{x} \equiv \bar{y}$, and such that $\bar{x}$ is never equivalent to gx. We further assume that X/R is an edge-free tree.

Suppose that there is an equivalence R which is maximal with respect to having these properties, and let X' be X/R. Then X' cannot have a fold along f'. For if it did we could lift the equivalence on E', which exists because there is a fold, to an equivalence S on E, which would strictly contain R, and which would satisfy all the relevant properties (X/S is the same as the result of folding X', and so it is a tree, by Lemma 36.).

Hence it is enough to show that there is such a maximal R, and we use Zorn's Lemma to prove this. So we take a collection of equivalences R_α, all satisfying the relevant conditions, such that for all α and β either $R_\alpha \subseteq R_\beta$ or $R_\beta \subseteq R_a$. Then $\cup R_\alpha$ is an equivalence, and all the relevant conditions obviously hold for $\cup R_\alpha$, except for the requirement that $X/\cup R_\alpha$ is a tree. We show that this condition also holds.

Since X is connected and maps onto $X/\cup R_\alpha$, it is immediate that $X/\cup R_\alpha$ is also connected. Suppose that we have a loop in $X/\cup R_\alpha$. Then we have edges $e_1,\ldots,e_n$ of X such that, for $1 \leq i \leq n$, $e_i\tau$ is equivalent to $e_{i+1}\sigma$ (where e_{n+1} means e_1) in the equivalence relation on V induced by $\cup R_\alpha$. This means that we have edges e_{ij} $(1 \leq j \leq m_i)$ such that $e_{i1}\tau = e_i\tau$, $e_{im_i}\tau = e_{i+1}\sigma$, and $e_{ij}\tau$ equivalent to $e_{i,j+1}\tau$ under $\cup R_\alpha$. Then there will be α_{ij} such that $e_{ij}\tau$ is equivalent to $e_{i,j+1}\tau$ under $R_{\alpha_{ij}}$. By the assumption on the family R_α, there will be some β such that $R_{\alpha_{ij}} \subseteq R_\beta$ for all i and j. It then follows that $e_1,\ldots,e_n$ gives a loop in X/R_β. Since X/R_β is a tree, this loop is reducible in X/R_β, and so our original loop, being the image of this in the map from X/R_β to $X/\cup R_\alpha$, will also be reducible.//

Lemma 38 *Let $f: X \to Y$ be a map of G-trees. Let X be edge-free and let there be no fold along f. Then either the induced map from $G\backslash X$ to $G\backslash Y$ is injective or else there is a vertex v of X, an edge e of Y, and some $g \in G$ such that $e\sigma = vf$ and g fixes v and e but does not fix all of Y.*

Proof Suppose that the conclusions do not hold. We will obtain a fold along f.

Suppose first that there are vertices v_1 and v_2 of X such that $v_2 \notin Gv_1$ but $v_2f = g(v_1f)$ for some g. Replacing v_1 by gv_1, we may assume that

$v_2f = v_1f$. Choose v_1 and v_2 with $v_1f = v_2f$ such that the irreducible path p (which must have positive length) from v_1 to v_2 has minimal length. Then pf is a closed path in Y. Since Y is a tree, it follows that we must be able to write p as $p_1e_1e_2p_2$ where e_1 and e_2 are edges such that $e_2f = \bar{e}_1f$.

If we had $e_2 \in Ge_1$, we would get $\bar{e}_1f \in \mathfrak{G}(e_1f)$. This cannot happen as G acts without inversions on Y. Since $e_2\sigma = \bar{e}_1\sigma$ we have a fold along f unless $e_2 \in G\bar{e}_1$.

So we may assume that there is some g such that $\bar{e}_1 = ge_2$. Thus $g(e_2f) = e_2f$ and $g(e_2\sigma) = e_2\sigma$. Since the second of the two conclusions was supposed false, it follows that g fixes all of Y; in particular, it fixes v_2f. Since $\bar{e}_1 = ge_2$, the path gp_2 starts where p_1 finishes, and so we have a path q which consists of p_1 followed by gp_2. This path starts at v_1 and finishes at gv_2, which are not in the same G-orbit. Also we have $v_1f = v_2f = g(v_2f) = (gv_2)f$. Since q is shorter than p, this contradicts our original choice of v_1 and v_2.

Now suppose that there are edges e_1 and e_2 such that $e_2 \notin Ge_1$ but $e_2f \in \mathfrak{G}(e_1f)$. As in the first part, we may assume that e_1 and e_2 are chosen so that $e_2f = e_1f$ and so that the irreducible path from $e_1\sigma$ to $e_2\sigma$ has minimal length. We get the same contradiction as before if this path has positive length. If this path has length 0, so that $e_1\sigma = e_2\sigma$, we have a fold along f, since $e_1f = e_2f$, $e_2 \notin Ge_1$, and $e_2 \notin G\bar{e}_1$, for the same reason as before.//

We define a homomorphism from the graph of groups $(\mathfrak{G},X)$ to the graph of groups $(\mathfrak{H},Y)$ to consist of a map of graphs $\varphi: X \to Y$, and homomorphisms from G_x to $H_{x\varphi}$ for each vertex or edge x of X such that, for every edge e of X, the homomorphisms $G_e \to G_{e\tau} \to H_{e\tau\varphi}$ and $G_e \to H_{e\varphi} \to H_{e\varphi\tau}$ are the same.

Lemma 39 *Let φ be a homomorphism from the graph of groups $(\mathfrak{G},X)$ to the graph of groups $(\mathfrak{H},Y)$. Suppose that φ maps a maximal tree T of X into a maximal tree S of Y. Then φ induces a homomorphism from $\pi(\mathfrak{G},X,T)$ to $\pi(\mathfrak{H},Y,S)$ whose restriction to G_v is the given homomorphism from G_v to $H_{v\varphi}$ for every vertex v of X. Further, let B be an orientation of Y, and let A be $B\varphi^{-1}$ (so that A is an orientation of X). Then φ induces a $\pi(G,X,T)$-map from the universal cover $\tilde{X}$ of $(\mathfrak{G},X)$ to the universal cover $\tilde{Y}$ of Y.*

Proof That φ induces a suitable homomorphism, which we still denote by φ, from $\pi(\mathfrak{G},X,T)$ to $\pi(\mathfrak{H},Y,S)$ is immediate from Lemma 20.

Let $\tilde{X}$ and $\tilde{Y}$ be the universal covers of $(\mathfrak{G},X)$ and $(\mathfrak{H},Y)$ corresponding to the maximal tree T and orientation A of X, and the maximal tree S and orientation B of Y, respectively. The homomorphism from $\pi(\mathfrak{G},X,T)$ to $\pi(\mathfrak{H},Y,S)$ makes $\tilde{Y}$ into a $\pi(\mathfrak{G},X,T)$-tree.

It is easy to check, first, that there is a well-defined function from $\tilde{X}$ to $\tilde{Y}$ sending $[g,x]$ to $[g\varphi,x\varphi]$ for all $g\in\pi(\mathfrak{G},X,T)$ and every vertex or edge x of X, and, second, that this function is a map of graphs which is a $\pi(\mathfrak{G},X,T)$-map.//

Theorem 40 *Let G_α and H_α be groups, and let φ_α be a homomorphism from G_α onto H_α (for α in some index set). Let Φ be the induced homomorphism from $*G_\alpha$ to $*H_\alpha$. Let A be a subgroup of $*G_\alpha$ such that $A\Phi = *H_\alpha$. Then there are subgroups A_α of A such that $A = *A_\alpha$, and $A_\alpha\Phi = H_\alpha$ for all α.*

Warning In general, $A_\alpha \neq A\cap H_\alpha$.

Proof Let Y be the graph with one vertex v_α for each α, one other vertex v, and one edge-pair $\{e_\alpha,\bar{e}_\alpha\}$ joining v to each v_α. Let $\mathfrak{G}$ and $\mathfrak{H}$ be the graphs of groups on Y with G_α (or H_α) being the vertex group at v_α, the vertex group at v and all the edge groups being trivial in both cases. Let X be the universal cover of $(\mathfrak{G},Y)$, and let $\tilde{Y}$ be the universal cover of $(\mathfrak{H},Y)$. Applying Lemma 39, and restrcting to the subgroup A, we have an A-map from X to $\tilde{Y}$. Because $A\Phi = *H_\alpha$, we find that $A\backslash\tilde{Y} = Y$. Since the action of $*H_\alpha$ on $\tilde{Y}$ is edge-free, an element of A which stabilises an edge of $\tilde{Y}$ must fix all of $\tilde{Y}$.

By Lemma 37, there is an edge-free A-tree Z and an A-map $f{:}Z\to\tilde{Y}$ such that f has no folds. Then Lemma 38, together with the above remarks about the action of A on $\tilde{Y}$, tell us that the induced map from $A\backslash Z$ to Y must be an injection. This map must also be a surjection, since f comes from a surjective map from X to $\tilde{Y}$ by factoring out an equivalence relation. Thus we may as well identify $A\backslash Z$ with Y.

Let $j{:}Y\to\tilde{Y}$ be the function considered in the definition of the universal cover. Suppose that we can find a function $k{:}Y\to Z$ such that $kf = j$. Then the Second Structure Theorem tells us that $A = *A_\alpha$, where $A_\alpha = \text{stab}(v_\alpha k)$. Also $A_\alpha\Phi \subseteq \text{stab}(v_\alpha kf) = H_\alpha$. Since Φ maps A onto $*H_\alpha$, it must map A_α onto H_α.

Now there will be a vertex u of Z such that $uf = [1,v]$, since there is a vertex of X which maps to $[1,v]$. There will be an edge z_α of Z starting at u whose image in Y is e_α. Then $z_\alpha f$ will be $[h,e_\alpha]$ for some $h\in *H_\alpha$. We have

$[1,v] = uf = z_\alpha \sigma f = z_\alpha f\sigma = [h,e_\alpha]\sigma = [h,v]$. Since the vertex group of $\mathfrak{H}$ at v is trivial, this gives $h = 1$. We may now define k by $vk = u$, $e_\alpha k = z_\alpha$, and, necessarily, $\bar{e}_\alpha k = \bar{z}_\alpha$ and $v_\alpha k = z_\alpha \tau$.//

Corollary 1 (Grushko's Theorem) *Let φ be a homomorphism from a free group F onto a free product $*H_\alpha$. Then F is the free product of subgroups F_α such that $F_\alpha\varphi = H_\alpha$ for all α.*

Proof Let $G_\alpha = H_\alpha\varphi^{-1}$, let φ_α be the restriction of φ to G_α, and let Φ be the induced map from $*G_\alpha$ to $*H_\alpha$. Because φ is onto, we have $G_\alpha\varphi = H_\alpha$, and so Φ is onto. Let $\vartheta{:}*G_\alpha \to F$ be the homomorphism which is inclusion on each G_α, so that $\Phi = \vartheta\varphi$. Since each G_α contains $\ker\varphi$, we see that $\mathrm{im}\vartheta \supseteq \ker\varphi$. From this it follows easily that ϑ maps onto F. Since F is free, there is a homomorphism $\psi{:}F \to *G_\alpha$ such that $\psi\vartheta$ is the identity homomorphism on F. Then ψ embeds F in $*G_\alpha$, and the result follows by applying the theorem to $F\psi$.//

Corollary 2 *Let A and B be finitely generated groups, whose minimum number of generators are m and n respectively. Then the minimum number of generators of $A*B$ is $m+n$.*

Proof Plainly $A*B$ can be generated by $m+n$ elements.

Let r be the minimum number of generators of $A*B$. Then there is a homomorphism from a free group F of rank r onto $A*B$. By Corollary 1, we can write F as F_1*F_2, where F_1 and F_2 map onto A and B respectively. Thus $\mathrm{rank}F_1 \geq m$ and $\mathrm{rank}F_2 \geq n$. Since $\mathrm{rank}F = \mathrm{rank}F_1 + \mathrm{rank}F_2$, we see that $r \geq m+n$, as required.//

We can generalise Corollary 1.

Corollary 3 *Let φ be a homomorphism from a free product $A = F**A_\lambda$ (for λ in some index set Λ) onto $H = *H_\alpha$ such that for each λ there are $h_\lambda \in H$ and an index α_λ such that $A_\lambda\varphi \subseteq h_\lambda H_{\alpha_\lambda} h_\lambda^{-1}$. Then A has subgroups G_α such that $A = *G_\alpha$ and $G_\alpha\varphi = H_\alpha$.*

Proof Let $G_{\alpha\lambda}$ be $(h_\lambda H_{\alpha_\lambda} h_\lambda^{-1})\varphi^{-1}$. Let 0 be a symbol not in Λ, and let $G_{\alpha 0}$ be $H_\alpha\varphi^{-1}$. Let G be the free product of the groups $G_{\alpha\mu}$ for all α and all μ in $\Lambda \cup \{0\}$. As in corollary 1, we have homomorphisms $\vartheta{:}G \to A$ and $\Phi{:}G \to *H_\alpha$ such that on each factor ϑ is inclusion and Φ is φ. As in Corollary 1, it is enough to define a homomorphism $\psi{:}A \to G$ such that $\psi\vartheta$ is the identity on A. On the factor A_λ we need only require ψ to be the inclusion of A_λ in $G_{\alpha_\lambda\lambda}$. Since F is free, we can define a suitable ψ on F if ϑ is onto. As before, this is immediate once we have seen that Φ is onto. But Φ certainly maps the groups $G_{\alpha 0}$ onto the corresponding H_α (this is why these extra groups were introduced), so Φ is onto.//

There are extended versions of these results, of which an example follows.

Proposition 41 *Suppose that we have homomorphisms from G_1 to H_1 and from G_2 to H_2 such that the induced homomorphism φ from $G_1 * G_2$ to $H_1 *_{H_0} H_2$ is onto. Let A be a subgroup of $G_1 * G_2$ such that $A\varphi = H_1 *_{H_0} H_2$. Then $A = A_1 * A_2$, where either $A_1\varphi \subseteq H_1$ and $A_2\varphi \subseteq H_2$ or else $A_1\varphi \cap H_0 \neq \{1\}$ and $A_1\varphi$ is contained in one of H_1 and H_2.*

Proof Let Y be the graph with two vertices v_1 and v_2 and with one edge-pair $\{e, \bar{e}\}$ joining v_1 and v_2. Let $\mathfrak{G}$ be the graph of groups on Y with G_i being the vertex group at v_i for $i = 1,2$, with the edge group being trivial. Let $\mathfrak{H}$ be the graph of groups on Y with H_i being the vertex group at v_i and the edge group being H_0. Let X be the universal cover of $(\mathfrak{G}, Y)$, and let $\tilde{Y}$ be the universal cover of $(\mathfrak{H}, Y)$. As in Theorem 40, there is an A-map from X onto $\tilde{Y}$, which, factoring by an equivalence relation, gives rise to an edge-free A-tree Z and an A-map $f: Z \to \tilde{Y}$.

As before, $A\backslash\tilde{Y} = Y$, and f induces a map from $A\backslash Z$ onto Y. If this map is injective then $A\backslash Z$ is isomorphic to Y. The Second Structure Theorem then shows, as before, that the first case in the proposition holds. If the map from $A\backslash Z$ is not injective then Lemma 38 shows that the second case of the proposition holds.//

The following corollary can be deduced from the proposition in a manner similar to the proof of Corollary 1 of Theorem 40 from the theorem. The details, and the formulation and proof of similar results for HNN extensions, will be left to the reader.

Corollary *Let φ be a homomorphism from a free group F onto $H_1 *_{H_0} H_2$. Then F is the free product of subgroups F_1 and F_2 such that either φ maps F_1 and F_2 into H_1 and H_2 respectively or $F_1\varphi \cap H_0 \neq \{1\}$ and $F_1\varphi$ is contained in one of H_1 and H_2.//*

Marshall Hall (1949b) proved that *any finitely generated subgroup of a free group F is contained in a subgroup of F whose index is finite.* His proof used Schreier transversals. Although he did not explicitly say so, his method proved a stronger result, which we specify below.

Definition A group G is called an *M.Hall group* if, given any finitely generated subgroup H of G and any finite number of elements $a_1,\dots,a_n$ of G not in H, there is a subgroup K of G whose index is finite and such that H is a free factor of K and $a_1,\dots,a_n \notin K$.

Then the stronger form of Hall's theorem is that any free group is an M.Hall group. This follows at once from the next theorem, which was first proved by Tretkoff (1975).

Plainly any finite group is an M. Hall group (we may take K to be H). Also the infinite cyclic group is an M. Hall group, since if H is non-trivial we may take K to be H, while if H is trivial we need only take K to be generated by a sufficiently large power of the generator of G.

We need to observe that if H is a finitely generated subgroup of an M. Hall group G and $b_1,\dots,b_n$ are elements of G no two of which lie in the same coset of H then there is a subgroup K of finite index containing H as a free factor and such that no two of $b_1,\dots,b_n$ lie in the same coset of K. We need only apply the definition to H and the elements $b_ib_j^{-1}$ for $i \neq j$.

Theorem 42 *The free product of M. Hall groups is an M. Hall group.*

Proof Suppose first that G is the free product of an infinite number of M. Hall groups. Let H be a finitely generated subgroup of G, and $a_1,\dots,a_n$ elements of $G - H$. Then we may write $G = P*Q$, where P is the free product of finitely many M. Hall groups, and P contains the subgroup H and the elements $a_1,\dots,a_n$.

Suppose that there is a subgroup K of P which has finite index, has H as a free factor, and contains none of $a_1,\dots,a_n$. Let φ be the homomorphism from G to P which is the identity on P and is trivial on Q. Then $K\varphi^{-1}$ has finite index in G. Since $K\varphi^{-1} \cap P = K$, we see that $K\varphi^{-1}$ contains none of $a_1,\dots,a_n$. Also the Kurosh Subgroup Theorem tells us that K is a free factor of $K\varphi^{-1}$, so that H is a free factor of $K\varphi^{-1}$.

It follows that it is enough to show that the free product of finitely many M. Hall groups is an M. Hall group. Then, inductively, it is enough to prove the result when $G = G_0*G_1$, where G_0 and G_1 are M. Hall groups.

In this case there is an edge-free G-tree X (namely, the universal cover of the graph of groups which has one edge-pair with trivial edge group, and two distinct vertices whose vertex groups are G_0 and G_1) with the following properties. X has an edge e and vertices v_0 and v_1 such that v_0 and v_1 are in different orbits, and have stabilisers G_0 and G_1 respectively, and such that $e\sigma = v_0$, $e\tau = v_1$, and every edge is in $Ge \cup G\bar{e}$ (and every vertex is in $Gv_0 \cup Gv_1$).

Let Y be $H\backslash X$, and let T be a maximal tree of Y. Let $j{:}Y \to X$ be the usual function. When convenient, we denote a vertex of Y by Hgv_0 or Hgv_1 and an edge by Hge or $Hg\bar{e}$. Give Y the orientation A consisting of all edges Hge.

Then $H = \pi(\mathfrak{H},Y,T)$, where the edge groups of $\mathfrak{H}$ are trivial and the vertex group at w is $H \cap \text{stab}wj$. Since H is finitely generated, and is the free product of $\pi_1(Y)$ and the vertex groups H_w, it follows that $H_w = \{1\}$ for all but finitely many w and that there are only finitely many edges of Y not in T.

Let S be the finite set consisting of the vertices Ha_iv_0, the vertices of the edges of $Y - T$, and all vertices w such that $H_w \neq \{1\}$. Let Z_0 be the finite subgraph of Y consisting of the edges of $Y - T$ and all those vertices and edges and their inverses which lie on the irreducible paths from Hv_0 to the vertices of S. Notice that $T \cap Z_0$ is connected, and so it is a maximal tree of Z_0.

Now take any vertex w of Y, and assume that w is of form Hgv_0. Consider $\{y \in A; y\sigma = w\}$, and let y_0 be an element of this set. Then $yj\sigma = wj = y_0j\sigma$. It follows that there is a bijection from the set of cosets of H_w in stabwj onto this set of edges, sending the coset H_wp onto that y such that $yj \in Hp(y_0j)$. Similarly, if w is of form Hgv_1 then there is a bijection from the set of cosets of H_w in stabwj onto $\{y \in A; y\tau = w\}$, where the coset H_wp maps to that y such that $\gamma_y^{-1}(yj) \varepsilon Hp\gamma_{y_0}^{-1}(y_0j)$, where γ_y is the element of $H = \pi(\mathfrak{H})$ corresponding to y

Take a vertex w of Z_0 of form Hgv_0. It follows that any edge starting at w is in A. Since Z_0 is connected, there is at least one edge starting at w. Choosing one such as the reference edge, there are finitely many elements $c_1 = 1, \ldots, c_r$ of stabwj in different cosets of H_w such that, in the bijection above, the cosets H_wc_i correspond to the edges of Z_0 starting at w.

Now stabwj is an M. Hall group, being isomorphic to G_0, and H_w is finitely generated, being a free factor of H. Thus there is a subgroup K_w of stabwj, containing H_w as a free factor, of finite index in stabwj, and with $c_1, \ldots, c_r$ in different cosets of K_w.

Let $c_{r+1}, \ldots, c_n$ be such that $c_1, \ldots, c_n$ form a transversal of K_w in stabwj. Add to Z_0 those edges of Y starting at w which correspond to the cosets H_wc_i for $r < i \leq n$, together with their inverses and their other end-points. Do this for all such vertices, and perform the similar construction for the vertices of form Hgv_1.

The result is a finite graph Z such that $Z_0 \subseteq Z \subseteq Y$. Let $\mathfrak{K}$ be the graph of groups on Z whose edge groups are trivial, whose group at a vertex w of Z_0 is K_w, and whose group at a new vertex w is stabwj. Then $T \cap Z$ is a maximal tree in Z, and $H = \pi(\mathfrak{H}|Z_0, Z_0, T \cap Z_0)$ is a free factor of $\pi(\mathfrak{K}, Z, T \cap Z)$, since H_w is a free factor of K_w for all vertices w of Z_0 and each edge of $Z - T$ is in Z_0.

By Lemma 20, there is a homomorphism $\varphi: \pi(\mathfrak{K}, Z, T \cap Z) \to G$ which is the inclusion on each K_w and which sends each z to the element γ_z of (the subgroup H of) G. The restriction of φ to H is just the inclusion of H in G.

Let $\tilde{Z}$ be the universal cover of the graph of groups $\mathfrak{K}$. Because φ maps K_w into stabwj, there is a well-defined function f from $V(\tilde{Z})$ to $V(X)$ sending $[k,w]$ to $(k\varphi)(wj)$. We may also define $f: E(\tilde{Z}) \to E(X)$ by $[k,z]f = (k\varphi)(zj)$ and $[k,\bar{z}]f = (k\varphi)(\bar{z}j)$ for all $z \in A \cap Z$. Then f is easily seen to be a map of trees.

We next show that the theorem holds provided that f is a bijection. For then, we see first that k acts trivially on $\tilde{Z}$ if $k\varphi = 1$, and so φ is one-one. It then follows, letting K be the image of φ, that H is a free factor of K. Second, letting w_0 be the vertex Hv_0 of Y and w_i the vertex Ha_iv_0, we see that w_i and w_0 are different vertices of Z. Hence the vertices $[1,w_0]$ and $[1,w_i]$ of $\tilde{Z}$ are in different orbits, and so w_0j and w_ij are in different K-orbits. Since $w_0j = v_0$ and $w_ij \in Ha_iv_0$, we see that $a_i \notin K$. Finally, there are only finitely many orbits of edges in $\tilde{Z}$, since Z is finite, and so there will be only finitely many K-orbits of edges of X. It follows that K has finite index in G.

To show that f is a bijection, it is enough, by Lemma 5, to show that it is locally bijective. It is plainly enough to show that it is locally bijective at all vertices of the form $[1,w]$. First let w be a vertex of Z_0 of form Hgv_0. We have a transversal $\{c_1, \ldots, c_n\}$ of K_w in stabwj, and the edges of Z starting at w are $y_1, \ldots, y_n$ where we may assume that $y_ij = c_i(y_1j)$ (multiplying each c_i by an element of H_w, if necessary). Then the edges of $\tilde{Z}$ starting at $[1,w]$ are all $[k, y_i]$ with $k \in K_w$ (these being all distinct). Now $[k, y_i]f = k(y_ij) = (kc_i)(y_1j)$. Since the elements kc_i are exactly the distinct elements of stabwj, we see that f provides a bijection from the edges of $\tilde{Z}$ starting at $[1,w]$ to the edges of X starting at wj, as required. When w is a vertex of Z_0 of form Hgv_1 a similar argument applies (but we need to take account of elements γ_y). When w is a new vertex similar but simpler arguments apply, since then $K_w =$ stabwj and there is only one edge of Z starting at w.//

Theorem 43 *Let G be a free product of non-trivial groups, let H be a finitely generated subgroup of G whose index is infinite, and let* $a_1,\ldots,a_n$ *be elements of G not in H. Then there is a subgroup K of G not containing any of* $a_1,\ldots,a_n$ *such that H is a proper free factor of K.*

Proof The proof is almost identical to that of the previous theorem.

We may write G as a free product $G_0 * G_1$. We construct Z_0 as before. We obtain Z by adding all edges (and their inverses and their endpoints) of Y which have one vertex in Z_0. We let K_w be H_w for any vertex w of Z_0, while K_w is to be stabwj for any new vertex w (stabwj is, of course, a conjugate of G_0 or G_1). The previous argument shows that we have a subgroup K of G of which H is a free factor and not containing any of $a_1,\ldots,a_n$. Since H has infinite index, Y will be infinite. As Z_0 is finite, there will be at least one vertex of Z not in Z_0. This shows that K is strictly larger than H.//

Proposition 44 *Let H be a finitely generated subgroup of a free product G of non-trivial groups. If H contains a non-trivial subnormal subgroup of G then H has finite index in G.*

Proof Suppose that H has infinite index. By Theorem 43, there is a subgroup K of G such that $K = H * L$ where $L \neq \{1\}$.

Let S be a subnormal subgroup of G contained in H. Then S is a subnormal subgroup of K. Hence there is a sequence $S_1 = K,\ldots,S_n = S$ of subgroups of K such that $S_{i+1} \triangleleft S_i$ for $i < n$. There will be some r such that S_{r+1} is contained in H but S_r is not contained in H. Take $k \in S_r - H$ and $s \in S_{r+1} - \{1\}$. Then $k^{-1}sk \in S_{r+1}$, since $S_{r+1} \triangleleft S_r$, but $k^{-1}sk \notin H$, since K has H as a proper free factor. This contradiction proves the result.//

We have already remarked that a different method of proof enables us to prove the following generalisation of Proposition 44. *If G is the fundamental group of a graph of groups with finite edge groups then any finitely generated subgroup containing an infinite subnormal subgroup will have finite index.*

This would follow, as in the proof of the proposition, if we proved the following generalisation of Theorem 43. *Let G be the fundamental group of a graph of groups whose edge groups are finite. Let H be a finitely generated subgroup of G, and let* $a_1,\ldots,a_n$ *be elements not in H. Then there is a subgroup* $K \neq H$ *of G not containing any of* $a_1,\ldots,a_n$ *such that K is the fundamental group of a graph of groups, one vertex group being H, and all edge groups being finite.*

The proof of this is a routine, but messy, generalisation of the proof of Theorem 43. It would be possible to generalise the definition of M.Hall groups so as to obtain a similar generalisation of Theorem 42.

8.6 *CONSTRUCTION OF TREES*

In order to apply the structure theorems, we need to be able to construct trees. Sometimes this can be done geometrically. For instance, Serre shows by such methods that certain linear groups are amalgamated free products.

It will obviously be of interest to know if a given set can be regarded as the set of vertices of a tree (or as the set of edges of a tree). It is easy to see that to any set A there is another set B such that A is the set of vertices of a tree and B is the set of edges, and also a set C which is the set of vertices of a tree for which A is the set of edges of an orientation.

The situation is very different when we are given a G-set A and wish to find a G-tree. For instance, it is easy to see that if $|A| > 2$ then A cannot be the set of vertices of a G-tree if G is doubly transitive, and (using Lemma 5.10) if A is finite then it cannot be the set of vertices of a G-tree if G is transitive. We give next some useful conditions which will ensure that A is the set of vertices (or edges) of a G-tree.

The next proposition is obvious.

Proposition 45 *Let T be a tree, and let $v_0 \in V(T)$. For each $v \in V(T)$, let $v\lambda$ be the length of the irreducible path from v_0 to v. Then $\lambda: V(T) \to \mathbb{N}$ has the following properties:* (*i*) *$v\lambda = 0$ iff $v = v_0$,* (*ii*) *if v and w are adjacent* (*that is, there is an edge joining v and w*) *then $v\lambda - w\lambda = \pm 1$,* (*iii*) *if $v \neq v_0$ then there is exactly one w with $w\lambda = v\lambda - 1$.//*

The converse of this is not difficult to prove.

Theorem 46 *Let V be a set with a symmetric relation called adjacency. Let $\lambda: V \to \mathbb{N}$ be a function satisfying* (*i*), (*ii*), *and* (*iii*) *of the previous proposition. Then there is a tree T such that $V(T) = V$, with an edge joining v and w iff they are adjacent, and such that the function of the previous proposition is the given function λ.*

Remark Let λ be a function satisfying (i) and (iii). If we are not given an adjacency relation, we could define v and w to be adjacent iff

$w\lambda - v\lambda = \pm 1$, and (ii) would automatically hold. But it is more useful to permit arbitrary adjacency relations; for instance, if V is a G-set then T will be a G-tree iff the adjacency relation is preserved under the action of G.

Proof There is a graph T whose vertex set is V and such that there is one edge-pair joining v and w for any two adjacent vertices v and w, and no edge-pair joining non-adjacent vertices.

By induction on $v\lambda$, there is for each vertex v a path joining v_0 to v.

Let $e_1, \ldots, e_n$ be a loop based at v_0, with $n > 0$. Then $n \neq 1$, by (ii). Let $v_i = e_i\tau$, and choose r so that $v_r\lambda = \max v_i\lambda$. Then, by (ii) and the choice of r, $v_{r-1}\lambda = v_r\lambda - 1 = v_{r+1}\lambda$. So, by (iii), $v_{r-1} = v_{r+1}$. Since there is at most one edge-pair joining any two vertices, this shows that $e_{r+1} = \bar{e}_r$. That is, all loops based at v_0 are reducible, and so T is a tree.

Suppose that $v\lambda > 1$, and let u be the vertex adjacent to v such that $u\lambda = v\lambda - 1$. Inductively, we may assume that the irreducible path from v_0 to u has length $u\lambda$, and that, when $u \neq v_0$, that the vertex w adjacent to u on this path has $w\lambda = u\lambda - 1$. Then there is a path from v_0 to v of length $v\lambda$, on which the vertex adjacent to v is u, and this path must be irreducible since $v \neq w$.//

It is slightly inconvenient to have one vertex playing a special role, and the next results avoid this.

Proposition 47 *Let T be a tree. For u and v in V, let $(u,v)d$ be the length of the irreducible path from u to v. Then d is an integer-valued metric on V (which is G-invariant if T is a G-tree). For any integer $r < (u,v)d$ there is exactly one vertex p such that both $(u,p)d = r$ and $(u,v)d = (u,p)d + (p,v)d$. If $(u,v)d = 1$ then, for any vertex w, $(u,w)d \neq (v,w)d$. If u,v, and w are vertices with $(u,v)d < (u,w)d + (w,v)d$ then there is a vertex q such that $(u,q)d < (u,w)d$ and $(q,v)d < (w,v)d$.*

Proof Suppose that there is an irreducible path $x_1, \ldots, x_n$ of length n from u to w, and an irreducible path $y_1, \ldots, y_m$ of length m from w to v. Then $x_1, \ldots, x_n, y_1, \ldots, y_m$ is a path from u to v. If this path is reducible, the irreducible path from u to v is obtained by reducing this path. Hence $(u,v)d \leq (u,w)d + (w,v)d$, which shows that d is a metric.

Suppose that $(u,v)d < (u,w)d + (w,v)d$. Then the path $x_1, \ldots, y_m$ will be reducible. This can only happen if $y_1 = \bar{x}_n$, and the required vertex q will then be $y_1\tau$.

Now suppose that $(u,v)d = 1$ and that $(u,w)d \neq (v,w)d + 1$. By the previous paragraph there will be a vertex q such that $(q,w)d < (v,w)d$ and $(u.q)d < (u,v)d$. Since $(u,v)d = 1$, this requires that $q = u$, and so $(u,w)d < (v,w)d$, as needed.

For any $r < (u,v)d$ there is, plainly, a unique vertex p on the irreducible path from u to v such that $(u,p)d = r$ and $(u,v)d = (u,p)d + (p,v)d$. Also the arguments of the first paragraph show that if w is a vertex such that $(u,v)d = (u,w)d + (w,v)d$ then w lies on the irreducible path from u to v.//

Theorem 48 *Let d be an integer-valued metric on a set V. Suppose that, for all $u \neq v$ there is a unique p such that $(u,p)d = 1$ and $(p,v)d = (u,v)d - 1$. Suppose also either that if $(u,v)d = 1$ then, for all w, $(u,w)d \neq (v,w)d$ or that for all u,v, and w such that $(u,v)d < (u,w)d + (w,v)d$ there is some q such that $(u,q)d < (u,w)d$ and $(q,v)d < (w,v)d$. Then V is the set of vertices of a tree in which there is an edge-pair joining u and v iff $(u,v)d = 1$, and, for all u and v, $(u,v)d$ is the length of the irreducible path from u to v. Also, if V is a G-set and $(gu,gv) = (u,v)d$ for all $g \in G$ and all $u,v \in V$ then this tree is a G-tree.*

Proof As in the proof of the previous proposition, the last condition on d implies the next-to-last condtion. Since d is an integer-valued metric, this in turn implies that if $(u,v)d = 1$ then $(u,w)d - (v,w)d = \pm 1$.

Take any vertex v_0, and define $v\lambda$ to be $(v_0,v)d$. Also define u and v to be adjacent iff $(u,v)d = 1$. Then the conditions of Theorem 46 are satisfied. Thus the result follows, except for identifying d with the length of the relevant irreducible path. However, Theorem 46 tells us that the length of the irreducible path from v_0 to v is $v\lambda$, which equals $(v_0,v)d$. Since v_0 is an arbitrary vertex, the proof is complete.//

We now look at edge-sets of trees.

Proposition 49 *Let T be a tree. For x and y in E write $x \le y$ iff the irreducible path from $x\sigma$ to $y\tau$ begins with x and ends with y. Then $\le$ is a partial order (which is preserved by G when T is a G-tree) satisfying the following conditions: (i) if $x \le y$ then there are only finitely many edges z such that $x \le z \le y$, (ii) for any x and y, at least one of $x \le y$, $x \le \bar{y}$, $\bar{x} \le y$, $\bar{x} \le \bar{y}$ holds, (iii) if $x \le y$ then $\bar{y} \le \bar{x}$, (iv) we never have $x \le y$ and $x \le \bar{y}$, (v) we never have $x \le y$ and $\bar{x} \le y$. Further, for $y \neq x \neq \bar{y}$, $x\tau = y\tau$ iff $x \le \bar{y}$ and there is no edge z such that $x \le z \le \bar{y}$ and $x \neq z \neq \bar{y}$.*

Proof Suppose that $x \leq y$ and $y \leq z$. Then there is an irreducible path beginning with x and ending with y, and an irreducible path beginning with y and ending with z. These paths fit together to give an irreducible path beginning with x and ending with z, so that $x \leq z$. Further, if $x \leq y$ and $y \leq x$ we must have $x = y$, since otherwise the paths fit together to give an irreducible path of length >1 beginning and ending with x, which is impossible as T is a tree. Thus $\leq$ is a partial order.

Suppose that $x \leq y$. Then the inverse of the irreducible path beginning with x and ending with y is an irreducible path beginning with $\bar{y}$ and ending with $\bar{x}$, so that $\bar{y} \leq \bar{x}$. Also, the unique irreducible path from $x\sigma$ to $\bar{y}\tau = y\sigma$ is obtained from the original path by deleting the edge y. In particular, its last edge cannot be $\bar{y}$, so that we do not have $x \leq \bar{y}$. Similarly we cannot have $\bar{x} \leq y$. (Alternatively, (v) is an immediate consequence of (iv) and (iii).)

If the irreducible path from $x\sigma$ to $y\tau$ does not end in y we ccan add the edge $\bar{y}$ at the end and still get an irreducible path. Similarly, if the path does not begin with x we can add the edge $\bar{x}$ at the beginning and still get an irreducible path. This proves (ii).

Suppose that $x \leq z \leq y$. Then, as already seen, there is an irreducible path beginning with x, ending with y, and containing z. Since there is only one irreducible path starting with x and ending with y, z must be one of the edges on this path, proving (i).

If $x\tau = y\tau$ and $x \neq y$ then $x, \bar{y}$ is an irreducible path, and so $x \leq \bar{y}$. Further, by the previous paragraph, there can be no edge z such that $x \leq z \leq \bar{y}$ and $x \neq z \neq \bar{y}$. Conversely, if $x \leq \bar{y}$ and there is no edge z such that $x \leq z \leq \bar{y}$ and $x \neq z \neq \bar{y}$ then either $x = \bar{y}$ or $x, \bar{y}$ is an irreducible path, so that $x\tau = \bar{y}\sigma = y\tau$.//

Theorem 50 *Let E be a set with an involution without fixed points (that is, with a map $^{-}$ such that $e \neq \bar{e}$ and $\bar{\bar{e}} \neq e$). Let $\leq$ be a partial order on E satisfying (i)-(iv) of Proposition 49. Then there is a tree T such that $E(T) = E$ with the given partial order on E being the same as the partial order obtained in Proposition 49. Further, when E is a G-set and $\leq$ and $^{-}$ are preserved by G then T is a G-tree.*

Proof Write $x < y$ if $x \leq y$ and $x \neq y$. Write $x \triangleleft y$ if $x < y$ but there is no z such that $x < z < y$. It is clear from (iii) that if $x \triangleleft y$ then $\bar{y} \triangleleft \bar{x}$.

Write $x \equiv y$ iff $x = y$ or $x \triangleleft \bar{y}$. We will show that $\equiv$ is an equivalence relation. When this has been shown, we obtain a graph T with $E(T) = E$ such that $V(T)$ is the set of equivalence classes, with $x\tau$ being the equivalence class of x (and $x\sigma$ being defined as $\bar{x}\tau$).

Plainly $\equiv$ is reflexive, and it is symmetric. Suppose that $x \equiv y$ and $y \equiv z$. We may assume that $x \neq y \neq z$ (since otherwise it is clear that $x \equiv z$), so that $x \lhd \bar{y}$ and $y \lhd \bar{z}$. Then we have $z \lhd \bar{y}$. Now $z \leq x$ requires $z = x$, since otherwise we cannot have $z \lhd \bar{y}$ and $x \lhd \bar{y}$. Also $\bar{z} \leq \bar{x}$ gives $x \leq z$, which again leads to $z = x$. We cannot have $\bar{z} \leq x$, since this would lead to $\bar{z} \leq \bar{y}$ (because $x \leq \bar{y}$) as well as the known $z \leq \bar{y}$, which contradicts (v) (as already remarked, (v) follows from (iii) and (iv)).

By (ii), the remaining possibility is that $z \leq \bar{x}$, and so $x \leq \bar{z}$. We want to show that $x \lhd \bar{z}$, so let w be such that $x \leq w \leq \bar{z}$. Now $w \leq y$ gives $x \leq y$, which contradicts (iv) since we are given $x \leq \bar{y}$, while $\bar{w} \leq y$ gives $\bar{w} \leq \bar{z}$ (as $y \leq \bar{z}$), which contradicts (v), since $w \leq \bar{z}$. If $\bar{w} \leq \bar{y}$ then $y \leq w$, and $y = w$ is impossible, as before. Thus $y < w$, which, with $y \lhd \bar{z}$ and $w \leq \bar{z}$, gives $w = \bar{z}$. Similarly, if $w \leq \bar{y}$ then we cannot have $w = \bar{y}$ (since then $\bar{y} \leq \bar{z}$, contradicting (v), as we are given $y \leq \bar{z}$), while $w < \bar{y}$ together with $x \leq w$ and $x \lhd \bar{y}$ leads to $w = x$. By (ii) these are the only possibilities for w, and so $x \lhd \bar{z}$, as required.

Thus we have constructed a graph T, which will be a G-graph if E is a G-set with $\leq$ and $^{-}$ being preserved by G.

Let $x \neq \bar{y}$. Then $x \lhd y$ iff $x \equiv \bar{y}$. By definition of τ this holds iff $x\tau = \bar{y}\tau$, which, by definition of σ, holds iff $x\tau = y\sigma$, which holds iff x,y is a path.

Suppose that $x \leq y$. Then either $x = y$ or $x \lhd y$ or there is an edge z such that $x < z < y$. By (i) there are only finitely many edges between x and y. By induction on the number of these edges we see that if $x < y$ then there are edges $z_1, \ldots, z_n$ such that $z_1 = x$, $z_n = y$, and $z_i \lhd z_{i+1}$ for $i < n$. Thus $z_1, \ldots, z_n$ is a path in T, which is irreducible since there is (by (iv)) no edge w such that $w \lhd \bar{w}$. Conversely, if there is an irreducible path whose first edge is x and whose last edge is y, then there are edges $z_1, \ldots, z_n$ such that $z_1 = x$, $z_n = y$, and $z_i \lhd z_{i+1}$ for $i < n$, and so $x \leq y$.

Now every vertex is the end of some edge. By (ii) and the preceding paragraph, it follows that T is connected.

As already remarked, we cannot have $w \lhd \bar{w}$, and, by definition, $w \neq \bar{w}$. Thus w is never equivalent to $\bar{w}$. Hence $w\tau \neq \bar{w}\tau$, and so $w\tau \neq w\sigma$, and there are no circles. Also there can be no irreducible loops $x_1, \ldots, x_n$ for $n > 1$, as then we would have $x_1 \lhd x_2 \lhd \ldots \lhd x_n \lhd x_1$, which is absurd. Hence T is a tree.//

We now look at another important way of constructing trees, and some of its consequences.

Let E be any set, and let A be a non-empty set. The set of all functions from E to A will be denoted by $[E,A]$. When E is a G-set we make $[E,A]$ into a G-set by defining $g\varphi$ by $e(g\varphi) = (g^{-1}e)\varphi$.

Let φ and ψ be in $[E,A]$. We say that φ is *almost equal to* ψ if $\{e;\ e\varphi \neq e\psi\}$ is finite. It is easy to see that almost equality is an equivalence relation; the equivalence classes are called *almost equality classes.* When E is a G-set, G also acts on the set of almost equality classes. A class will be called *G-stable* if it is fixed by G.

The following theorem, due to Dicks and Dunwoody (1989), is long and difficult to prove, and I will not give the proof here. However, it is fairly easy to use, and we will obtain an important application.

Theorem 51 (Dicks-Dunwoody Almost Stability Theorem) *Let E be a G-set such that each element of E has finite stabiliser, and let A be a non-empty set. Let V be a G-stable almost equality class of $[E,A]$. Then there is a G-tree with vertex set V whose edges have finite stabilisers.//*

We can identify $[E,\mathbb{Z}_2]$ with the set of subsets of E, identifying φ with $\{e;\ e\varphi = 1\}$. Thus we may refer to two subsets of E as being almost equal. For φ and ψ in $[E,\mathbb{Z}_2]$ we may define $\varphi+\psi$ by $e(\varphi+\psi) = e\varphi + e\psi$. If P and Q are the subsets of E corresponding to φ and ψ then the subset corresponding to $\varphi+\psi$ is the symmetric difference of P and Q (that is, the set of elements belonging to exactly one of P and Q).

Let H be a subgroup of the group G, and let E be an H-set. On $G\times E$ there is an equivalence relation given by $(g_1,e_1) \equiv (g_2,e_2)$ iff $\exists h \in H$ such that $e_2 = he_1$ and $g_2 = g_1h^{-1}$. The class of (g,e) is denoted by $g\times e$, and the set of equivalence classes by $G\times_H E$; it has a G-action defined by $g(g'\times e) = (gg')\times e$. We can regard E as a subset of $G\times_H E$ by regarding e as $1\times e$.

Lemma 52 *Let H be a subgroup of G with $|G{:}H| < \infty$, and let E be an H-set. To each H-stable almost equality class V of $[E,\mathbb{Z}_2]$ there is a G-stable almost equality class W of $[G\times_H E,\mathbb{Z}_2]$ whose restriction to E is V.*

Proof Let $g_1 = 1,\dots,g_n$ be a transversal of H in G. Then $G\times_H E$ can be regarded as the disjoint union of the sets g_iE for $i = 1,\dots,n$. Let v be an element of the class V; we regard v as a subset of E. Let w be $g_1v\cup\dots\cup g_nv$. Then w is a subset of $G\times_H E$, which we regard as an element of $[G\times_H E,\mathbb{Z}_2]$, and let W be the almost equality class of w. By construction, the restriction of W to E is V. We need to show that W is G-stable; that is, we need to show that gw is almost equal to w for all $g\in G$.

Now there exists a permutation $j_1,\dots,j_n$ of $1,\dots,n$, and elements $h_1,\dots,h_n$ of H such that $gg_i = g_{j_i}h_i$. Thus $gw = \cup gg_iv = \cup g_{j_i}h_iv$. Since h_iv is almost equal to v, this union almost equals $\cup g_{j_i}v = w$, as required.//

Let T be a tree, v a vertex of T, and e an edge of T. We say that *e points towards v* if the irreducible path from $e\sigma$ to v begins with e. To each v we take the subset of those edges pointing towards v. Thus we may regard v as an element of $[E,\mathbb{Z}_2]$. Let w be another vertex of T. Unless e or $\bar{e}$ is in the irreducible path from v to w, e either points to both of v and w or to neither of them. Hence v and w are almost equal elements of $[E,\mathbb{Z}_2]$. Thus $V(T)$ lies in one almost equality class of $[E,\mathbb{Z}_2]$, which we denote by $\mathfrak{B}$. When T is a G-tree it is clear that $\mathfrak{B}$ is a G-stable almost equality class.

There are elements of $\mathfrak{B}$ which do not lie in $V(T)$. We proceed to construct a G-map from $\mathfrak{B}$ to $V(T)$ which is the identity on $V(T)$.

Fix some vertex v of T. Take any $u \in \mathfrak{B}$. Then, since u and v are almost equal, the element $u+v$ of $[E,\mathbb{Z}_2]$, regarded as a subset of E, is finite. Let it be $\{e_1,\ldots,e_n\}$.Take $g \in \text{stab}u$. Since $gu = u$, we have $u+gu = 0$, and so $v+gv = (u+v)+(gu+gv)$ contains only edges in $\{e_1,\ldots,e_n, ge_1,\ldots,ge_n\}$. We have already remarked that an edge e occurs in $v+gv$ iff either e or $\bar{e}$ is on the irreducible path from v to gv. It follows that the distance from v to gv is at most $2n$. Consequently the orbit $(\text{stab}u)v$ is bounded. It follows from Proposition 30 that stab$u$ fixes some vertex of T. Now choose one element u from each G-orbit of $\mathfrak{B}$, and a corresponding element of $V(T)$ fixed by stab$u$, taking the element to be u itself when $u \in V$. We may now extend this function to a G-map.

Lemma 53 *Let H be a subgroup of G with $|G:H| < \infty$, and let T be an H-tree. Then there is a G-stable almost equality class W of $[G\times_H E,\mathbb{Z}_2]$ such that there is an H-map from W to $V(T)$.*

Proof Let $\mathfrak{B}$ be the almost equality class of $[E,\mathbb{Z}_2]$ containing $V(T)$. By lemma 52 we have a G-stable almost equality class W of $[G\times_H E,\mathbb{Z}_2]$ whose restriction to E is $\mathfrak{B}$. In particular, there is an H-map from W to $\mathfrak{B}$.. By the previous remarks there is also an H-map from $\mathfrak{B}$ to $V(T)$, and so we get an H-map from W to $V(T)$.//

We can now prove some interesting theorems.

Theorem 54 *Let H be a subgroup of G with $|G:H| < \infty$. Suppose that there is an H-tree T whose edges have finite stabilisers. Then there is a G-tree $\tilde{T}$ whose edges have finite stabilisers and such that there is an H-map from $V(\tilde{T})$ to $V(T)$.*

Proof All elements of the G-set $G\times_H E(T)$ have finite stabilisers, since $g\times e = 1\times e$ iff g is in H and stabilises e. By the previous lemma, there is a G-stable almost equality class W of $[G\times_H E(T),\mathbb{Z}_2]$ with an H-map from W to $V(T)$. By the Almost Stability Theorem, there is a G–tree $\tilde{T}$ whose edges have finite stabilisers and whose vertex set is W.//

Corollary *Let H be a subgroup of G with $|G{:}H|<\infty$. If there is an H-tree all of whose vertices have finite stabilisers then there is a G-tree all of whose vertices have finite stabilisers. Also, if the vertex stabilisers of the H-tree all have order $<k$ then the vertex stabilisers of the G-tree all have order $<k|G{:}H|$.*

Proof Let T be the given H-tree. Plainly every edge of T has finite stabiliser. Let $\tilde{T}$ be the tree obtained in the theorem. Let v be a vertex of $\tilde{T}$. Then $|\mathrm{stab}v : H\cap \mathrm{stab}v| < |G{:}H|$. Since we have an H-map from $V(\tilde{T})$ to $V(T)$, $H\cap\mathrm{stab}v$ lies in the stabiliser of a vertex of T, and the result follows.//

Theorem 55 *The following are equivalent : (i) G acts on a tree with the vertex stabilisers being finite and all of order $<k$ for some k, (ii) G is the fundamental group of a graph of groups whose vertex groups are finite and all of order $<k$ for some k, (iii) G has a free subgroup of finite index.*

Corollary *A torsion-free group with a free subgroup of finite index is free.//*

Proof The equivalence of (i) and (ii) is immediate from the Structure Theorems.

Suppose that G has a free subgroup H of finite index. Then H acts on a tree with all vertex stabilisers being trivial. Thus (i) holds by the corollary to the previous theorem.

The next proposition shows that (iii) follows from (ii).//

We now look at free actions of a group G on a set; that is, actions for which all elements of the set have trivial stabilisers.

Let S be any set. Then G acts on $G\times S$ by $g(g_1,s)=(gg_1,s)$, and this is plainly a free action. Hence any set X bijective with $G\times S$ can be given a free G-action.

Conversely, let G act freely on X. Let S be a transversal for the action; that is, S is a subset of X with exactly one element in each G-orbit. Then the function from $G\times S$ to X which sends (g,s) to gs is easily seen to be a G-map which is a bijection.

It follows that G can act freely on X iff either X and G are finite and $|G|$ divides $|X|$ or (using known properties of infinite sets) if X is infinite and $|G| \le |X|$.

Suppose that we also have another free G-action on X, which we denote by $g{\circ}x$. Let S_1 be a transversal for this action. Then the map sending (g,s_1) to $g{\circ}s_1$ is also a G-map which is a bijection. If $|S| = |S_1|$ we have a bijection from S to S_1 which will extend to a G-map from $G\times S$ to $G\times S_1$ which is a bijection. It follows that there will be a bijection $\alpha: X \to X$ such that $(gx)\alpha = g{\circ}(x\alpha)$ for all x and g. We may regard the G-actions as homomorphisms φ and φ_1 from G to Sym$X$, the group of permutations of X. We then have $\alpha \in \mathrm{Sym}X$, and $g\varphi_1 = \alpha^{-1}(g\varphi)\alpha$ for all g.

When X is infinite and $|G| = |X|$ we may not have $|S| = |S_1|$. However, we do have $|S| = |S_1|$ when X is finite and $|G|$ divides $|X|$ and when X is infinite and $|G| < |X|$ (using properties of infinite sets).

We now apply these facts to the study of graphs of groups.

Proposition 56 *Let $\mathfrak{G}$ be a graph of groups. Let X be a set such that either X is infinite and $|X| > |G_v|$ for all vertices v or else X is finite and $|G_v|$ divides $|X|$ for all vertices v. Then there is a homomorphism from $\pi(\mathfrak{G})$ to SymX such that the maps $G_v \to \pi(\mathfrak{G}) \to$ SymX are one-one for all vertices v. The kernel of the homomorphism from $\pi(\mathfrak{G})$ to SymX will be free.*

Remarks For any $\mathfrak{G}$ there exists a set X such that $|X| > |G_v|$ for all v (by properties of infinite sets). Thus the proposition provides another proof of the fact that the maps from G_v to $\pi(\mathfrak{G})$ are one-one.

There is plainly a finite set X such that $|G_v|$ divides $|X|$ for all v iff the groups G_v are finite and there is a bound on their order. This completes the proof of Theorem 55.

Proof The conditions ensure that X has a free G_v-action for all v. Thus we have a one-one homomorphism from G_v to Sym$X$ for all v.

Let e be an edge. Then we have two free actions of G_e on X, the first coming from embedding G_e in $G_{e\sigma}$, and the second coming from embedding G_e in $G_{e\tau}$. Since $|G_e|$ divides $|G_v|$ when G_e is finite, we see that, for any edge e, we have $|G_e| < |X|$ when X is infinite and $|G_e|$ divides $|X|$ when X is finite.

By the discussion above, it follows that the two maps from G_e to Sym$X$ are conjugate. Lemma 20 now shows the existence of a homomorphism from $\pi(\mathfrak{G})$ to Sym X such that $G_v \to \pi(\mathfrak{G}) \to \mathrm{Sym}X$ is one-one for all v. Further, Corollary 2 to Theorem 27 tells us that the kernel of the map from $\pi(\mathfrak{G})$ to Sym$X$ is free.//

For those readers who know about cohomology of groups, I mention that the structure of groups of cohomological dimension one can be obtained from the Almost Stability Theorem by similar techniques; *a group has cohomological dimension one over a ring R iff it is the fundamental group of a graph of groups in which all vertex groups are finite and the order of every vertex group is invertible in R.*

We say that the group G *has more than one end* if there is a subset S of G, infinite and with infinite complement, such that gS is almost equal to S for all $g \in G$. (This notion has a geometric significance. It is possible to define the number of ends of G in a way which essentially measures "the number of ways of going to infinity in G"; this number is always 0 (which holds iff G is finite), 1, 2, or ∞.) Another similar proof yields the following result. *The following are equivalent:* (*i*) *G has more than one end*, (*ii*) *there is a G-tree whose edges have finite stabilisers and such that G stabilises no vertex*, (*iii*) *G is either an amalgamated free product* $A *_C B$ *with C finite and* $A \neq C \neq B$ *or an HNN extenstion* $\langle A, t;\ t^{-1}Ct = D\rangle$ *with C finite or a locally finite countably infinite group.*

9 DECISION PROBLEMS

9.1 *DECISION PROBLEMS IN GENERAL*

Can we tell whether or not an element of a free group is in the commutator subgroup? Whether or not it is a commutator? Given a finite presentation, can we tell whether or not an element of the free group is 1 in the group presented? Can we tell whether or not two finite presentations present isomorphic groups?

These questions, and other similar ones, are obviously of interest. We shall see later that the first two questions have the answer "Yes", while the other two have the answer "No". The techniques used are those of combinatorial group theory, but have no specific connection with the topological approach. I include this chapter because I find the material particularly interesting.

In order to make these questions precise, we need to be clearer about what is meant by "We can tell ...". This should mean that we can tell by some purely mechanical process, not requiring thought. In other words, we want to be able to feed the data to a computer and have the computer arrive at the answer as to whether or not the required property holds for the given data.

Readers might think that a computer's ability to arrive at the answer for given data would depend very much on the computer. This turns out not to be so, though the speed and efficiency of the computer's answer will depend on the computer.

More precisely, there is a class of functions (called *partial recursive functions*) with the following properties. In many different models of computation the computable functions turn out to be exactly the partial recursive functions, while for other models of computation the computable functions are a subset of the partial recursive functions. No-one has suggested a model of computation in which there are computable functions which are not partial recursive. The techniques used in proving that in certain models the computable functions are partial recursive are sufficiently general that it appears likely that any model suggested in future can be treated by similar methods so as to show that the functions computable in that model are partial recursive. Finally, the partial recursive functions do appear to be intuitively computable. (Here I say "do appear to be" rather than "are" because I cannot claim to know what someone else's intuition is; however, anyone who accepts that certain fairly simple processes applied to intuitively computable functions yield intuitively computable functions will find that the partial recursive functions are intuitively computable.)

This identification of the intuitively computable functions with the partial recursive functions is known as *Church's Thesis*. For the most part, we will work with intuitively computable functions, and I hope that the reader will find various claims made to be intuitively acceptable. It would be possible to take a formal approach, but often the technicalities would obscure the ideas. At a few crucial points we will have to look at some of the precise definitions. For the general background and the detailed formalism, the reader should look at books on the theory of computation, for instance Cohen (1987).

Certain specialised classes of computers, such as the finite-state automata and the push-down automata, do not have the full generality we need. They compensate for this, however, by being simpler and faster. A new theory of automata and groups has been developed in the past few years, and is discussed in a book-length paper in preparation by Cannon, Epstein, and others; a little more will be said about this in Chapter 10.

We think of our computers as operating in discrete steps. For convenience, to begin with we think of them as working with elements of $\mathbb{N}$, the natural numbers, though in the later development it may be convenient to think of them as working with elements of some other countable set, for instance a free monoid $M(X)$, where X is finite or countable. On some inputs the computer will stop after a finite number of steps, giving us an output. On other inputs the computer may continue computing for ever. We do not permit our computers to stop without giving an output. (Real computers may stop giving an error message, but we could think of this as the computer continuing for ever saying to itself "Something's wrong".) Thus our computers enable us to compute functions from $\mathbb{N}$ to $\mathbb{N}$ or, more generally, from $\mathbb{N}^k$ to $\mathbb{N}$ for some k. However it should be noted that these functions are *partial functions*. That is, they need not be functions from $\mathbb{N}$ to $\mathbb{N}$ but rather from a subset of $\mathbb{N}$ to $\mathbb{N}$, since on some inputs the computer runs for ever and gives no output. It might be thought that this situation could be avoided by a careful choice of our computer programs, but it turns out that it is an unavoidable feature of any theory of computation. When we are looking at partial functions we will generally use the word 'partial';

a function in the ordinary sense will sometimes be called a *total function* for emphasis (in Cohen (1987) and other books on computability, though, the word 'function' usually means 'partial function').

We shall need to refer to *algorithms*. An algorithm is a description, in language which may be at various levels of formality, of a precise mathematical procedure to be followed. The euclidean algorithm for finding the greatest common divisor of two integers is a typical example. A computer program is an example of an algorithm; it is written in a programming language suitable for the computer being considered. As before, our algorithms operate one step at a time, although this step may consist of performing a number of simpler processes. For instance, we might have an algorithm one step of which involved finding the greatest common divisor of two integers or finding the maximum common part of two words. As before, on a given input, an algorithm may either halt after a finite number of steps, giving an output, or it may continue for ever, in which case it gives no output. We sometimes use the phrase '*partial algorithm*' to empasise that for certain inputs the process may not halt.

Definition Let A be a subset of $\mathbb{N}$. Then A is called *recursive* if there is an algorithm which on input n has output "Yes" if $n \in A$ and has output "No" if $n \notin A$. Recursive sets are sometimes called *decidabe*.

Definition Let A be a subset of $\mathbb{N}$. Then A is called *recursively enumerable* (usually abbreviated to *r.e.*) if there is a partial algorithm which on input n has output "Yes" iff $n \in A$. For $n \notin A$ the partial algorithm may or may not halt.

When A is r.e. we can always modify the partial algorithm so that if $n \notin A$ then it does not halt. All that is needed is to add further instructions such as "If the algorithm halts and its output is not "Yes" then go through all the members of $\mathbb{N}$ in order starting at 0."

Note that any recursive set is r.e., since the algorithm for deciding membership of A is a suitable partial algorithm. Also, if A is recursive so is $\mathbb{N} - A$, simply interchanging "Yes" and "No".

Any finite set is recursive. If the set is $\{n_1, \ldots, n_k\}$ then a suitable algorithm is "Check whether $n = n_1$. If so answer "Yes". If not check whether $n = n_2$. ... Check whether $n = n_k$. If so answer "Yes". If none of these checks have given answer "Yes" then answer "No"." It should be reasonably clear that such sets as the set of all even numbers, the set of all squares, the set of all powers of 3, and the set of all primes, are recursive.

It is not immediately obvious that there are sets which are r.e. but not recursive. The proof of this fact, which we shall need, may be found in Chapter 2 of Cohen (1987) and elsewhere. It is not difficult, but the details would take us too far afield.

Suppose that we are given a set A, and we have a computable total function $f:\mathbb{N}\to\mathbb{N}$ such that $A=\mathbb{N}f$. We can regard f as enabling us to find a list of the elements of A, namely $0f$, $1f$, $2f$, ... Consequently sets of this kind are sometimes called *listable*; we also regard the empty set as listable. We shall see that listable sets are the same as r.e. sets.

One might at first sight believe that a list of the elements of A would enable us to decide membership of A. This is not so, because we did not require the list to give the elements of A in any sensible order. If the integer 1 has not appeared in the first 193674 terms of the list, for instance, this may be because 1 is not on the list at all, but it could be that its first appearance on the list is at the 193675 th term.

There is an important case where a list of members of A enables us to decide membership of A. This occurs when we can tell for any n how far along the list we need to check before n will occur if it is ever going to occur. More formally, we have $A=\mathbb{N}f$, where f is computable and total, and we also have a computable total function g such that if $n\in A$ then $n=mf$ for some $m\leq ng$. Then, given n, we have the answer "No" if n does not occur among $0f$, $1f$, ..., $(ng)f$, and "Yes" if it does appear among these elements.

In particular, if the list is given in strictly increasing order then A is recursive. For we then have $mf>n$ for all $m>n$, so we may take ng to be n.

If the list is given in increasing order which is not strictly increasing, it is still true that A is recursive. Here, though, we run into a subtle point which it is easy to get confused about; similar points will also occur later. When we claim that the set A is recursive we are simply saying that we can show that there must be some algorithm that will work. We do not need to exhibit the algorithm explicitly. A similar point in a less confusing form often occurs in analytical and topological situations; for instance, Brouwer's Fixed Point Theorem (Theorem 4.7) tells us that any continuous map from a disk to itself must have a fixed point, but it does not tell us any way to find the fixed point.

So let $A=\mathbb{N}f$, where f is an increasing computable total function. Then A must be either finite or infinite, though we may not know which. If A is finite then it is recursive. Let A be infinite. Given n, compute $0f$, $1f$,... until (which must happen at some point) we find m with $mf\geq n$. Then $n\in A$ iff it occurs among $0f, \ldots, mf$. However this algorithm will not decide membership of A when A is finite, since it continues for ever when n is larger than all elements of A. So we know that some algorithm must work to enable us to decide membership of A, but we do not know what algorithm works.

Proposition 1 *Let* $A \subseteq \mathbb{N}$. *Then the following are equivalent:* (*i*) *A is r.e.,* (*ii*) *A is listable,* (*iii*) $A = \mathbb{N}f$, *where f is a computable partial function,* (*iv*) *The function which is* 0 *on A and undefined elsewhere is computable,* (*v*) *A is the domain of some computable partial function.*

Proof Suppose that (ii) holds. Since $\emptyset$ is r.e. (the partial algorithm which never answers "Yes" shows this), we may assume that $A = \mathbb{N}f$, where f is a computable total function. Then a partial algorithm for membership of A is "Given n, test $0f$, $1f$, and so on until, if ever, we find m with $n = mf$."

Suppose that (i) holds. Then a computable partial function f is given by the following (partial) algorithm. "Given n, apply the partial algorithm for membership of A to n. If this partial algorithm halts, then output n and halt." Plainly $\mathbb{N}f$ is A, which proves (iii). Replacing "output n and halt" by "output 0 and halt", we obtain a partial algorithm for the function in (iv), so this function is computable. This also shows that (v) holds.

Suppose that (v) holds. Take a partial algorithm for computing this function f. Then a partial algorithm for membership of A is given by "On input n, run the partial algorithm for computing f, and output "Yes" if this halts."

Suppose that (iii) holds. If $A = \emptyset$ then A is listable. So we may assume $A \neq \emptyset$. Hence there is some $a_0 \in A$. Here we again encounter the subtlety previously mentioned. We do not know a_0, but we are sure that such an a_0 exists. In fact, the problem really arises at an earlier stage. Given a partial algorithm for f, we cannot tell whether A is empty or non-empty, but we know it must be one or the other, and we use different arguments to show that A is listable in the two cases. It is quite easy (by a variant of the technique below) to obtain from a partial algorithm for f a partial algorithm which never halts if A is empty and outputs an element of A if A is not empty, and we could then use the element obtained as our choice of a_0.

From a partial algorithm for f we obtain, when $A \neq \emptyset$, by the procedure below a total algorithm for a function $G:\mathbb{N}^2 \to \mathbb{N}$, so that G is a computable total function.

On input (n,k) run the partial algorithm for f with input n for (at most) k steps. Output the output of the algorithm for f if it has stopped in this time; otherwise output a_0. Denote the output by $(n,k)G$.

Then $(n,k)G$ is either nf or a_0, and, if nf is defined then there is some k such that $(n,k)G = nf$. Thus $\mathbb{N}^2 G = A$.

Now we know that $\mathbb{N}^2$ is bijective with $\mathbb{N}$. It is easy to choose a specific bijection J for which the formula shows that J is computable. We can also find an algorithm for computing J^{-1}. Thus the function $J^{-1}G$ is computable and total, and its range is A, which shows that (ii) holds.//

Proposition 2 (*i*) *Let A and B be recursive subsets of $\mathbb{N}$, and let $f{:}\mathbb{N} \to \mathbb{N}$ be a computable total function. Then $A \cup B$, $A \cap B$, and Af^{-1} are recursive.*

(*ii*) *Let C and D be r.e. subsets of $\mathbb{N}$, and let $g{:}\mathbb{N} \to \mathbb{N}$ be a computable partial function. Then $C \cup D$, $C \cap D$, Cg^{-1}, and Cg are r.e.*

Remark Note that Af is r.e. by (ii), since recursive sets are r.e. However, Af need not be recursive, since Proposition 1 tells us that any non-empty r.e. set can be written as $\mathbb{N}f$ for some computable total f.

Proof (i) We have total algorithms for deciding membership of A and of B. On input n, apply first one algorithm and then the other. Then $n \in A \cup B$ iff at least one algorithm answers "Yes", while $n \in A \cap B$ iff both algorithms answer "Yes".

The algorithm for deciding membership of Af^{-1} is "Apply an algorithm which computes f to the input n. When this stops, take the output and apply the algorithm for A to this."

(ii) Take a partial algorithm for C which does not halt on input n if $n \notin C$, and similarly for D. The partial algorithms for $C \cap D$ and Cg^{-1} are obtained in the same way as the corresponding ones in (i), but they will now halt for some values of n but not for others.

We cannot use an algorithm similar to (i) for $C \cup D$. For this algorithm will not halt if $n \notin C$ and $n \in D$, whereas we want the answer "Yes" in this case. However, instead of running the two algorithms consecutively, we may run them simultaneously. More precisely, we take alternately a step of one algorithm and then a step of the other (in terms of computers, we could feed the same input n to both computers). As soon as one algorithm halts on input n, we output the answer "Yes". This is the required partial algorithm for $C \cup D$.

Alternatively, we could (unless C or D is empty, when the result is obvious) find computable total functions φ and ψ such that $C = \mathbb{N}\varphi$ and $D = \mathbb{N}\psi$. Define ϑ by $(2n)\vartheta = n\varphi$ and $(2n+1)\vartheta = n\psi$. It is easy to see that ϑ is computable, total, and with range $C \cup D$.

Finally, $Cg = \mathbb{N}\varphi g$. Plainly, φg is a computable partial function, so (ii) of Proposition 1 shows that Cg is r.e.

We could also use the following partial algorithm for Cg. "On input n, search sytematically through all pairs (m,k) until, if ever, we find a pair such that the partial algorithm for computing g, when given input m produces output n in at most k steps. Then output "Yes"."

To complete this argument, we have to make precise the notion of 'searching sytematically'. This can be done by successively testing $0J^{-1}$, $1J^{-1}$, and so on.//

Notice that this notion of a systematic search of the elements of $\mathbb{N}^2$ can be extended to $\mathbb{N}^k$, using the bijection J to obtain bijections from $\mathbb{N}^k$ to $\mathbb{N}$.

While a list of the elements of A does not make A recursive, two lists, one of the elements of A and the other of the elements of $\mathbb{N} - A$, combine to show that A is recursive. Essentially, given n we check the two lists alternately until we see which list n occurs in. This is made precise by the proposition below.

Proposition 3 *A is recursive iff both A and $\mathbb{N} - A$ are r.e.*

Proof If A is recursive then so is $\mathbb{N} - A$. Hence both A and $\mathbb{N} - A$ are r.e., since all recursive sets are r.e.

Now suppose that A and $\mathbb{N} - A$ are r.e. Take partial algorithms for the two sets. On any input one of these two must halt with answer "Yes". Hence the following is an algorithm for deciding membership of A. "Run both algorithms simultaneously on the input n until one of them stops with answer "Yes". If it is the partial algorithm for A which has given answer "Yes" then output "Yes"; otherwise output "No"."

Alternatively, take total computable φ and ψ such that $A = \mathbb{N}\varphi$ and $\mathbb{N} - A = \mathbb{N}\psi$. Define a computable total function ϑ by $(2m)\vartheta = m\varphi$ and $(2m+1)\vartheta = m\psi$. Plainly $\mathbb{N}\vartheta = \mathbb{N}$, and $k\vartheta \in A$ iff k is even. Thus we have the following algorithm for deciding membership of A. "On input n, compute 0ϑ, 1ϑ, ... until (which must happen sooner or later) we find k with $n = k\vartheta$. Output "Yes" if k is even, and "No" if k is odd."//

I have already mentioned that the next result is true, but will not be proved here. It is important enough to be stated as a proposition.

Proposition 4 *There is a set which is r.e. but not recursive.//*

For applications to group theory we need to extend these concepts to countable sets other than $\mathbb{N}$. This is easy to do. We simply take a bijection φ from S to $\mathbb{N}$, and then we call a subset T of S recursive (or r.e.) if $T\varphi$ is recursive (or r.e.). All the propositions above lead at once to corresponding results for this situation.

Thus the notion of recursive (or r.e.) subsets of S depends on the bijection φ chosen. In practice this does not matter. There will usually be a natural choice of one or more bijections, all of which give the same recursive and r.e. subsets.

It is possible to be more general, which is often useful. A *Gödel numbering* of a countable set S is a one-one function φ from S onto a recursive subset of $\mathbb{N}$. It turns out (see Chapter 8 of Cohen (1987)) that we may satisfactorily define a subset T of S to be recursive (or r.e.) if its image $T\varphi$ is recursive (or r.e.), where φ is a Gödel numbering which need not map onto $\mathbb{N}$.

Similarly, let S_1 be another countable set with a Gödel numbering φ_1. Given a partial function $f:S \to S_1$, we may define f to be computable if the partial function $\varphi^{-1}f\varphi_1$ from $\mathbb{N}$ to $\mathbb{N}$ is computable.

Let X be a countable set. Choosing a bijection from X to $\mathbb{N}$, we may denote the elements of X by x_n for all $n \in \mathbb{N}$. This leads to the following Gödel numbering φ of the free monoid $M(X)$.

Let $w \in M(X)$ be $x_{i_1} \dots x_{i_k}$. We define $w\varphi$ to be $2^k \Pi_{r=1}^{r=k} p_r^{\,i_r}$, where p_n are the prime numbers in increasing order (so $p_0 = 2$, $p_1 = 3$, $p_2 = 5$, and so on). It is easy to check that φ is one-one and has recursive image.

When X is finite we may choose a one-one map from X into $\mathbb{N}$, which enables us to denote the elements of X by x_n for some finite set of values of n. Then the previous formula for φ still gives a Gödel numbering for $M(X)$. Also when X is countable we may, if we wish, use the same formula but starting from a Gödel numbering of X rather than from a bijection of X with $\mathbb{N}$.

When X is finite all choices of one-one maps from X into $\mathbb{N}$ turn out to leave the recursive and r.e. subsets of $M(X)$ unchanged. This will not be true in general when X is infinite, but they will be left unchanged if the map is changed in a sufficiently straightforward way. Also, when X is finite we may obtain a Gödel numbering of $M(X)$ by taking any bijection of X with $\{1,\dots,n\}$, with x_i corresponding to i, say, and mapping $x_{i_1}\ldots x_{i_k}$ to $\Sigma i_r n^{r-1}$, or, if preferred, to $\Sigma i_r n^{k-r}$; all these give the same recursive and r.e. subsets of $M(X)$ as the previous Gödel numberings.

Any homomorphism from $M(X)$ to $M(Y)$ is determined by its values on X. For finite X we find that any such homomorphism is computable. For infinite X the homomorphism will be computable provided that the map from X to $M(Y)$ is conputable. Both of these results are fairly easy to see, though their precise proof requires some messy technical details.

When considering the free group $F(X)$ we need to look at the free monoid $M(X \cup X^{-1})$, where X^{-1} is a set bijective with X and disjoint from it. Thus we need a one-one map α from $X \cup X^{-1}$ into $\mathbb{N}$ with recursive image. Given a one-one map from X to $\mathbb{N}$ with recursive image, and denoting the element (if any) which maps to n by x_n, we may define α by $x_n\alpha = 2n$ and $x_n^{-1}\alpha = 2n+1$. The details of these maps and the corresponding Gödel numberings of $M(X)$ and $M(X \cup X^{-1})$ are not important, but we need to know that such maps exist and have the expected properties.

When we have a set which consists of all elements with some property, it is often convenient to say "We can decide whether or not an element has the property" when we mean that the set is recursive. Note that the words "whether or not" are used; if we simply said "We can decide whether an element has the property" we might only mean that the set is r.e.

Exercise 1 Find a bijection fron $\mathbb{N}^2$ to $\mathbb{N}$ which is computable and for which (the coordinate functions of) its inverse are also computable. (One possibility is to arrange the elements of $\mathbb{N}^2$ in an infinite square array and obtain a bijection by moving succesively along the finite diagonals. This gives the bijection J defined by $(m,n)J = (m+n)(m+n+1)/2+m$.)

Exercise 2 Suppose that we have an algorithm for a partial function $f:\mathbb{N} \to \mathbb{N}$. Show how to obtain an algorithm (on no inputs!) which runs for ever if f is not defined anywhere and which halts after a finite amount of time with output an element of $\mathbb{N}f$ if f is defined somewhere.

9.2 *SOME EASY DECISION PROBLEMS IN GROUPS*

When we look at free monoids or free groups, we shall always assume that they are countable; that is, that the basis X is finite or countable. This is because the notion of a (partial) algorithm requires us to be working in a countable set. We can sometimes work in uncountable groups by restricting attention to countable parts of them.

In the monoid $M(X \cup X^{-1})$ we can decide whether or not a word is reduced. We need only look through the word to see whether or not some x and the corresponding x^{-1} are next to each other or not. It should be clear that this is an algorithm.

The function λ sending each word to its reduced form is computable. This can be shown using the canonical reduction approach to Theorem 1.4. This method lets us compute (inductively on the length of u) the reduced words $u\lambda$ for all the initial segments u of w until we finally obtain $w\lambda$ itself. Alternatively, we may observe that leftmost reduction (as discussed in Exercise 1.2) of a word which is not reduced is a computable partial function, and $w\lambda$ can be obtained by repeatedly performing leftmost reduction, checking at eeach stage before we perform the reduction whether or not we have a reduced word, until we reach a reduced word.

We can then decide whether or not two words give the same element of $F(X)$, since we need only compute their reduced forms and see whether or not these are the same.

Let S be any set of reduced words, and let S' be the set of those words whose reduced form is in S. By Proposition 2 S is r.e. iff S' is r.e. and S' is recursive if S is recursive. Also S is recursive if S' is recursive, since S is the intersection of S' with the the recursive set of all reduced words.

It follows that, when we want to ask whether certain subsets of the free group $F(X)$ are recursive (or r.e.), it does not matter whether we consider all words giving members of this subset or if we only consider reduced words. We shall use whichever form is most convenient.

The cyclic reduction of a given (reduced) word is computable. For we need only reduce the word, and then successively remove pairs of elements x^{ε} from the beginning and $x^{-\varepsilon}$ from the end, until this can no longer be done.

It follows that we can decide whether or not two elements of $F(X)$ are conjugate. By Proposition 1.9, we need only take the cyclic reductions of the two words, and then check whether or not one can be obtained from the other by a cyclic permutation.

Proposition 5 *The commutator subgroup of a free group is a recursive subset.*

Proof Let w be the word $x_{i_1}^{\varepsilon_1} \ldots x_{i_n}^{\varepsilon_n}$. We define the *exponent-sum over* x_k of w, written $w\sigma_k$, to be $\Sigma\varepsilon_r$, the sum being taken over those r such that $i_r = k$. (Sometimes it is more convenient to denote x_k by a single letter; if y is x_k we then write σ_y rather than σ_k.) Plainly σ_k is a computable function.

It is now clear that w is in the commutator subgroup iff $w\sigma_k = 0$ for every k such that x_k occurs in w. We can plainly check whether or not this condition holds.//

Proposition 6 *In a free group the set of commutators is recursive.*

Remark The set of all words (not necessarily reduced) of form $uvu^{-1}v^{-1}$ is plainly recursive. The set of commutators is the image of this set under the map which sends each word to its reduced form. But this argument only shows that the set of commutators is r.e.

Proof It is enough to show that the reduced word w is a commutator iff $w = abca^{-1}b^{-1}c^{-1}$, where the right-hand side is reduced as written (we permit one of a,b, and c to be 1). For suppose we have shown this. Then w is not a commutator if its length is odd. If w has even length, there are only finitely many possibilities for words a, b, and c such that abc is reduced as written and is the first half of w. We can find all these possibilities, and for each possibility check whether or not w is $abca^{-1}b^{-1}c^{-1}$, reduced as written.

Now $abca^{-1}b^{-1}c^{-1} = (ab)(ca^{-1})(ab)^{-1}(ca^{-1})^{-1}$, and so it is a commutator.

Conversely, let w be a commutator. Then we can certainly write w as $abca^{-1}b^{-1}c^{-1}$, where the right-hand side is not necessarily reduced (for instance, by taking c to be 1). Among all ways of writing w in this form, choose one for which $|a| + |b| + |c|$ is as small as possible. We shall show this expression must be reduced as written.

Suppose not. Then (other cases are possible, but they are dealt with similarly) we may write a as ux^{ε} and b as $x^{-\varepsilon}v$. Thus $w = uvcx^{-\varepsilon}u^{-1}v^{-1}x^{\varepsilon}c^{-1}$. This is of the same form as before, this time using the three elements u, v, and d, where d is the reduced form of $cx^{-\varepsilon}$. Since $|u| + |v| + |d| < |a| + |b| + |c|$, this contradicts our choice of a,b, and c.//

We can decide whether or not a finite subset of a free group is Nielsen reduced (see section 8.2). If not, we can perform an elementary Nielsen reduction. Continuing this process for as long as necessary, given a finite subset A of a free group, we can find a finite subset B such that B is Nielsen reduced and $\langle A\rangle = \langle B\rangle$.

There is a slight problem in this. If a set is not Nielsen reduced, there is usually more than one way of reducing it. We can either allow our algorithms to have choices, so that the final result of the computation is just one possible Nielsen reduced set. Alternatively, we can reorganise the algorithm so that a specific choice is always made. For instance, we could insist that the elements of our set are given in some specific order, and we could always make the first Nielsen reduction possible (having specified how the reductions are ordered).

It is not clear from the definition that we can tell whether or not a finite subset S of $F(X)$ is weakly reduced, since the definition requires us to look at all products of elements of $S \cup S^{-1}$. However, Lemma 8.17 tells us that we need only look at products of at most $2^{|S|}$ elements, and so it is possible to tell whether or not S is weakly reduced.

Proposition 7 *Given an element w of a free group $F(X)$ and a finite subset A of the group, we can decide whether or not $w \in \langle A \rangle$.*

Proof Begin by computing a Nielsen reduced set B such that $\langle A \rangle = \langle B \rangle$.

Let w have length n in the basis X. By Corollary 1 to Lemma 8.7, we know that if $w \in \langle B \rangle$ then w is the product of at most n elements of $B \cup B^{-1}$. Hence, to decide whether or not $w \in \langle A \rangle$, we need only construct the finite set of products of at most n elements of $B \cup B^{-1}$, and see whether or not w is in this set.//

9.3 *THE WORD PROBLEM*

Let $\langle X;R \rangle$ be a presentation of a group G. Let N be $\langle R \rangle^{F(X)}$. We say the presentation $\langle X;R \rangle$ ***has solvable word problem*** if N is recursive. We also say that $\langle X;R \rangle$ has *solvable conjugacy problem* if $\{(u,v);\ u, v \in F(X)$, and Nu and Nv are conjugate in $G\}$ is recursive. If $\langle X;R \rangle$ has solvable conjugacy problem then it also has solvable word problem, since u is in N iff in G the element Nu is conjugate to $N1$. In particular, the presentation $\langle X;\emptyset \rangle$ of $F(X)$ has solvable word and conjugacy problem. Also, as already remarked, when considering the word and conjugacy problems for G, we can look at reduced words only or at all words, whichever is most convenient.

We shall see later that when G is finitely generated and one finitely generated presentation of G has solvable word problem then every finite presentation of G has solvable word problem. We then say that the group G has solvable word problem. The same applies to the conjugacy problem.

This does not hold for groups which are not finitely generated. For instance, let A be a set which is r.e. but not recursive. Let X be $\{x_n;\ n \in \mathbb{N}\}$, and let R be $\{x_n;\ n \in A\}$. Then $x_n \in N$ iff $n \in A$. Since A is not recursive, it follows from Proposition 2 that N cannot be recursive. However, G is just a free group of countable rank, so it has a presentation $\langle Y;\emptyset\rangle$ whose word problem is solvable. We can write Y as $\{y_n;\ n \in \mathbb{N}\}$, and we may take a function $f{:}\mathbb{N} \to \mathbb{N}$ with $\mathbb{N}f = \mathbb{N} - A$. This gives rise to a function from Y into X, and so to a homomorphism from $F(Y)$ to $F(X)$. However, these functions cannot be computable. For if one of them were computable then they would all be computable. But, by Proposition 3, $\mathbb{N} - A$ is not r.e. (as A is r.e. but not recursive) so it cannot be the image of a computable function.

Lemma 8 *Let $\langle Y;S\rangle$ be a presentation of a group H. Suppose that there is a computable homomorphism $\alpha{:}F(Y) \to F(X)$ which induces a monomorphism from H to G. Then $\langle Y;S\rangle$ has solvable word problem if $\langle X;R\rangle$ has solvable word problem. Also $\langle S\rangle^{F(Y)}$ is r.e. if N is r.e.*

Remark We have already seen that α will be computable if Y is finite, and also if Y is countable and α comes from a computable map from Y to $F(X)$.

Proof Let w be in $F(Y)$. Because α induces a monomorphism, we see that $w \in \langle S\rangle^{F(Y)}$ iff $w\alpha \in N$. The result follows from Proposition 2.//

Letting α induce the identity (or inclusion), the following corolllary is immediate.

Corollary *Let the presentation $\langle X;R\rangle$ of G have solvable word problem, and let X be finite. Then any finitely generated presentation of G has solvable word problem. More generally, any finitely generated presentation of a subgroup of G has solvable word problem.//*

Lemma 9 *Let R be recursive. Then N is r.e.*

Proof We may assume that $R = R^{-1}$, since, if not, we may replace R by $R \cup R^{-1}$, which is also recursive, by Proposition 2.

Let N_1 be the set of all (not necessarily reduced) words of the form $u_1^{-1}r_1u_1 \ldots u_n^{-1}r_nu_n$ for some n, some $r_i \in R$, and some words u_i, where no reductions have been made in this product. We can plainly determine all the ways of writing a word w in the form $u_1^{-1}r_1u_1 \dots u_n^{-1}r_nu_n$ for some n (this would not be true if we were looking at the reduced form of such a word); then, since R is recursive, for each such way we can decide whether or not every r_i is in R. Hence N_1 is recursive.

As N is the image of N_1 under the map which assigns to each word the corresponding reduced word, we see from Proposition 2 that N is r.e.//

Note that, in particular, N is r.e. if R is finite.

Lemma 10 *Let R be r.e. Then G has a presentation $\langle X_1;R_1\rangle$ such that R_1 is recursive, and X is a recursive subset of X_1.*

Proof Write the elements of $F(X)$ as $\{w_n;\ n\in\mathbb{N}\}$. By Proposition 1, there is a computable total function $f:\mathbb{N}\to\mathbb{N}$ such that R is $\{w_{nf};\ n\in\mathbb{N}\}$. (This holds unless $R=\emptyset$, and the result is immediate in this case).

Let X_1 be $X\cup\{y\}$, where y is some new symbol. Thus X is a recursive subset of X_1.

Let R_1 be $\{y\}\cup\{w_{nf}y^n;\ n\in\mathbb{N}\}$. Plainly $\langle X_1;R_1\rangle$ presents G. Also, given any element of $F(X_1)$, we can decide whether or not it is of the form wy^n for some n and some $w\in F(X)$. If so, we can decide whether or not w is w_{nf}, since f is computable and n is known. Hence R_1 is recursive.//

This procedure, and similar procedures in logic and elsewhere, is known as Craig's Trick.

When R is recursive, we naturally refer to $\langle X;R\rangle$ as a *recursive presentation* of G. A group with a recursive presentation is called *recursively presentable*. We have just seen that a group with a presentation in which R is r.e. is recursively presentable.

Further, G is recursively presentable iff there is, for some X, a homomorphism from $F(X)$ onto G whose kernel is r.e. For Lemma 9 tells us that the kernel is r.e. if the set of defining relators is recursive. Conversely, if the kernel is r.e. then we have a presentation whose set of defining relators is r.e. (namely, take every element of the kernel to be a defining relator) and Lemma 10 then tells us that there is another presentation whose set of defining relators is recursive.

Let G be finitely generated and recursively presentable. Then Lemmas 8, 9, and 10 tell us that for every finitely generated presentation $\langle Y;S\rangle$ of G the group $\langle S\rangle^{F(Y)}$ is r.e.

This does not hold for infinitely generated groups. For instance, take G to be free of countable rank, and let Y be $\{y_n;\ n \in \mathbb{N}\}$, and let S be $\{y_n;\ n \in A\}$, where A is a set which is not r.e. However Lemma 8 tells us that, even for infinitely generated groups, the subgroup $\langle S \rangle^{F(Y)}$ will be r.e. if the map from Y to $F(X)$ is computable.

Lemma 8 also tells us that any finitely generated subgroup of a finitely presented group is recursively presentable (and that the same holds for any finitely generated subgroup of any recursively presentable group). Higman's Embedding Theorem, which will be proved later, is the converse of this; that is, *any finitely generated recursively presentable group is a subgroup of some fintely presented group.*

Lemma 11 *Let R be r.e. Then N is r.e.*

Proof There are two possible proofs. With the notation of Lemma 10, the inclusion of X in X_1 is computable, because X is a recursive subset of X_1 (the algorithm for computing the inclusion is "Take an element x_1 of X_1. Determine whether or not it is in X. If it is, output x_1; if not, continue computing for ever.")

As R_1 is recursive, $\langle R_1 \rangle^{F(X_1)}$ is r.e. Now apply Lemma 8.

Alternatively, let $F(X)$ be $\{w_n;\ n \in \mathbb{N}\}$, and let R be $\{w_{nf};\ n \in \mathbb{N}\}$, where $f: \mathbb{N} \to \mathbb{N}$ is a computable total function. We may assume that $R = R^{-1}$, since otherwise we may replace R by $R \cup R^{-1}$, which is also r.e. by Proposition 2. We have the following partial algorithm for membership of N.

"At stage k, consider all words $u_1^{-1} r_1 u_1 \ldots u_n^{-1} r_n u_n$ satisfying the following conditions. (i) $n \leq k$, (ii) $u_i \in \{w_m;\ m \leq k\}$ for all $i \leq n$, (iii) $r_i \in \{w_{mf}; m \leq k\}$ for all $i \leq n$ (note that all such words can be explictly constructed). Take the reduced forms of the finitely many such words. Then the word w is in N iff it appears among the reduced words obtained at some stage." //

We now look at some kinds of finitely presented groups whose word problem is solvable. The corollary to Lemma 8 tells us that it is enough to find one presentation which has solvable word problem.

Any finitely generated free group has solvable word problem. This is clear, taking a free presentation of the group.

Any finite group G has solvable word problem. We take a multiplication table presentation. This may be written (using a bijection of G with $\{1, \ldots, n\}$ for suitable n) with generators $X = \{x_1, \ldots, x_n\}$ and defining relations $x_i x_j = x_k$; here there is one relation for each pair i, j, and, for each fixed j, the function sending x_i to the corresponding x_k is a permutation of X which is the identity when $j = 1$. It is then easy to compute for each $w \in F(X)$ an integer r such that $w \in Nx_r$. In particular, $w \in N$ iff $r = 1$, and we can check whether or not this holds.

Any finitely generated abelian group has solvable word problem. For it is known, from the theory of abelian groups, that any finitely generated abelian group G has a presentation with generators $x_1, \ldots, x_n$ for some n, and defining relators $[x_i, x_j]$ for all i and j together with $x_i^{d_i}$ for $i \leq k$ for some k, where the d_i are positive integers. Letting $w\sigma_i$ be the exponent sum of x_i in w, we see that $w \in N$ iff $w\sigma_i = 0$ for $i > k$ and d_i divides $w\sigma_i$ for $i \leq k$. Since σ_i is a computable function, we can decide whether or not these conditions hold.

Suppose that we have a different finite presentation of the same group G, with generators $y_1, \ldots, y_m$ and defining relators including all the commutators $[y_i, y_j]$. The general theory of this section tells us that this presentation must have a solvable word problem. In principle, we know how to solve the word problem for this presentation, using the homomorphism from $F(Y)$ to $F(X)$ which induces the identity on G. This is not a very practical method until we know this homomorphism. However, the general theory of abelian groups enables us to compute explicitly from this presentation a presentation of the form in the previous paragraph, and it also lets us obtain functions from Y into $F(X)$ which induce the identity on G. Then, to determine whether or not an element of $F(Y)$ is 1 in G we need only take its image in $F(X)$ and check whether or not this is in N.

Note that there is no reason to suppose that we can decide of a finite presentation $\langle X; R\rangle$ whether or not the group it presents is abelian (or finite, or free). We shall see later that these problems are not decidable. However, there is plainly a class of finite presentations of an abelian group from which we can tell that the group is abelian, namely, those presentations for which every commutator of two distinct generators is one of the defining relators.

Similarly, there is at least one finite presentation of a finite group from which we can tell that it is finite, and at least one presentation of a free group from which we can tell that the group is free.

By contrast, in the next two results, not only are we unable to decide of a finite presentation whether or not the group has the relevant property, but there is no known class of presentations such that we can tell from the presentation that the group has the property.

Proposition 12 *Finitely presentable residually finite groups have solvable word problem.*

Proof Let the residually finite group G have finite presentation $\langle X;R\rangle$. We know that N is r.e. Hence, by Proposition 3, it is enough to prove that $\{w;\ w \notin N\}$ is also r.e.

If $w \notin N$ then w and 1 are different in G, and so there is a homomorphism from G to a finite group such that the image of w is not 1. Such a homomorphism can be regarded as a homomorphism from $F(X)$ which sends R to 1, and homomorphisms from $F(X)$ to a group can be identified with functions from X to the group.

We cannot consider all finite groups, since there are too many of them (and, in formal set theory, there is no set whose members are exactly the finite groups). But we need only consider one group in each isomorphism class. Now a finite group H of order n is bijective with $\{1,\ldots,n\}$. Using this bijection to transform the multiplication on H into a multiplication on $\{1,\ldots,n\}$, we have an isomorphism from H to some group whose underlying set is $\{1,\ldots n\}$.

This gives the following partial algorithm for membership of $F(X) - N$.

"Given $w \in F(X)$, at stage n perform the following construction. First take all the (finitely many) multiplication tables on $\{1,\ldots,n\}$ and determine which of them make $\{1,\ldots,n\}$ a group. For each such group, take all the (finitely many) functions from X to $\{1,\ldots,n\}$. Each of these can be regarded as a homomorphism from $F(X)$, and for each one we determine whether or not R is mapped to the identity. For those maps with this property, we determine whether or not w is mapped to the identity. If there is one such map for which w is not mapped to the identity, we answer "Yes"."//

Proposition 13 *Countably generated recursively presented simple groups have solvable word problem.*

Proof If G is trivial, it certainly has solvable word problem. Hence we need only consider the case when G is non-trivial. (In general, we cannot tell from the presentation whether G is trivial or not. Hence we cannot give an algorithm for solving the word problem; however, we show that such an algorithm must exist.)

As before, it is enough to show that $\{w;\ w \notin N\}$ is r.e.

Since G is not trivial, there is some word a not in N. (As usual, the fact that we may not be able to find such an element a only means that we cannot explicitly find an algorithm, but the existence of an algorithm follows from the existence of the word a.) If $w \notin N$ then w is not 1 in G. Thus we get the trivial group if we cancel the element w; that is, $\langle X; R \cup \{w\}\rangle$ presents the trivial group. In particular, $a \in \langle R \cup \{w\}\rangle^{F(X)}$. If $w \in N$ then $\langle R \cup \{w\}\rangle^{F(X)}$ is just N, and so, by our choice of a, we have $a \notin \langle R \cup \{w\}\rangle^{F(X)}$.

Also, since we are assuming that R is recursive, the subgroup $\langle R \cup \{w\}\rangle^{F(X)}$ is r.e., and, by Proposition 1, we can make a list of its elements. Thus we have the following partial algorithm for membership of $F(X) - N$.

"Given $w \in F(X)$, write down the list of elements of $\langle R \cup \{w\}\rangle^{F(X)}$. If we find the word a in the list then answer "Yes"."

There is an alternative algorithm which applies to all non-trivial finitely generated simple groups (but we still cannot tell whether or not the group is trivial). This algorithm for membership of $F(X) - N$ is

"Given $w \in F(X)$, write down the list of elements of $\langle R \cup \{w\}\rangle^{F(X)}$. If at some stage we have found all the members of X in the list then answer "Yes"."//

Proposition 14 *Let X be finite and let R be recursive. Suppose that every $w \in N - \{1\}$ contains more than half of some member of R. Then the presentation $\langle X;R\rangle$ has solvable word problem.*

Remark The hypothesis is that there are words a, b, u, and v such that $|u| > |v|$ such that $w = aub$, reduced as written, and uv is reduced as written and is in R.

Certain of the small cancellation groups discussed in Chapter 10 satisfy this condition.

Proof Let $U = \{u;\ u$ is more than half of some member of $R\}$. Then U is recursive. For, given u, we need only write down all the finitely many words v with $|v| < |u|$ and check for each of them whether or not uv is reduced as written and lies in R.

We also have a computable function α with domain U such that $u\alpha$ is some element v of the kind considered. For instance, we could choose, among the finitely many possibilities for v given u, that one with the smallest Godel number.

Given a word w, we can obtain all the finitely many ways of writing it as aub, reduced as written, and see whether or not any of these ways has u in U. Thus the set S of all words containing more than half of a member of R is recursive. Also the only reduced word in N but not in S is 1.

We can define a computable function β on S by letting $w\beta$ be the reduced form of $a(u\alpha)^{-1}b$. For definiteness, we may require that among all possibilities for u we choose that one with the smallest Godel number, and if this u occurs more than once we take its leftmost occurrence. It is clear that, for $w \in S$, we have $w \in N$ iff $w\beta \in N$. Also $w\beta$ is shorter than w, by the choice of u and $u\alpha$.

This gives the following algorithm for deciding whether or not a reduced word is in N.

"Given w, determine whether or not w is in S, and, if so, replace w by $w\beta$. Now determine whether or not $w\beta$ is in S, and, if so, replace $w\beta$ by $(w\beta)\beta$. Repeat this process as often as necessary until we reach a reduced word w' which is not in S. Then w is in N iff w' is 1."//

Plainly, when the above condition holds, if $|w| = n$ then after at most n replacements we reach a word w' which is not in S. Thus we can say that the solution to the word problem is linear (in terms of how many steps it takes to make the decision, as function of the length of the word). Book (1988) shows that the solution is linear in a more precise sense.

Proposition 15 *Let G and H have solvable word problems. Then their free product $G*H$ also has solvable word problem.*

Proof Let X generate G, and let Y generate H, where $X \cap Y = \emptyset$, so that $X \cup Y$ generates $G*H$.

Any $w \in F(X \cup Y)$ can be written as $u_1v_1 \ldots u_nv_n$, where $u_i \in F(X)$, $v_i \in F(Y)$, and u_1 and v_n may be trivial, but no other syllable u_i or v_i is trivial.

Let S be the set of such words for which some non-trivial u_i equals 1 in G or else some non-trivial v_i equals 1 in H. Since G and H have solvable word problems, the set S is recursive. Further, a word w equals 1 in $G*H$ only if it is trivial or lies in S, by the Normal Form Theorem for free products.

For $w \in S$, let $w\alpha$ be the word obtained by deleting the first syllable which equals 1 in G or H. Then α is a computable partial function, and $w\alpha$ is shorter than w. Also $w = 1$ in $G*H$ iff $w\alpha = 1$ in $G*H$.

This give rise to the following algorithm for solving the word problem in $G * H$.

"Given a word $w \in F(X \cup Y)$ decide whther or not it is in S. If so, compute $w\alpha$. Decide whether or not $w\alpha$ is in S, and if so compute $(w\alpha)\alpha$. Continue this process until it stops. We will end with a word w' which is not in S. Answer "Yes" if w' is trivial and "No" otherwise."//

If we want to extend this result to amalgamated free products or HNN extensions we have to consider the positions of the amalgamated or associated subgroups in the larger groups. This leads to the *membership problem* discussed below.

Let G be $\langle X;R\rangle^{\varphi}$. Let Y be a subset of $F(X)$, and let A be $\langle Y\varphi\rangle$. We say that *the membership problem for A in G* is solvable if we can decide whether or not an element of $F(X)$ maps by φ to an element of A. (Strictly speaking, we should refer to the membership problem for the presentation $\langle X;R\rangle$ and the subset Y; for finite X and Y the solvability or otherwise of the problem depends only on G and A, but in general it may depend on the generators chosen.)

Suppose that the word problem for $\langle X;R\rangle$ is solvable, the membership problem for A in G is solvable, and that Y is an r.e. subset of $F(X)$. Then, for any $w \in F(X)$ such that $w\varphi \in A$ we can compute an element $v \in \langle Y\rangle$ such that $w\varphi = v\varphi$. For we can list all the elements of $\langle Y\rangle$, since Y is r.e. For each such v we can decide whether or not wv^{-1} equals 1 in G, since G has a solvable word problem. As our element w was chosen such that $w\varphi \in A$ (and, by hypothesis, for any w we can decide whether or not this holds) there must be some element of the list with this property, and we just take the first such element. In fact, this procedure gives an explicit expression for $w\varphi$ as a product of elements of $(Y \cup Y^{-1})\varphi$.

Let Y be a subset of a countable set X. It is easy to see that the membership problem for $F(Y)$ in $F(X)$ is solvable iff Y is a recursive subset of X.

Lemma 16 *Let $K \leq H \leq G$. If the membership problems for H in G and for K in H are solvable then the membership problem for K in G is solvable.*

Proof Let G be $\langle X;R\rangle$ and let H be $\langle Y;S\rangle$. Take a word $w \in F(X)$. Then w represents an element of K only if it represents an element of H. By hypothesis, we can decide whether or not this holds. If it does, we have seen above that we can compute an element v of $F(Y)$ such that w equals in G the element of H represented by v. By hypothesis, we can tell whether or not v represents an element of K.//

Lemma 17 *Let G and G_1 be groups with solvable word problems. Let H and H_1 be subgroups of G and G_1 respectively whose membership problems are solvable. Then the membership problem for $H * H_1$ in $G * G_1$ is solvable.*

Proof Let X and X_1 generate G and G_1, where $X \cap X_1 = \emptyset$. An element w of $F(X \cup X_1)$ may be written as $u_1 v_1 \ldots u_n v_n$ for some n, where $u_i \in F(X)$, $v_i \in F(X_1)$, and u_1 and v_n may be trivial but all other u_i and v_i are non-trivial. Since the word problems for G and G_1 are solvable we can determine whether any non-trivial syllable represents 1 in G or G_1. If so, we can obtain a shorter word by omitting such a syllable. Thus we can compute from w another word w' whose non-trivial syllables do not represent 1 in G or G_1. We may as well assume that w itself has this property.

It then follows that w represents an element of $H * H_1$ iff each u_i represents an element of H and each v_i represents an element of H_1. By hypothesis, we can decide whether or not these hold.//

Proposition 18 *Let $G = \langle X;R \rangle$ have solvable word problem. Let A_{-1} and A_1 be subgroups of G whose membership problem is solvable. Suppose that there is an isomorphism from A_1 to A_{-1} which is induced by a computable bijection from a subset Y_1 of $F(X)$ to a subset Y_{-1}. Then the HNN extension $H = \langle G,p;\ p^{-1}A_1p = A_{-1} \rangle$ has solvable word problem.*

Remark As Y_1 is the domain of a computable function it is r.e. Then the inverse bijection from Y_{-1} to Y_1 is also a computable partial function. To compute the element of Y_1 corresponding to y_{-1} we need only list the elements of Y_1, computing their images as we list the elements, and see which one has image y_{-1}.

Further, there is an induced computable map from $\langle Y_1 \rangle$ to $\langle Y_{-1} \rangle$, and conversely. For given any word w, we can list the products of elements of $Y_1 \cup Y_1^{-1}$ until the first time, if ever, we find a product which, when reduced, equals w, and then take the corresponding product of elements of $Y_{-1} \cup Y_{-1}^{-1}$.

Proof H is generated by $X \cup \{p\}$. Any $w \in F(X \cup \{p\})$ can be written as $u_0 p^{\varepsilon_0} u_1 p^{\varepsilon_1} \ldots u_{n-1} p^{\varepsilon_{n-1}} u_n$ where $n \geq 0$, $\varepsilon_i = \pm 1$, and $u_i \in F(X)$.

Let S be the set of those elements for which, for some i, either $\varepsilon_i = 1 = -\varepsilon_{i-1}$ and u_i represents an element of A_1 or else $\varepsilon_{i-1} = 1 = -\varepsilon_i$ and u_i represents an element of A_{-1}. By Britton's Lemma any w which equals 1 in H must either lie in S or else have $n = 0$. Further, if $n = 0$ then $w = 1$ in H iff $w = 1$ in G, and we can decide whether or not this holds, since the word problem for G is solvable.

It is enough to show that S is recursive and that there is a computable function α with domain S such that $w\alpha$ contains fewer occurrences of p than w. For then the required algorithm for solving the word problem for H is a very slight modification of the algorithm of Proposition 15.

Since the membership problems for both A_1 and A_{-1} are solvable, it is easy to see that S is recursive.

For $w \in S$, take the smallest possible i satisfying the relevant conditions. For convenience, assume $\varepsilon_i = 1$, so that u_i represents an element of A_1 (the other case is similar). As we have seen, since the word problem for G is solvable, we can explicitly find a product of elements of $Y_1 \cup Y_1^{-1}$ which equals u_i (when reduced). Let v_i be the reduced form of the product of the corresponding elements of $Y_{-1} \cup Y_{-1}^{-1}$. Let $w\alpha$ be obtained from w by replacing $p^{-1} u_i p$ by v_i. Using earlier remarks, we see easily that α is a computable function with the required properties.//

A similar result holds for amalgamated free products.

The result extends at once to a finite number of stable letters.

Suppose that we have a countable number of stable letters and associated subgroups A_i and A_{-i}, with generating sets Y_i and Y_{-i}. Then it is not enough for the relevant conditions to hold for each i, as the algorithms for varying i might have no connection with one another.

Instead, we need algorithms which are uniform in i. More precisely, we need an algorithm which decides for any word w and number i whether or not w represents an element of A_i (and similarly for A_{-i}). We also need a computable partial function f of two variables w and i such that $(w,i)f$ is defined iff $w \in Y_i$, and such that, for any fixed i, the function sending w to $(w,i)f$ is a bijection from Y_i onto Y_{-i}. If these conditions are satisfied and the word problem for G is solvable, the argument extends at once to show that the HNN extension also has solvable word problem.

A group is called *computable* if it has a Gödel numbering for which there is a computable partial function $\mu:\mathbb{N}^2 \to \mathbb{N}$ such that, when m and n are the Gödel numbers of g and h respectively, $(m,n)\mu$ is the Gödel number of gh. These groups were discussed by Rabin (1960). Some of their properties are given in the exercises below.

Exercise 3 Let G be a computable group. Show that there is a partial computable function $\iota:\mathbb{N} \to \mathbb{N}$ such that $n\iota$ is the Gödel number of g^{-1} when n is the Gödel number of g.

Exercise 4 Let φ and ψ be Gödel numberings of G. Show that if $\varphi\psi^{-1}$ is a computable partial function then so is $\psi\varphi^{-1}$.

Exercise 5 Let φ and ψ be Gödel numberings of G such that $\varphi\psi^{-1}$ is computable. Show that if G is computable when the Gödel numbering φ is used then it is also computable when the Gödel numbering ψ is used.

Exercise 6 Show that the free group $F(X)$ is computable, when the Gödel numbering of $F(X)$ comes from a Gödel numbering of X.

Exercise 7 Let G be a computable group. Let X be a recursive set of generators of G. Show that G has solvable word problem in these generators. (In particular, if G is finitely generated, we find that G has solvable word problem).

Exercise 8 Let M be a subgroup of $F(X)$ whose membership problem is solvable (with respect to the generators X), and let φ be the Godel numbering of $F(X)$ corresponding to some Gödel numbering of X. Let $w\alpha$ be the smallest member of $\{u\varphi;\ u \in Mw\}$. Show that α is computable.

Exercise 9 Using the previous exercise, show that if G has solvable word problem in some set of generators X then G is computable for a Gödel numbering such that the map from X to G is computable.

Exercise 10 Formulate and prove a result about amalgamated free products similar to Proposition 18.

9.4 *MODULAR MACHINES AND UNSOLVABLE WORD PROBLEMS*

A *modular machine* M consists of the following data: an integer $m > 1$, and a number of quadruples of form (a,b,c,R) and (a,b,c,L) where a, b, and c are integers with $0 \le a < m$, $0 \le b < m$, $0 \le c < m^2$, and R and L are two special symbols (they stand for 'right' and 'left'). At most one quadruple is to begin with a given pair a,b. (A further integer n is required if we want to define the function computed by M, but this will not concern us.)

The modular machine M operates on $\mathbb{N}^2$ as follows. Take any $(\alpha,\beta) \in \mathbb{N}^2$, and write $\alpha = um + a$, $\beta = vm + b$, where $0 \le a < m$ and $0 \le b < m$. If no quadruple begins with a,b, we say that (α,β) is *terminal*. If there is a quadruple (a,b,c,R) then (α,β) *yields* $(um^2 + c, v)$, while if there is a quadruple (a,b,c,L) then (α,β) yields $(u, vm^2 + c)$.

The *computation of M starting with* (α_0,β_0) is the finite or infinite sequence (α_i,β_i) of elements of $\mathbb{N}^2$ such that $(\alpha_{i-1},\beta_{i-1})$ yields (α_i,β_i) for all $i > 0$, and such that the last element is terminal when the sequence is finite. When $(0,0)$ is terminal for M we define $H_0(M)$ to be $\{(\alpha,\beta)$; the computation of M starting at $(\alpha.\beta)$ is finite and ends with $(0,0)$ $\}$.

This definition is not at all intuitive. When we discuss Higman's Embedding Theorem in a later section, we will look at a class of machines called Turing machines. These will be a fairly intuitive class of machines, and they are capable of computing any computable function. We shall then be able to indicate that modular machines are essentially a way of obtaining a numerical coding for Turing machines, in a fashion which is particularly convenient for group theory and some aspects of the theory of computable functions. Modular machines are discussed in detail in Cohen (1987).

Let A be a set which is r.e. but not recursive. Then the function which is 0 on A and not defined outside A is computable. We can then find a modular machine which computes this function. It turns out that $H_0(M)$ cannot be recursive, because A is not recursive.

This indicates (but we cannot give a proper proof here) that the following result holds.

Proposition 19 *There is a modular machine M such that $H_0(M)$ is not recursive.//*

From this machine we shall be able to construct a finitely presented group with unsolvable word problem. Another construction of such a group can be made using Higman's Embedding Theorem. All such constructions use in some form or another the fact that there is a set which is r.e. but not recursive.

We shall begin with the group $K = \langle x, y, t;\ xy = yx \rangle$. This group can be regarded as the free product $\langle x, y;\ xy = yx \rangle * \langle t \rangle$ of a free abelian group of rank two and an infinite cyclic group. It can also be regarded as the HNN extension of the free group $\langle x,t \rangle$ of rank two with the stable letter y and both associated subgroups being $\langle x \rangle$.

Let $t(r,s)$ be $y^{-s}x^{-r}tx^ry^s$, and let T be the subgroup $\langle t(r,s);\text{ all } r \text{ and } s\rangle$ of K. Then T is easily seen to be the normal closure $\langle t\rangle^K$ of $\langle t\rangle$ in K. Also T is free with basis $\{t(r,s)\}$. This can be proved using the Kurosh subgroup theorem, but it is more easily proved directly, by taking a reduced word in the letters $t(r,s)$ and expanding it out.

For any i, j, m, and n, the subgroup $T\cap\langle t(i,j),\ x^m,\ y^n\rangle$ has basis $\{t(r,s);\ r\equiv i \bmod m,\ s\equiv j \bmod n\}$. For these elements are plainly in this subgroup. Also any element of $\langle t(i,j),\ x^m,\ y^n\rangle$ is easily seen to be of form $ux^{pm}y^{qn}$ for some integers p and q and some element u of $\langle t(r,s);\ r\equiv i \bmod m,\ s\equiv j \bmod n\rangle$, from which the result follows.

The subgroups $\langle t(i,j),\ x^m,\ y^n\rangle$ for any i and j and any non-zero m and n are all isomorphic. For such a subgroup is conjugate to the subgroup $\langle t,\ x^m,\ y^n\rangle$, and this is the free product of $\langle t\rangle$ and the subgroup $\langle x^m,\ y^n\rangle$. Since $\langle x,y\rangle$ is free abelian of rank two, we see that our subgroup is isomorphic to K.

Let i, j, i_1, and j_1 be arbitrary integers, and let m, n, m_1, and n_1 be arbitrary non-zero integers. It follows from the paragraphs above that the map sending $t(i,j)$ to $t(i_1,j_1)$, x^m to x^{m_1}, and y^n to y^{n_1} induces an isomorphism of $\langle t(i,j),\ x^m,\ y^n\rangle$ with $\langle t(i_1,j_1),\ x^{m_1},\ y^{n_1}\rangle$. It also induces an isomorphism of $T\cap\langle t(i,j),\ x^m,\ y^n\rangle$ with $T\cap\langle t(i_1,j_1),\ x^{m_1},\ y^{n_1}\rangle$ which sends $t(um+i,vn+j)$ to $t(um_1+i_1,vn_1+j_1)$.

Now take a modular machine M with quadruples $(a_i,b_i\,c_i,R)$ and (a_j,b_j,c_j,L) for i in some index set I and j in some index set J. Let K_M be the group with generators x, y, t, r_i (all $i\in I$), and l_j (all $j\in J$) and defining relations

$$xy=yx,\ r_i^{-1}t(a_i,b_i)r_i=t(c_i,0),\ r_i^{-1}x^mr_i=x^{m^2},\ r_i^{-1}y^mr_i=y,\ l_j^{-1}t(a_j,b_j)l_j=t(0,c_j),$$

$$l_j^{-1}x^ml_j=x,\ l_j^{-1}y^ml_j=y^{m^2}\ (\text{for } i\in I \text{ and } j\in J).$$

Then K_M is just the HNN extension of K with stable letters r_i and l_j, the subgroups $\langle t(a_i,b_i),\ x^m,\ y^m\rangle$ and $\langle t(c_i,0),\ x^{m^2},\ y\rangle$ being associated with r_i, and similarly for l_j.

We define T_M to be the subgroup of T generated by $\{t(\alpha,\beta);\ (\alpha,\beta)\in H_0(M)\}$. Because T has basis $\{t(i,j),\text{ all } i \text{ and } j\}$, we see that $t(\alpha,\beta)\in T_M$ iff $(\alpha,\beta)\in H_0(M)$.

Take a quadruple (a,b,c,R) with corresponding stable letter r, and take a pair (α,β) such that $\alpha\equiv a \bmod m$ and $\beta\equiv b \bmod m$. Then the isomorphism of associated subgroups sends $t(\alpha,\beta)$ to $t(\alpha',\beta')$, where (α,β) yields (α',β'). In particular, this isomorphism maps $T_M\cap\langle t(a,b),\ x^m,\ y^m\rangle$ onto

$T_M \cap \langle t(c,0),\ x^{m^2},\ y\rangle$. A similar result holds for a quadruple (a,b,c,L).

It now follows from Proposition 1.34 that

$K \cap \langle T_M,\ r_i \text{ (all } i \in I),\ l_j \text{ (all } j \in J)\rangle = T_M$.

Finally, we show that $\langle T_M,\ r_i,\ l_j \text{ (all } i,\ j)\rangle = \langle t, r_i, l_j\rangle$.We need only prove that $t(\alpha,\beta) \in \langle t, r_i, l_j\rangle$ for all $(\alpha,\beta) \in H_0(M)$. We use induction on the length of the computation. If the computation has length 0 then we must have $\alpha = 0 = \beta$, since (α,β) is then terminal and in $H_0(M)$, so the result is immediate. Otherwise, let (α,β) yield (α',β'). We may suppose, without loss of generality, that the quadruple (a,b,c,R) is the one which applies to (α,β), and that the corresponding stable letter is r. Then $r^{-1}t(\alpha,\beta)r = t(\alpha',\beta')$. Inductively, we have $t(\alpha',\beta') \in \langle t, r_i, l_j\rangle$, and so we also have $t(\alpha,\beta) \in \langle t, r_i, l_j\rangle$.

Proposition 20 *Let M be a modular machine. Let* G_M *be* $\langle K_M,\ k;\ kt = tk,\ kr_i = r_i k,\ kl_j = l_j k \text{ (all } i \text{ and } j)\rangle$. *Then G is a finitely presented grouk such that* $(\alpha,\beta) \in H_0(M)$ *iff* $kt(\alpha,\beta) = t(\alpha,\beta)k$. *In particular,* $H_0(M)$ *is recursive if* G_M *has solvable word problem.*

Proof Plainly G_M is finitely presented. Also, G_M is an HNN extension of K_M, the stable letter being k and both associated subgroups being $\langle t, r_i, l_j\rangle$. Hence $kt(\alpha,\beta) = t(\alpha,\beta)k$ iff $t(\alpha,\beta) \in \langle t, r_i, l_j\rangle$. We have seen that this holds iff $(\alpha,\beta) \in H_0(M)$.

Since the function sending (α,β) to $k^{-1}t(\alpha,\beta)kt(\alpha,\beta)^{-1}$ is plainly computable, it follows from Proposition 2 that $H_0(M)$ is recursive if G_M has solvable word problem.//

Combining the purely group-theoretical Proposition 20 with the machine-theoretical Proposition 19, the following theorem is immediate.

Theorem 21 *There is a modular machine M such that the corresponding finitely presented group* G_M *has unsolvable word problem.//*

9.5 *SOME OTHER UNSOLVABLE PROBLEMS*

By arguments similar to those in the previous section, but more complicated, we can find a group whose word problem is solvable but whose conjugacy problem is unsolvable.

It is possible to classify unsolvable problems according to their degrees of difficulty. It is also possible to classify an interesting subclass of solvable problems according to how hard it is to solve them. In both cases it is possible to find a finitely presented group whose word problem is of a specified degree of difficulty and whose conjugacy problem is of any harder degree of difficulty. We shall not pursue these aspects further, as they would take us too far into the theory of computable functions, and they are also rather technical genralisations of the construction we made previously.

From the unsolvability of the word problem we can show that other interesting questions about elements of a group are also unsolvable.

Proposition 22 *There is a finite presentation* $\langle X;R\rangle$ *of a group* G *such that it is not possible to decide whether or not an element of* $F(X)$ *represents in* G *an element of any of the following kinds*: (*i*) *an element of the centre of* G, (*ii*) *an* n *th power, where* $n>1$, (*iii*) *an element with only finitely many conjugates*, (*iv*) *a commutator*, (*v*) *an element of finite order* (*not* 1). *Also there is* $g\in G$ *such that we cannot decide whether or not an element of* $F(X)$ *represents an element of the centraliser of* g.

Proof We start with a finitely presented group $H=\langle Y;S\rangle$ with unsolvable word problem.. Let A be the group $\langle a,b\colon b^3\rangle$. The group $A*H$ will be the required group. It has presentation $\langle X;R\rangle$, where $X=Y\cup\{a,b\}$ and $R=S\cup\{b^3\}$.

Take any $w\in F(Y)$. Let W be the commutator $[a,w]$ in $F(X)$. In G, W represents a non-trivial commutator of elements of A and H if $w\neq1$ in H, while $W=1$ in G if $w=1$ in H.

Since a non-trivial free product has trivial centre, W represents an element in the centre of G iff $w=1$ in H. Also, the element represented by W commutes with the image of b iff $w=1$ in H, and it lies in A iff $w=1$ in H, since otherwise it has length (in the free product sense) four.

Further, since the element represented is not in A when $w\neq1$ in H, it cannot commute with any element of A, and so it has infinitely many conjugates (namely, the conjugates by the powers of a). Since the element has length 4, it could only be a proper power if it were $(ch)^2$ for some $c\in A$ and $h\in H$. We would then have in G both $c=a$ and $c^{-1}=a$, which is impossible, as a has infinite order.

Thus we could decide whether or not $w=1$ in H if we could decide of W whether or not the element it represented was in the centre of G, commuted with the image of b, lay in A, had only finitely many conjugates, or was a proper power. Since the word problem for H is unsolvable, none of these problems can be decided.

Similarly, bw represents an element of order 3 if $w = 1$ in H and an element of infinite order (since its length is 2) if $w \neq 1$ in H. Also, $[a,b]w$ represents a commutator if $w = 1$ in H, while if $w \neq 1$ in H then its length in G is 2. However, as in Proposition 6, we can show that any commutator in a free product which is not in a conjugate of one or the factors may be written as $pqrp^{-1}q^{-1}r^{-1}$, where there is no cancellation between adjacent pairs, though there may be coalescence. Analysing the possibilities for coalescence it is not hard to see that commutators in a free product but not in a conjugate of one of the factors have length at least 4.//

By contrast, for any finitely presented group $G = \langle X;R \rangle$ we can decide whether or not an element of $F(X)$ represents an element of the commutator subgroup G. For this holds iff it represents 1 in G/G', and this latter group has solvable word problem, being finitely generated abelian.

We now look at decision problems for properties of subgroups of a given group, and of presentations of groups. We need to have some idea of Gödel numberings for these objects if we are to look at such problems. For definiteness, I give an indication of this below, although we shall never need to use the details.

Suppose that we have a Gödel numbering of a set S, so that the elements of S can be written as s_n for certain n. We know how to find a Gödel numbering for the finite sequences of elements of S (or, equivalently, for the elements of the free monoid $M(S)$). We can regard this as giving a Gödel numbering of the finite subsets of S, regarding a subset as being a sequence $s_{i_1}, \ldots, s_{i_k}$ where $i_1 < i_2 < \ldots < i_k$.

If we want to look at finite presentations we cannot possibly consider all $\langle X;R \rangle$ where X is an arbitrary finite set, as there are set-theoretical difficulties. But, of course, we need only consider presentations $\langle X;R \rangle$ where X is a finite subset of some fixed countable set $\mathfrak{X}$. Such a presentation gives rise to a pair of numbers m and n, where m is the number corresponding to the finite subset X of $\mathfrak{X}$, while n is the number corresponding to the finite subset R of $F(\mathfrak{X})$. We let $J(m,n)$ correspond to $\langle X;R \rangle$, where J is a computable bijection from $\mathbb{N}^2$ to $\mathbb{N}$. Given m and n, we can tell whether or not they correspond to some X and R, and, if so, we can tell whether or not R is a subset of $F(X)$ and not just a subset of $F(\mathfrak{X})$. Thus we get a Gödel numbering of finite presentations.

Proposition 23 *The set of pairs $\langle X;R \rangle$ and $\langle Y;S \rangle$ of finite presentations such that the corresponding groups are isomorphic forms an r.e. set.*

Proof First, replacing Y by another subset of $\mathfrak{X}$ bijective with it if necessary, we may assume that $X \cap Y = \emptyset$.

Applying Tietze transformations of type I, the group with presentation $\langle X;R\rangle$ also has presentation $\langle X \cup Y;R_1\rangle$, where $R_1 = R \cup \{yw_y$, all $y \in Y\}$, where w_y can be any element of $F(X)$. Applying Tietze transformations of type II to this, we get a presentation $\langle X \cup Y;T\rangle$, where T comes from R_1 by adding certain products of conjugates of elements of R_1 and their inverses.

It was shown in Theorem 1.15 that $\langle X;R\rangle$and $\langle Y;S\rangle$ present isomorphic groups iff it is possible to find T such that the presentation $\langle X \cup Y;T\rangle$ also comes from $\langle Y;S\rangle$ by the same two processes.

This leads to the following partial algorithm for determining whether the two presentations are isomorphic.

"At stage n, do the following. First take all R_1 of the form $R \cup \{yw_y$, all $y \in Y\}$ for which the elements w_y in $F(X)$ have length $\leq n$ (there are only finitely many of them). For each such R_1, take all the finitely many T which are obtained from R_1 by adding some products of conjugates of elements of R_1 and their inverses, where each product has at most n factors, and each conjugating element in $F(X \cup Y)$ has length $\leq n$. Perform the similar construction on $\langle Y;S\rangle$, and see whether or not some presentation occurs in both constructions at this stage. If so, answer "Yes", and if not go on to the next stage."//

We now look at properties of groups, and decision problems connected with them. The only properties we consider are those such that whenever a group has the property then so does any isomorphic group. Thus we do not consider the property of being a subgroup of a group of matrices, but we could consider the property of having a monomorphism into a group of matrices. We shall always assume that there are some finitely presented groups having the property and other finitely presented groups not having the property.

Proposition 24 *Suppose that there is an integer k such that no free group of rank $\geq k$ has the property $\mathfrak{P}$. Then there is a finitely presented group $G_{\mathfrak{P}} = \langle Y;S\rangle$ such that we cannot decide of a subset Z of $F(Y)$ whether or not the subgroup of $G_{\mathfrak{P}}$ generated by Z has the property $\mathfrak{P}$.*

Remark Among the properties included in the proposition are those of being trivial, finite, cyclic, abelian, and simple, while, looking at the complement of $\mathfrak{P}$ rather than $\mathfrak{P}$ itself, the property of being free is also covered. Readers can find many other such properties for themselves.

Proof Let P be a finitely presented group having the property $\mathfrak{P}$. We may assume that P has k generators $p_1, \ldots, p_k$, since we are permitted to increase both k (from its definition) and the number of generators of P (adding suitable relators).

Let $G = \langle X;R \rangle$ be the group considered in Proposition 22, and use the notation of that proposition. Define $G_{\mathfrak{P}}$ to be $P * G$.

For any $w \in F(Y)$, let u be $[a,w]$ and let v be $[b,w]$. For $i \leq k$, let z_i be $p_i v^{-i} u v^i$, and consider the subgroup of $G_{\mathfrak{P}}$ generated by $\{z_1, \ldots, z_k\}$. If $w = 1$ in H then $z_i = p_i$, and so this subgroup is P, which has the property $\mathfrak{P}$. We shall show that this subgroup is free of rank k when $w \neq 1$ in H. Consequently the subgroup has $\mathfrak{P}$ iff $w = 1$ in H. Since the word problem for H is unsolvable, we cannot decide whether or not this subgroup has $\mathfrak{P}$.

It is enough, by Corollary 1 to Proposition 1.8, to show that the elements $v^{-i} u v^i$, regarded as elements of G, are a free basis for the subgroup they generate. For this, it is enough to show that u and v are a free basis for the subgroup of G which they generate.

Since $G = A * H$, by the Kurosh Subgroup Theorem if the subgroup $\langle u,v \rangle$ were not free it would have a non-trivial intersection with some conjugate of A or H. This is not possible, as we see by mapping G onto one factor. We are left with the possibility that this subbgroup, though free, only has rank 1. But that would require both u and v to be powers of the same element. Since they both have length 4, and they are not equal or inverses of each other, this is plainly impossible.//

Theorem 25 *Suppose that there is a finitely presented group Q which is not a subgroup of any finitely presented group having the property $\mathfrak{P}$. Then we cannot decide of a finite presentation whether or not it presents a group having $\mathfrak{P}$.*

Remark The easiest way to ensure this condition is to assume that $\mathfrak{P}$ is hereditary; that is, that if a group has $\mathfrak{P}$ then all its subgroups have $\mathfrak{P}$. The slightly weaker condition that whenever a finitely presented group has $\mathfrak{P}$ then all its finitely presented subgroups have $\mathfrak{P}$ is also enough.

In particular, $\mathfrak{P}$ could be any of the following properties: being trivial, finite, abelian, cyclic, free, having a faithful representation in some matrix group $GL_n(K)$, or being residually finite.

By Lemma 8 $\mathfrak{P}$ could also be the property of having a solvable word problem.

Also, by Lemma 8 and Proposition 13, a finitely presented group with unsolvable word problem cannot be embedded in a finitely presented simple group. Thus $\mathfrak{P}$ could also be the property of being embeddable in a finitely presented simple group.

It follows from this theorem that *the isomorphism problem is unsolvable; that is, it is impossible to decide of two finite presentations whether or not they present isomorphic groups.* Indeed, we have the stronger result that it is impossible to decide whether or not a finite presentation presents the same group as the trivial presentation $\langle \emptyset;\emptyset \rangle$, which presents the trivial group.

Proof Let P be a finitely presented group having the property $\mathfrak{P}$. Let U be a finitely presented group with unsolvable word problem. Let A be the finitely presented group $U * Q$, and let $\langle X;R \rangle$, where $X = \{x_1, \ldots, x_n\}$, be a presentation of A. Then A has unsolvable word problem, since its subgroup U has unsolvable word problem.

In the free product $A * \langle b \rangle$ the elements $x_i b^2$ have infinite order. Hence the group B with presentation

$$\langle X, b, c_1, \ldots, c_n, c_{n+1};\ c_i^{-1} x_i b^2 c_i = (x_i b^2)^2 \text{ for } i \le n,\ c_{n+1}^{-1} b c_{n+1} = b^2 \rangle$$

is an HNN extension of $A * \langle b \rangle$ with stable letters $c_1, \ldots, c_{n+1}$.

In particular, $\{c_1, \ldots, c_{n+1}\}$ is a basis for a free subgroup of B. It follows that there is an isomorphism of $\langle c_1, \ldots, c_{n+1} \rangle$ to $\langle c_1^2, \ldots, c_{n+1}^2 \rangle$ sending c_i to c_i^2 for all i.

Hence we can form the HNN extension $C = \langle B, d;\ d^{-1} c_i d = c_i^2, \text{ all } i \le n+1 \rangle$ of B, and this will also contain $A * \langle b \rangle$, and hence also Q, as a subgroup.

The group $Y = \langle y, z;\ z^{-1} y z = y^2 \rangle$ is an HNN extension of the infinite cyclic group $\langle y \rangle$ with stable letter z. In particular, $\langle z \rangle$ is an infinite cyclic subgroup of Y, and we may form the HNN extension $T = \langle Y, t;\ t^{-1} z t = z^2 \rangle$.

In Y, non-trivial powers of y and z cannot be equal, by Britton's Lemma. It follows, using Britton's Lemma again, that $\{y, t\}$ is a basis for the subgroup of T which it generates.

Now take any word $w \in F(X)$. If w does not equal 1 in A then $[w,b]$ is not 1 in $A * \langle b \rangle$. Further, using Britton's Lemma, we then see that $[w,b]$ is not equal in B to any word in $\{c_1, \ldots, c_{n+1}\}$. By Britton's Lemma again, it follows that $\{[w,b], d\}$ is a basis for the subgroup of C which it generates.

Let H_w be obtained from the free product $C * T$ by adding the relations $[w,b] = t$ and $d = y$. Let G_w be $H_w * P$.

Thus we have obtained for each $w \in F(X)$ a finite presentation of a group G_w, and this construction is computable. Also, if w does not equal 1 in A then H_w is just an amalgamated free product of C and T. Thus, if $w \neq 1$ in A then G_w contains Q, and so, by the definition of Q, the group G_w does not have the property $\mathfrak{P}$.

On the other hand, if $w = 1$ in A, it is easy to check that the relations of H_w tell us that all the generators of H_w equal 1 in H_w. Hence G_w has the property $\mathfrak{P}$ if $w = 1$ in A, since it is then just P.

That is, $w = 1$ in A iff the presentation constructed presents a group with the property $\mathfrak{P}$.

Since we cannot decide whether or nor $w = 1$ in A, it follows that we cannot decide whether or not a finitely presented group has the property $\mathfrak{P}$.//

9.6 *HIGMAN'S EMBEDDING THEOREM*

We have proved in Lemmas 8 and 9 that any finitely generated subgroup of a finitely presented group is recursively presentable (and in Lemma 10 that a group with an r.e. presentation also has a recursive presentation). In this section we prove the converse, and deduce some important consequences.

Theorem 26 (Higman's Embedding Theorem) *A finitely generated group can be embedded in a finitely presented group iff it is recursively presentable.*

Before proving the theorem, we shall obtain some propositions that follow from it.

Corollary *Let G be a countably generated recursively presentable group. Then G can be embedded in a finitely presented group.*

Proof Let G have presentation $\langle X;R \rangle$, where X is countable and R is an r.e. subset of $F(X)$.

We showed in Proposition 1.36 that G can be embedded in a group H with two generators. The defining relators of H are obtained in a computable way from the members of R, and so they form an r.e. set, being the image of an r.e. set by a computable mapping. By the theorem, H can be embedded in a finitely presented group K.//

Proposition 27 *Let H be a subgroup of the finitely presented group* $\langle X;R\rangle$. *Then H is the image of an r.e. subset of* $F(X)$ *iff there is a fintely presented group K containing G and a finitely generated subgroup L of K such that* $G \cap L = H$.

Remark Let H be the image of the r.e. subset S of $F(X)$. Then H is also the image of $\langle S\rangle$, which is an r.e. subgroup of $F(X)$, and the counter-image of H in $F(X)$ is $\langle S \cup R^{F(X)}\rangle$, which is again r.e. Thus there are several different but equivalent ways of expressing the property considered.

Proof Suppose there are such groups K and L. Let K be $\langle Y;T\rangle^{\psi}$ and let L be $\langle Z\rangle^{\psi}$, where Y, Z, and T are finite. There is a homomorphism from $F(X)$ into $F(Y)$ which induces the inclusion of G into K. The counter-image of H in $F(X)$ is the counter-image under this map, which is a recursive function, of the subgroup $\langle Z\rangle T^{F(Y)}\rangle$, which is r.e. Hence the counter-image of H is itself r.e.

Conversely, suppose that H is the image of an r.e. subset S. Let G_1 be a group isomorphic to G, with H_1 being the subgroup corresponding to H, and take the amalgamated free product $M = G *_{H=H_1} G_1$. Let G have the finite presentation $\langle X;R\rangle$. Then M is finitely generated by $X \cup X_1$, and has as defining relators $R \cup R_1 \cup \{s^{-1}s_1$, all $s \in S\}$, which is an r.e. set. Thus, by the Embedding Theorem, M can be embedded in a finitely presented group K. Since $G \cap G_1 = H$, by construction, the required subgroup L is just G_1.//

Let A be a subset of $\mathbb{N}$ which is r.e. but not recursive. Then the group $G = \langle x_n$, all n; $x_n, n \in A\rangle$ has an r.e. presentation and an unsolvable word problem (for this presentation, though, being free, it has a solvable word problem when a suitable different generating set is taken) . By the corollary above, this group can be embedded in a finitely presented group K (with generators Z), by first embedding it in a finitely generated group H (with generators Y). Now, by the construction of H, there is a computable function sending n to an element of $F(Y)$ which is equal in H to x_n. Since Y and Z are finite, we saw towards the end of section 1 that any map from $F(Y)$ to $F(Z)$ is computable. Thus we have a computable function sending n to an element of $F(Z)$ which is equal in K to x_n. By Lemma 8, K has unsolvable word problem, since it contains G. This provides another proof that there is a finitely presented group with unsolvable word problem.

Proposition 28 *There is a finitely presented group which contains a copy of every recursively presentable group.*

Proof We may restrict attention to presentations with a countable generating set and an r.e. set of relators, since we may always enlarge a finite set of generators to a countable one and enlarge the defining relators by making each new generator a relator. Thus we need only look at presentations $\langle X;R\rangle$ where X is a fixed countable set $\{x_n, n \in \mathbb{N}\}$, and R is an r.e. subset of $F(X)$.

Letting the elements of $F(X)$ be $\{w_n, n \in \mathbb{N}\}$, which is more convenient than taking a general Gödel numbering, we know that there is a computable partial function $f:\mathbb{N} \to \mathbb{N}$ such that R is $\{w_n, n \in \mathbb{N}f\}$, and that any such f gives rise to an r.e. presentation.

Now it may be shown that there is a computable partial function $\Phi:\mathbb{N}^2 \to \mathbb{N}$ such that for any partial computable $f:\mathbb{N} \to \mathbb{N}$ there is some k with $nf = (n,k)\Phi$ for all n. The proof of this requires detailed knowledge of computable functions, but an indication can be given here. The heart of the argument is that computer programs are just words in some language, and we can tell whether or not a given word is a program. Thus it makes sense to define Φ by requiring $(n,k)\Phi$ to be the result of applying the kth program to the input n. We then find that Φ is computable and satisfies the required property. Note also that (by a version of Cantor's diagonal argument which is used in showing that $\mathbb{R}$ is uncountable) it is not hard to prove that $\{n; (n,n)\Phi \text{ is defined}\}$ is r.e. but not recursive.

We may then replace X by other countable sets, and we find that for each fixed k the presentation $\langle x_{nk}, n \in \mathbb{N};\ w_{nk}, \text{ where } n = (m,k)\Phi \text{ for some } m\rangle$ is an r.e. presentation (where w_{nk} comes from w_n by replacing each x_i by the corresponding x_{ik}), and that each group with an r.e. presentation has a presentation of this kind.

It follows that the group G with generators x_{nk}, all n and k, and with defining relators w_{nk} for all n and k with $n = (m,k)\Phi$ for some m contains copies of every recursively presentable group. Since Φ is a computable function, this presentation of G is itself an r.e. presentation. By the corollary to the Embedding Theorem, G embeds in a finitely presented group, as required.//

Theorem 29 *A finitely generated group has solvable word problem iff it can be embedded in a simple subgroup of a finitely presented group.*

Proof Suppose that $G \leq P \leq H$, where P is simple and H has finite presentation $\langle Y;S\rangle$. Let G have presentation $\langle X;R\rangle$ with X finite. Let

$\varphi:F(X) \to F(Y)$ induce the inclusion of G in P. Let N be $\langle R\rangle^{F(X)}$.

If P is trivial then so is G, and so G has solvable word problem. Otherwise let $u \in F(Y)$ represent in H a non-trivial element of P.

As H is finitely presented, by Lemmas 8 and 9 N is r.e. Hence we need only show that $F(X) - N$ is also r.e.

Take $w \in F(X)$. If $w \in N$ then $w\varphi \in \langle S\rangle^{F(Y)}$, and, as $u \notin \langle S\rangle^{F(Y)}$ by definition, we have $u \notin \langle S \cup \{w\varphi\}\rangle^{F(Y)}$.

If $w \notin N$, then $w\varphi$ represents a non-identity element of P. Since P is simple, we see that $u \in \langle S \cup \{w\varphi\}\rangle^{F(Y)}$.

Thus a partial algorithm for membership of $F(X) - N$ is to list the members of $\langle S \cup \{w\varphi\}\rangle^{F(Y)}$ and answer "Yes" if u occurs in this list.

Note that this argument is almost the same as that in Proposition 13. Note also that X can be infinite provided that the map φ from X to $F(Y)$ is computable.

To prove the converse, take a countably generated group G whose word problem with respect to some countable set of generators X is solvable. We shall assume that we have a Gödel numbering of X which is not necessarily a bijection. It is not enough to assume that G is finitely generated, because our construction below will lead to countably generated groups.

The Gödel numbering of X leads to a Gödel numbering of $F(X)$, as usual, and so also to a Gödel numbering of all pairs of elements (u,v) of $F(X)$. Because G has a solvable word problem, there is a recursive subset A of $\mathbb{N}$ such that the pairs (u_i,v_i) for $i \in A$ are all the pairs (u,v) such that $u \neq 1 \neq v$ in G.

We may now form the HNN extension G_1 of $G * \langle y\rangle$ with stable letters t_i for $i \in A$, with $t_i^{-1}u_iy^{-1}u_iyt_i = v_iy^{-1}u_iy$. This is an HNN extension because the relevant elements of $G * \langle y\rangle$ are cyclically reduced of length 4, and so both have infinite order.

Evidently we have a Gödel numbering of the generators of G_1. Also, given a word w in $F(X \cup \{y\})$ and a number i, we can first decide whether or not $i \in A$, and, if so, whether or not w equals in $G * \langle y\rangle$ a power of $u_iy^{-1}u_iy$ (or of $v_iy^{-1}u_iy$). For this holds iff w freely equals $w_1y^{-1}w_2y \ldots w_{2n-1}y^{-1}w_{2n}y$ (or the inverse of such an element) for some n, where each w_r equals u_i in G. Because G has solvable word problem, we can decide whether or not this holds.

It follows from Proposition 18 and the comments following it that G_1 also has solvable word problem.

Since G_1 is recursively generated and has solvable word problem, we may apply the same construction to embed G_1 in a group G_2. We may continue this process, getting groups G_k for all k, and we let P be $\bigcup G_k$.

Let p and q be non-identity elements of P. They are both in G_k for some k. By construction, there is a stable letter t and a generator y_k such that $t^{-1}py_k^{-1}py_kt = qy_k^{-1}py_k$ in G_{k+1}. Hence $q \in \langle p \rangle^P$, which shows that P is simple.

By the corollary to the Embedding Theorem, it is enough to show that P is recursively presented. Let us assume that the original Gödel numbering of X maps into the even integers. Let $J:\mathbb{N}^2 \to \mathbb{N}$ be a computable bijection. In going from G_k to G_{k+1} we add a new generator y_k and new generators t_{ik} where i is in some recursive set A_k. We get a Gödel numbering of the generators of P by mapping y_k to $2(0,k)J+1$ and t_{ik} to $2(i,k)J+1$. It is not difficult to check that the set of defining relations is r.e., as required.//

Before proving Higman's theorem, we need to look at the computers called Turing machines, and their relationship with modular machines. The discussion of Turing machines will be informal, using terms from the everyday world.

Consider a tape, infinite in both directions, divided into squares. On each square is written a letter from some alphabet. One letter, called "blank", plays a special role, in that all but finitely many squares are required to have the blank letter written on them; we simply say that such a square is blank.

In order to convey information to and from the tape, we need a read-write head, which at any instant is positioned on some square of the tape (consider a cassette recorder). The machine may be in any one of a finite number of states (for instance, position of switches, whether or not a capacitor is charged, etc.). At any instant what the machine does will depend on what state it is in and on what symbol it is scanning; it may do one or all of changing the symbol on the scanned square, changing its state, or moving the head one square right or left. Instructions detailing which actions are to be performed are given to us. Thus, started in a configuration which consists of some symbols on the tape, a particular square being scanned, and a particular state, the instructions will successively cause it to pass through various configurations. It may compute for ever, or it may reach a configuration for which no instructions are given about what to do when in that state scanning that symbol. If this happens the machine stops.

There is only a notational change if we assume that the alphabet consists of the numbers $0,1,\ldots,n$ for some n, with 0 being the blank, and that the states are also numbers (at most one of which can be a letter, else technical problems arise). Take m greater than any letter or state. Then any configuration can be described by four numbers; the state q, the letter a scanned, the number u whose m-ary representation is the sequence of letters to the left of the scanned square (taken from left to right, so the letter to the left of the scanned square is the coefficient of m^0), and the number v whose m-ary representation is the sequence of letters to the right of the scanned square (taken from right to left, so the letter to the right of the scanned square is the coefficient of m^0).

These four numbers can be conveniently combined into two, namely either $(um + a, vm + q)$ or $(um + q, vm + a)$, whichever turns out to be more convenient (sometimes one is better, sometimes the other). If we analyse what an element of $\mathbb{N}^2$ corresponding to a configuration changes to when we perform the next step in the computation, we find that the change is exactly that of some modular machine. It is this that leads to the definition of modular machines. For more about this topic, see the book by Cohen (1987) and the papers by Aanderaa and Cohen (1980a,b).

Turing machines can be used to compute numerical functions, and, in addition, because of their definition, they can be conveniently used to discuss questions of recursiveness for subsets of a free monoid without using a Godel numbering. In particular, let S be an r.e. subset of the free monoid on the finite set X. Then there is a Turing machine T which, started in a special starting state with an element w of the monoid written on the tape, the scanned square being the last letter of w, will halt iff $w \in S$, and if $w \in S$ will halt on a blank tape with its state being the starting state.

In particular, let C be a recursively presented group. Thus C is $\langle c_1, \ldots, c_n; S\rangle$, where S is an r.e. subset of the free group $\langle c_1, \ldots, c_n\rangle$. We can regard a word as being an element of the free monoid on $\{c_1, \ldots, c_{2n}\}$, where $c_{n+i} = c_i^{-1}$. We take the Turing machine of the last paragraph, and the corresponding modular machine M, and the integer m which is part of the definition of M.

To every word w there is a number α whose m-ary representation is w (regarding the element c_i for $i \le 2n$ as the integer i). Let I be the set of those $\alpha \in \mathbb{N}$ which correspond to some word. For $\alpha \in I$, let $w_\alpha(c)$ be the corresponding word. In particular, $w_0(c) = 1$, and $w_i(c) = c_i$ for $i \le 2n$, while $w_{\alpha m+i}(c) = w_\alpha(c)c_i$ for all α. We shall also write $w_\alpha(b)$ and $w_\alpha(bc)$ for the words obtained from $w_\alpha(c)$ by replacing each $c_j^{\pm1}$ (for $j \le n$) by $b_j^{\pm1}$ and $(b_jc_j)^{\pm1}$ respectively, where the b_j are new letters.

From the discussion above, when all the details are filled in, we find that, for $\alpha \in I$, $w_\alpha(c) \in S$ iff $(\alpha,0) \in H_0(M)$. It is this fact that we shall use in proving Higman's theorem.

Let K_M be the group contructed in the discussion of Proposition 20. Let U be the subset of K_M consisting of t and all the r-symbols and

l-symbols. We will write t_α instead of $t(\alpha,0)$.

Let $H_1 = K_M * (C \times \langle b_1, \ldots, b_n \rangle) * \langle d \rangle$, and let $b_{n+i} = b_i^{-1}$.

We know that $\{t_\alpha;\ \alpha \in I\}$ is a basis for the subgroup it generates. We may map H_1 to K_M by mapping K_M identically and mapping the other generators to 1. Then, by Corollary 1 to Proposition 1.8, it follows that $\{t_\alpha w_\alpha(b)d;\ \alpha \in I\}$ is also a basis for the subgroup it generates.

Hence we may form the HNN extension $H_2 = \langle H_1, p;\ p^{-1}t_\alpha p = t_\alpha w_\alpha(b)d,\ \text{all } \alpha \in I \rangle$. In particular $C \subseteq H_2$. We note for future reference that one of the relations of H_2 is $p^{-1}tp = td$.

Let A be the subgroup $\langle t, x, d, b_j\ (1 \le j \le n), p \rangle$ and, for $1 \le i \le 2n$, let A_i be the subgroup $\langle t_i, x^m, b_i d, b_j\ (1 \le j \le n), p \rangle$. Then A is just the HNN extension of the free group F with basis $\{t, x, d, b_j\ (1 \le j \le n)\}$ with stable letter p, the relations being $p^{-1}t_\alpha p = t_\alpha w_\alpha(b)d$ for all $\alpha \in I$.

We know that $\langle t_i, x^m \rangle \cap \langle t_\alpha\ (\text{all } \alpha \in I) \rangle$ is just $\langle t_\beta\ (\text{all } \beta \in I \text{ with } \beta \equiv i \bmod m \rangle$. Mapping F onto the free group with basis $\{t, x\}$ by sending d and b_j to 1, we see that $\langle t_i, x^m, d, b_j\ (1 \le j \le n) \rangle \cap \langle t_\alpha\ (\text{all } \alpha \in I) \rangle$ is also $\langle t_\beta\ (\text{all } \beta \in I \text{ with } \beta \equiv i \bmod m) \rangle$. The group $\langle t_i, x^m, d, b_j\ (1 \le j \le n) \rangle \cap \langle t_\alpha w_\alpha(b)d\ (\text{all } \alpha \in I) \rangle$ is, for the same reason, $\langle t_\beta w_\beta(b)d\ (\text{all } \beta \in I \text{ with } \beta \equiv i \bmod m) \rangle$. In particular, Proposition 1.34 tells us that A_i is the HNN extension of the free group $\langle t_i, x^m, b_i d, b_j\ (1 \le j \le n) \rangle$ with stable letter p, the HNN relations being $p^{-1}t_\beta p = t_\beta w_\beta(b)d$ for all $\beta \in I$ with $\beta \equiv i \bmod m$. Since $w_{\alpha m + i}(b) = w_\alpha(b)b_i$, it is easy to see from this that there is an isomorphism from A to A_i which maps each of the stated generators of A to the corresponding stated generator of A_i.

Now let A_+ be the subgroup $\langle U, d, b_j\ (1 \le j \le n), p \rangle$ and let A_- be the subgroup $\langle U, d, b_j c_j\ (1 \le j \le n), p \rangle$. We will show that there is an isomorphism of A_+ onto A_- under which the stated generators correspond.

There is a homomorphism of H_2 to itself which maps each c_j to 1 and maps every other generator to itself. This induces a homomorphism from A_- to A_+ for which the relevant generators correspond.

We know that $\langle U \rangle \cap K = \langle t(\alpha,\beta),\ \text{all } (\alpha,B) \in H_0(M) \rangle$, so that $\langle U \rangle \cap \langle t_\alpha\ (\text{all } \alpha \in I) \rangle = \langle t(\alpha,\beta),\ \text{all } (\alpha,\beta) \in H_0(M) \rangle \cap \langle t_\alpha,\ \text{all } \alpha \in I \rangle =$ $\langle t_\alpha,\ \text{all } \alpha \in I \text{ with } (\alpha,0) \in H_0(M) \rangle$. As above, it follows that $\langle U, d, b_j\ (1 \le i \le n) \rangle \cap \langle t_\alpha,\ \text{all } \alpha \in I \rangle = \langle t_\alpha,\ \text{all } \alpha \in I \text{ with } (\alpha,0) \in H_0(M) \rangle$, and that $\langle U, d, b_j\ (1 \le j \le n) \rangle \cap \langle t_\alpha w_\alpha(b)d,\ \text{all } \alpha \in I \rangle = \langle t_\alpha w_\alpha(b)d,\ \text{all } \alpha \in I \text{ with } (\alpha,0) \in H_0(M) \rangle$. In particular, by Proposition 1.34, A_+ is just the HNN extension of the free product of U and the free group with basis $\{d, b_j (1 \le j \le n)\}$ with stable letter p, the HNN relations being $p^{-1}t_\alpha p = t_\alpha w_\alpha(b)d$ for all $\alpha \in I$ with $(\alpha,0) \in H_0(M)$. Thus, to show that the given correspondence of generators provides a homomorphism of

A_+ to A_- (which will then be an isomorphism, since we have found a homomorphism which is its inverse), it is enough, by von Dyck's theorem, to show that $p^{-1}t_\alpha p = t_\alpha w_\alpha(bc)d$ for all $\alpha \in I$ with $(\alpha,0) \in H_0(M)$. Because each b_j commutes with each c_i, we have $w_\alpha(bc) = w_\alpha(b)w_\alpha(c)$ for all α. By the definition of M, we have $w_\alpha(c) = 1$ in C when $(\alpha,0) \in H_0(M)$. Since $p^{-1}t_\alpha p = t_\alpha w_\alpha(b)d$ for all α, the required relation does hold.

It follows that there is an HNN extension H_3 of H_2 whose stable letters are $a_i, 1 \le i \le 2n$, and k, the subgroups associated with a_i being A and A_i, and the subgroups associated with k being A_+ and A_-.

Then H_3 contains C, and it is plainly finitely generated. We will show that it is finitely related, which will complete the proof of Higman's theorem.

The relations of H_3 consist of the following: (I) the finitely many relations of K_M, the relations $b_ic_j = c_jb_i$ for $1 \le i, j \le n$, the relation $p^{-1}tp = td$, and the finitely many relations involving the stable letters of H_3, (II) the relations $p^{-1}t_\alpha p = t_\alpha w_\alpha(b)d$ for all $\alpha \in I$ with $\alpha \neq 0$, and (III) the relations corresponding to the relators in S, that is to say the relations $w_\alpha(c) = 1$ for all $\alpha \in I$ with $(\alpha,0) \in H_0(M)$.

We know that one consequence of the relations of K_M is that if $(\alpha,0) \in H_0(M)$ then t_α can be written as a product of elements of $U \cup U^{-1}$. Then, as a consequence of the relations involving k, we have $k^{-1}t_\alpha k = t_\alpha$ if $(\alpha,0) \in H_0(M)$. Since $k^{-1}pk = p$ and $k^{-1}dk = d$, we have, as consequences of the relations (I) and (II), the relations $k^{-1}w_\alpha(b)k = w_\alpha(b)$ for all α with $\alpha \in I$ and $(\alpha,0) \in H_0(M)$. However, the relations in (I) give $k^{-1}w_\alpha(b)k = w_\alpha(bc) = w_\alpha(b)w_\alpha(c)$. Thus the relations (III) are consequences of (I) and (II).

Regard a word $w_\alpha(b)$ as a word involving only positive powers of b_i for $1 \le i \le 2n$ (recall that b_{n+i} is b_i^{-1}). Let $w_\alpha(a)$ be obtained from the word $w_\alpha(b)$ by replacing each b_i by a_i for $1 \le i \le 2n$. It is easy to check, by induction on the length of the word, that $w_\alpha(a)^{-1}tw_\alpha(a) = t_\alpha$ and $w_\alpha(a)^{-1}dw_\alpha(a) = w_\alpha(b)d$ are consequences of the relations in (I). Since the relations $p^{-1}tp = td$ and $a_i^{-1}pa_i = p$ for all i are in (I), we see that all the relations in (II) are consequences of the relations in (I). Since (I) is finite, we have proved Higman's Embedding Theorem.

9.7 *GROUPS WITH ONE RELATOR*

Groups with only one defining relator have nice properties. In particular, their word problem is solvable. This class of groups includes the fundamental groups of compact 2-manifolds, which are either

$\langle a_1, b_1, \ldots, a_n, b_n;\ \Pi[a_i, b_i]\rangle$ or $\langle c_1, \ldots, c_m;\ \Pi c_i^{\,2}\rangle$. The word problem for these was originally solved by Dehn (1912a,b) using geometrical methods.

Theorem 30 (Freiheitssatz) *Let G be $\langle X;r\rangle$, where r is cyclically reduced. Let $Y \subset X$ be such that r involves a generator in $X-Y$. Then the subgroup of G generated by Y is free with basis Y.*

Proof We use induction on the length of r. Notice first that the result is obvious when r involves only one generator (no matter what the length of r is), as in this case G is just the free product of a finite cyclic group and a free group.

So we assume that r involves at least two generators. First suppose that some generator t occurs in r with exponent sum 0. Let $a, b, \ldots$ be the remaining generators, where a occurs in r. Taking a cyclic conjugate of r if necessary, we may assume that r begins with $a^{\pm 1}$. Note that a could be any generator occurring in r other than t.

Write a_i for $t^{-i}at^i$, etc. Because t occurs with exponent sum 0 in r, we can write r as a word s in a_i, $b_i, \ldots$, and s will have shorter length than r. For instance, if r is $a^2t^{-1}b^3t$ then s is $a_0^{\,2}b_1^{\,3}$. Notice that our assumption on r tells us that s begins with $a_0^{\,\pm 1}$.

Let m be the smallest index such that a_i occurs in s, and let M be the largest index such that a_i occurs in s. Thus $m \le 0 \le M$. It is easy to check, using Tietze transformations, that G has the presentation

$\langle t,\ a_i\ (m \le i \le M),\ b_i\ (\text{all } i), \ldots;\ s,\ t^{-1}a_it = a_{i+1}\ (m \le i \le M-1),\ t^{-1}b_it = b_{i+1}\ (\text{all } i), \ldots\rangle$.

Here we may have $m = M$, in which case the relators involving a_i are omitted.

Let H be the group $\langle a_i\ (m \le i \le M),\ b_i\ (\text{all } i), \ldots;\ s\rangle$. Since s is shorter than r, we may assume (inductively) that the subgroups $\langle a_i\ (m \le i \le M-1),\ b_i\ (\text{all } i), \ldots\rangle$ and $\langle a_{i+1}\ (m \le i \le M-1),\ b_{i+1}\ (\text{all } i), \ldots\rangle$ are free with the given generators as bases, since they do not involve all the generators occurring in s. Note that if $m = M$ we omit the generators a_i from these subgroups. The presentation we have obtained for G then shows that G is an HNN extension of H.

Now let Y be a subset of X which does not contain every generator occurring in r. Suppose first that $t \notin Y$. Then Y is just a subset of H, regarding $a, b, \ldots$ as $a_0, b_0, \ldots$. Since t does occur in r, some generator of H which is not in Y must occur in s. The result now follows by induction.

Since a could be any element of X occurring in r other than t, if $t \in Y$ we may assume that $a \notin Y$. Let $w \in F(X)$ be a non-trivial reduced word not involving a. If the exponent sum of t in w is not 0 then w cannot equal 1 in G. If the exponent sum of t in w is 0 then we can write w as a non-trivial reduced word in $b_i, \ldots$. Inductively, such a word, as it does not involve the generator a_0 which occurs in s, cannot be 1 in H, and so cannot be 1 in G.

Now t in the above argument could have been any generator which occurs with exponent sum 0 in r. Hence the result is proved unless every generator occurring in r does so with non-zero exponent sum. In this case we choose t to be any generator occurring in r but not in Y, and we let a be any other generator occurring in r.

Let the exponent sums of t and a in r be τ and α (both non-zero, by assumption). Let G_1 be the group $\langle X,p;\ r,\ p^{\tau}a^{-1}\rangle$. This is a one-relator group, since we can simply omit the generator a and the second relator. Further, we may define q to be tp^{α}, and we have a presentation of G_1 generated by $X_1 = \{p,q\} \cup (X - \{t,a\})$, and with a single relator r_1. It is easy to check that with this generating set the exponent sum of p in r_1 is 0, by our choice of p and q. Thus the construction above gives G_1 as an HNN extension of a group H_1 with one relator s_1.

Now r_1 may be longer than r, because various powers of p have been inserted. But the length of s_1 is the total number of occurrences of generators other than p in r_1, and this is the same as the number of occurrences of generators other than a in r. Hence s_1 is shorter than r. It follows by the argument of the first case that the subgroup of G_1 generated by $p \cup (Y - \{a\})$ is freely generated by this set.

Now let w be a non-trivial reduced word of $F(X)$ which only involves the generators in Y. Let w' be obtained from w by replacing a by p^{τ}. Then $w = w'$ in G_1. Since w' is a non-trivial reduced word in $\{p\} \cup (Y - \{a\})$, we see that $w' \neq 1$ in G_1, and so $w \neq 1$ in G, as required.//

When no generator occurs in r with exponent sum 0, the group G_1 is just the amalgamated free product of G and the infinite cyclic group $\langle p\rangle$, with a being identified with p^{τ}. This is immediate, once we have see (by the Freiheitssatz) that a has infinite order in G.

We give a similar inductive proof that the word problem for one-relator groups is solvable. In order to give an inductive proof, we need to prove a stronger result.

Theorem 31 *Let G be $\langle X;r\rangle$, where X is finite or countable. Let Y be a recursive subset of X. Then the membership problem for the subgroup $\langle Y\rangle$ of G is solvable. In particular, the word problem for G is solvable.*

Proof We first note that it is enough to prove that the result holds when Y is obtained from X by deleting one generator which occurs in r.

For suppose we have shown this. If we are given a recursive set Z which omits at least one such generator, then we can regard Z as a recursive subset of such a Y. Now, by the Freiheitssatz, the subgroup $\langle Y\rangle$ of G is free with basis Y. Thus the membership problem for $\langle Z\rangle$ in $\langle Y\rangle$ is solvable, and as we are assuming that the membership problem for $\langle Y\rangle$ in G is solvable, Lemma 16 tells us that the membership problem for $\langle Z\rangle$ in G is solvable.

As a particular case, taking Z to be empty, we see that the word problem for G is solvable.

On the other hand, suppose that Z contains all the generators occurring in r. We can write G as the free product of a group G' with a finite set X' of generators and one relator r and a free group with basis X'', say. Now the word problem for G' will be solvable (by the argument above). Since $\langle Z\rangle = G' * \langle Z \cap X''\rangle$, we see that $\langle Z\rangle$ has solvable membership problem in G by Lemma 17.

So we are left to prove the result when Y is obtained from X by deleting a generator occurring in r. If R only involves one generator then G is the free product of a finite cyclic group and a free group, and the result is then obvious.

Suppose first that some generator t occurs in r with exponent sum 0. We use the notation of the Freiheitssatz. Thus G is an HNN extension of a one-relator group H whose relator is shorter than r. Inductively, the membership problems for the two associated subgroups of H are solvable. Now we may regard any reduced word w in $F(X)$ as a word in $t, a_0, b_0, \ldots$ .As in Proposition 18 we may compute from this another word v in these latter generators such that v has no pinches and $w = v$ in G.

Let Y be $X - \{t\}$. Then $\langle Y\rangle = \langle Y_0\rangle \subseteq H$, where $Y_0 = \{a_0, b_0, \ldots\}$. Then w represents in G an element of $\langle Y\rangle$ iff, first, t does not occur in v and, if so, v represents in H an element of $\langle Y_0\rangle$, and this is decidable, by induction.

Now let Y be $X - \{a\}$. Let the exponent sum of t in w be τ. We may assume that $\tau = 0$, since w represents an element of $\langle Y\rangle$ iff $wt^{-\tau}$ represents an element of $\langle Y\rangle$. Then the exponent sum of t in the corresponding word v is also 0. Also, if $w = u$ in G, where $u \in \langle X - \{a\}\rangle$, then t has exponent sum 0 in u (since t has exponent sum 0 in r and wu^{-1} is a consequence of r).

It follows that u can be written as a word in b_i (all i) etc. (not involving t or any a_i). Since v has no pinches and t does not occur in u, we can only have $vu^{-1} = 1$ in G if t does not occur in v and $v = u$ in H.

Thus, to decide whether or not w represents in G an element of $\langle X - \{a\}\rangle$ we first find v, and when t does not occur in v decide (which is possible by induction) whether or not v represents in H an element of $\langle b_i, \ldots\rangle$.

We are left with the case when no generator occurs with exponent sum 0 in r. We use the same notation as before, and we assume that Y is $X - \{t\}$. We have seen that G_1 is the amalgamated free product of G and an infinite cyclic group. Consequently it is enough to decide whether or not a word $w \in F(X)$ represents in G_1 an element of $\langle Y\rangle$. Now we may compute an element w^* in the generators X_1 of G_1 such that $w = w^*$ in G_1 (simply replacing t and a by $qp^{-\alpha}$ and p^{τ}). Thus it is enough to show that the membership problem for $\langle Y\rangle$ in G_1 is solvable.

As in the proof of the Freiheitssatz, we may write G_1 as an HNN extension of a one-relator group whose relator is shorter than r. The first part of the current proof now tells us that the membership problem for the subgroup $\langle Y_1\rangle$ of G_1, where $Y_1 = \{p\} \cup (Y - \{a\})$, is solvable. Also $\langle Y_1\rangle$ is free with basis Y_1. The membership problem for $\langle Y\rangle$ in $\langle Y_1\rangle$ is then also solvable, by Lemma 17, since Y is just $\{p^{\tau}, b, \ldots\}$. The result follows by Lemma 16.//

Exercise 11 Let G be $\langle X; s^n\rangle$, where s is not a proper power. Show that the only elements of finite order in G are the conjugates of powers of s.

Exercise 12 Let G be $\langle a,b;\ a^{-1}b^2a = b^3\rangle$, and let $c = b^2$. Show that G is generated by $\{a,c\}$, and find a presentation in terms of these generators. Show that there is an endomorphism of G onto itself which sends a to a and b to c. Show, using Britton's Lemma, that this endomorphism is not one-one.

10 FURTHER TOPICS

In this chapter we will consider a few other topics in combinatorial group theory where the methods have a topological flavour.

10.1 *SMALL CANCELLATION THEORY*

Small cancellation theory is one of the major aspects of combinatorial group theory. The methods are somewhat more geometrical than topological (insofar as it is possible to make such a distinction). I have something of a blind spot in this area, so I only summarise the results. For details see Lyndon and Schupp (1977). The paper by Greendlinger and Greendlinger (1984) simplifies one of the proofs given there.

Suppose that, in $F(X)$, we have $w = \Pi_{i=1}^{i=m} u_i r_i u_i^{-1}$, where w and each u_i and r_i are reduced. Then it is possible to make a diagram in the plane, composed of regions, edges, and vertices, with each edge being given a label from $X \cup X^{-1}$, in such a way that the boundary of the whole diagram is a sequence of edges whose label (up to cyclic permutation) is w, while there are m regions, whose boundaries consist of sequences of edges whose labels are (up to cyclic permutation) the r_i.

Let R be a subset of $F = F(X)$, and let $N = \langle R\rangle^F$. We will assume that R is symmetrised; that is, that if $r \in R$ then all cyclic permutations of r and of r^{-1} are in R.

We define a *piece* (of R) to be an element u of F such that there are distinct r_1 and r_2 in R with $r_1 = uv_1$ and $r_2 = uv_2$, both reduced as written (r_2 is permitted to be r_1^{-1} or a cyclic permutation of r_1 or r_1^{-1}). For instance, if r_1 is a^3b then r_2 could be a^2ba, so that a^2b is a piece.

We say that R satisfies the condition $C(n)$ if no element of R is the product of fewer than n pieces, and we say that R satisfies the condition $C'(\lambda)$ if for any $r \in R$ and any piece p of r we have $|p| < \lambda|r|$.

Another condition $T(n)$, $n \geq 3$, is also useful. We say that R satisfies $T(n)$ if whenever we have $r_1, \ldots, r_k \in R$ for some $k < n$ then either at least one of the products $r_1r_2, r_2r_3, \ldots, r_{k-1}r_k, r_kr_1$ is reduced as written or at least one of them is the identity.

The conditions $C(n)$ and $T(n)$ give conditions on the number of boundary edges of a region and the number of edges at a vertex, respectively. There is a well-known formula due to Euler which relates the number of vertices, edges, and regions of a map (this formula is used in showing that there are only five regular polyhedra). A detailed analysis of this formula, and related properties, gives a proof of the following propositions.

Proposition 1 *Let R be a symmetrised set satisfying either* $C'(1/6)$ *or both* $C'(1/4)$ *and* $T(4)$. *Then any non-trivial element of N contains more than half of some element of R.//*

As already remarked in Proposition 9.14, this allows us to solve the word problem for the group $\langle X;R\rangle$ when R is finite and satisfies the above condition. It is also clear that the solution of the word problem is linear, in the sense that if $w \in N$ then w is the product of at most $|w|$ conjugates of elements of R, and we can show that each conjugating element has length at most $|w|$. As remarked after Proposition 9.14, the solution is linear in another sense.

Proposition 2 *There is a number K such that for any symmetrised R satisfying either* $C(6)$, *or* $C(4)$ *and* $T(4)$, *or* $C(3)$ *and* $T(6)$ *the minimal diagram for an element* $w \in N$ *has at most* $K|w|^2$ *regions.//*

Thus, under these conditions, w is the product of m conjugates of elements of R, where $m \leq K|w|^2$. Also, when R is finite, there is a bound (in terms of m and $|w|$, and hence in terms of $|w|$) for the lengths of the conjugating elements. Thus we can solve the word problem for such finitely presented groups by simply checking whether or not w is one of the finitely many products of conjugates of elements of R allowed by these conditions.

It is also possible to give similar conditions under which the conjugacy problem can be solved.

These results have a strong geometric flavour. Proposition 2 is closely related to the *isoperimetric inequality*, which states that any set in $\mathbb{R}^2$ whose boundary has a given length L has area $\leq L^2/4\pi$, the maximum occurring only for a disk. This result was known from classical Greek times (with a legend relating to the foundation of the city of Carthage), but a full proof did not come until modern times.

Versions of the isoperimetric inequality hold not just in $\mathbb{R}^2$ but, more generally, in surfaces of non-positive curvature (and in surfaces of strictly negative curvature, there is a linear inequality, rather than a quadratic one; Proposition 1 is connected with this fact). The various conditions imposed on R are essentially ways of regarding the diagrams as surfaces of non-positive curvature.

The theory can be extended without essential difficulty (but with some complication in the details) to the situation where R is a subset of a free product, rather than of a free group. Length will, of course, now refer to length in the free product. We now define a piece to be a reduced word u in the free product such that there are reduced words v_1 and v_2 and elements r_1 and r_2 of R such that $r_1 = uv_1$ and $r_2 = uv_2$, where these products are without cancellation but may have coalescence. The following is one interesting application due to Schupp (1971).

Proposition 3 *Let X and Y be non-trivial groups, not both of order 2. Then every countable group is isomorphic to a subgroup of some quotient group of X*Y.//*

As remarked, the theory of small cancellation groups is closely related to the theory of spaces of non-positive curvature. The various conditions given in Proposition 2 are enough to show that a certain space has non-positive curvature. Juhasz (1986, 1987a,b) has given variants of these conditions, which are not uniform (in the sense that different conditions apply for different regions and vertices of the diagram, while in Proposition 2 there is no distinction between one region or vertex and another), from which similar conclusions can be drawn. He has recently (unpublished, even as preprint, at present) shown that any one-relator group satisfies one of a number of such conditions, from which he has shown that the conjugacy problem for one-relator groups is solvable.

Gromov has developed a theory of hyperbolic groups, closely related to small cancellation theory. Similar isoperimetric inequalities hold for these groups, and so the word problem can be solved for them. A survey of this theory is given in an article by Gromov in the book edited by Gersten (1987).

As already mentioned, there is a theory of automatic groups developed in a long paper (of which preprints are available from Epstein) by Cannc Epstein, Holt, Paterson and Thurston. If this theory only told of how and when we could compute quickly in groups, it would be of practical importance but perhaps not of great theoretical importance. However, it turns out that small cancellation groups, hyperbolic groups, and other important groups such as some knot groups, are all automatic groups. The theory of automatic groups is moving very rapidly at present, and the generalisations of small cancellation theory and related results are very interesting.

10.2 *OTHER TOPICS*

Whitehead (1936a,b) used topological methods to prove some results about automorphisms of free groups. His results were given an algebraic form by Higgins and Lyndon (1974), and Hoare (1979) gave a proof using some aspects of graph theory.

Let F be free on the finite set X. A *Whitehead automorphism* α is an automorphism which is either a permutation of $X \cup X^{-1}$ or is such that, for some fixed $a \in X \cup X^{-1}$, for all x $x\alpha$ is one of x, xa, $a^{-1}x$, or $a^{-1}xa$, and $a\alpha = a$. Plainly there are only finitely many Whitehead automorphisms.

We shall consider *cyclic words*; that is, cyclically reduced words, with a cyclic permutation of a word being identified with the word itself. Two cyclic words are *equivalent* if there is an automorphism which sends one to the other. A cyclic word w is *minimal* if $|w| \leq |w\alpha|$ for all automorphisms α.

Theorem 4 *Let u and v be equivalent cyclic words, and let v be minimal. Then there are Whitehead automorphisms $\alpha_1, \ldots, \alpha_n$ (for some n) such that $v = u\alpha_1 \ldots \alpha_n$ and $|u\alpha_1 \ldots \alpha_i| \leq |u|$ for all i. Further, if u is not minimal then these inequalities are all strict.//*

It follows at once that we can determine for all u an equivalent minimal cyclic word. For either u is itself minimal or its length can be reduced by applying one of the finite set of Whitehead automorphisms. After performing the latter operation a number of times, we will reach (and recognise) a minimal cyclic word equivalent to u.

We can also decide for all pairs of cyclic words (and therefore for all pair of words, since two words which give the same cyclic word come from each other by an inner automorphism) whether or not they are equivalent. Given any two cyclic words u and v, we can find minimal cyclic word u_0 and v_0 which are equivalent to u and v, resepctively. Then u and v are equivalent iff u_0 and v_0 are equivalent. If this happens, there is a sequence of distinct cyclic words $w_0, \ldots, w_n$ for some n, all of the same length, with $w_0 = u_0$ and $w_n = v_0$, such that w_{i+1} comes from w_i by a Whitehead automorphism. Since the words are distinct and all of the same length, there is a computable upper bound for n, and we simply have to see whether or not v_0 can be obtained from u_0 by applying a sequence of at most this number of Whitehead automorphisms.

More detailed analysis enabled McCool (1975b) to prove the following.

Proposition 5 *The subgroup of the automorphism group of F which fixes a finite set of words (or cyclic words) is finitely presentable; further, there is a procedure to determine a finite presentation of this subgroup.*

McCool (1974) was also able to obtain explicitly a presentation of Aut$F$, and later (1975a) to obtain by Tietze transformations another presentation which was originally obtained using difficult methods by Nielsen.

Generalisations of the theory were obtained by Gersten (1984) who was able to prove that *we can decide, given two subgroups H and K, whether or not there is an automorphism of F which sends H to K.*

Collins (in an article in the book by Gersten and Stallings (1987) and elsewhere) has extended the theory of Whitehead automorphissm to free products.

Howie (1981) uses CW-complexes (the topological form of the complexes we have discussed) to obtain some results on equations over groups. This is just one of many places where deeper aspects of topology than those we have looked at enable us to prove results in group theory.

According to Propositions 8.47 and 8.48 a set V is the set of vertices of a tree iff it is a metric space with an integer-valued metric d such that *for any a and b in V the function sending x to* $(a,x)d$ *is a bijection from* $\{x;\ (a,x)d+(x,b)d=(a,b)d\}$ *to the interval* $[0,(a,b)d]$. If we removed the condition that the metric is integer-valued we would get a space which is called an $\mathbb{R}$-*tree*. More generally, we could require the function d to map into any ordered group Λ, and we would obtain a Λ-tree. The theory of these parallels some topological results in an important way, in which not only does the topology influence the group theory, but the group theory feeds back into the topology. For more about this subject see an article by Shalen called "Dendrology of groups" in the book edited by Gersten (1988).

NOTES AND REFERENCES

These notes and references are highly selective. For some topics I have listed all or most relevant papers, while for others I have only listed those I have actually used. In particular, the work of Bernard Neumann and of Roger Lyndon is so influential in combinatorial group theory that, rather than making references to their work on almost every page, I have cited their papers only occasionally.

CHAPTER 1

Van der Waerden's method for proving the Normal Form Theorem for free groups appears in van der Waerden (1948). The Diamond Lemma approach, in the form given in the exercises, appears in Newman (1942). See also Bergman (1978) for more about this interesting topic, which comes up in many situations requiring normal forms.

Von Dyck's theorem is one of the earliest in combinatorial group theory, dating from 1882.

Britton's Lemma was first proved in Britton (1963). The argument that proves both Britton's Lemma and the results on residual finiteness of HNN extensions appears in Cohen (1977) and in Baumslag and Tretkoff (1978).

The general theory of HNN extensions was begun by Higman, Neumann, and Neumann (1949), in which theorems 36-39 occurr. The finitely generated infinite simple group was obtained by Higman (1951), while Schupp (1971) gives a proof that any countable group embeds in a quotient of the group G of Theorem 40.

Miller and Schupp (1973) give a proof of Britton's Lemma using small cancellation theory (see Chapter 10), and this does not require the Axiom of Choice.

Further general results on combinatorial group theory can be found in the books by Magnus, Karrass, and Solitar (1966) and by Lyndon and Schupp (1977), and the historical account of Chandler and Magnus (1982). For more about the topological connections see the long articles by Collins and Zieschang (to appear) and by Scott and Wall (1979), as well as some of the books cited.

CHAPTER 2

For more about spaces, paths, etc., see books on topology, in particular Brown (1968, 1989) and Massey (1967).

CHAPTER 3

The book by Higgins (1971) gives a full account of the theory of groupoids and its applications (many of which are algebraic forms of our later geometric theorems).

CHAPTER 4

Van Kampen's theorem appears in van Kampen (1933). A very similar result had been obtained earlier by Seifert (1931), and the theorem is sometimes referred to as the Seifert-van-Kampen theorem. The proof using groupoids is due to Brown (1967).

The example of two contractible spaces whose join is not simply connected comes from Griffiths (1954), but has been modified so that the calculation of the fundamental group of the Hawaiian earring in Griffiths (1956) is not needed. Morgan and Morrison (1986) show that Griffiths' proof (1956) is wrong, and they provide a correct proof of his result. Lemma 25 and Proposition 26 come from their 1986 paper, with the proofs changed slightly to tidy up some minor difficulties in their argument.

CHAPTER 5

The result that Fixα is finitely generated when α is an automorphism of a finitely generated free group has proofs by Culler and Vogtmann (1986), Cooper (1987), and Gersten (1987), as well as the proof by Goldstein and Turner (1986) which I have followed. The results on the rank of Fixα come from M. Cohen and Lustig (1988). Bestvina and Handel (1989) show that Fixα has rank at most rank$F$.

CHAPTER 7

The Schreier and Reidemeister-Schreier theorems were proved in Schreier's paper (1927) and Reidemeister's book (1932, 1950). The Kurosh subgroup theorem appears first in Kurosh (1934).

The covering space argument for the Kurosh subgroup theorem appears in Baer and Levi ((1936). The algebraic proof of the Reidemeister-Schreier and Kurosh theorems is given in the form due to Weir (1956).

Further applications of covering spaces to group theory may be found in papers by myself (1963) and by Tretkoff (1973a, 1973b, 1975).

CHAPTER 8

The general results of Bass-Serre theory may be found in the book by Serre and its translation (1977, 1980).

Theorem 6 was proved by Serre; essentially the same proof, but without the details, can be found in the book by Reidemeister (1932, 1950).

Nielsen (1921, more easily available in the 1986 collection of his papers) proved his theorem for finitely generated subgroups. The general case is due to Federer and Jonsson (1950).

The discussion of weakly reduced sets comes from Hoare (1976, and also 1981).

The theory of abstract length functions on groups was started by Lyndon (1963). This has been developed in many ways, including the cited papers by Hoare and papers by Chiswell (1976a, 1979a, 1981).

In the proof of the First Structure Theorem I have followed the approach of Chiswell (1979b).

A detailed proof of the subgroup theorems for amalgamated free products and HNN extensions using Bass-Serre theory was first given by Cohen (1974). Earlier purely algebraic proofs were given by Karrass and Solitar (1970, 1971), improved by Karrass, Pietrowski, and Solitar (1974). Their method was extended to arbitrary tree products by Fischer (1975).

The version of the Kurosh Subgroup Theorem obtained from Theorem 27 is slightly weaker than that in Chapter 7, but is almost always enough for any uses we need to make of it. To obtain the precise form we must see how a Kurosh system gives rise to a representative tree.

Let $G = *G_\alpha$. Then G is the fundamental group of a graph of groups $(\mathfrak{G},X)$, where X has vertices x_α at which the vertex group is G_α, and another vertex x_0 which has trivial vertex group, with one edge pair joining x_0 and x_α. The universal cover of $\mathfrak{G}$ is a G-tree Y, which has a vertex y_0 with trivial stabiliser, vertices y_α with stabiliser G_α, an edge-pair $\{e_\alpha,\bar{e}_\alpha\}$ joining y_0 and y_α, with every vertex and edge being in the G-orbit of one of these.

Suppose that we are given a Kurosh system. Let S be the subgraph of Y with vertices uy_α and uy_0 for all α-representatives in the large and all α, together with the edges joining uy_0 and uy_α and the edges joining uy_0 and uy_β whenever u ends in G_β. The conditions on a Kurosh system ensure that S is connected.

Choose some index, which we denote by 1. Obtain T from S by adding those vertices gy_0 and the edges joining gy_0 to gy_1 for all g such that g is a 1-representative for which $Hgy_0 \cap S = \emptyset$.

Unless there are representatives in the large u and v with $Hu = Hv$, it is easy to check that T is a representative tree for H, and the detailed application of the Second Structure Theorem will give the Kurosh Subgroup Theorem in the precise form obtained in Chapter 7.

It is part of the algebraic proof of the theorem that we must have $Hu \neq Hv$ for distinct representatives in the large u and v. This should be easy to prove directly, but I am not sure how. It is easy, however, to find a Kurosh system which obviously satisfies this condition. We need only construct the Kurosh system as in Chapter 7 by induction on the length of the shortest element of the double coset, and, for the double cosets of a given length, construct the representatives in the large in some order (for instance, using a well-ordering of the index set), ensuring that at each step we choose our representative among the representatives already chosen, if this is possible.

The results on finitely generated subgroups are due to Cohen (1974, 1976), based on algebraic arguments of Burns (1972, 1973). The proof of Grushko's theorem is based on that of Stallings (1987). The original proof is due to Grushko (1940), with additions and extensions from Burns and Chau (1984), Chiswell (1976b), Higgins (1966), Neumann (1943), Stallings (1965), and Wagner (1957). M. Hall's theorem first appears in Hall (1949b); see also Burns (1971b). Tretkoff (1975) first proved that the free product of M. Hall groups is an M. Hall group. The proof given here is new, but is essentially no more than a translation into Bass-Serre theory of the proofs of Tretkoff and of Imrich (1977). Howson's theorem was proved in Howson (1954); the graphical proof was first published by Imrich (1977). An improvement of the bound was given in an algebraic proof by Burns (1971a), which has been given a graphical form by Nickolas (1985). These improvements are still of the order of magnitude $2(\text{rank}A-1)(\text{rank}B-1)$ rather than the conjectured bound (which we know could not be improved) of $(\text{rank}A-1)(\text{rank}B-1)+1$.

The methods of Imrich (1977) are graphical but rather different from those of Bass-Serre theory, and are an interesting alternative.

The Almost Stability Theorem appears in the book by Dicks and Dunwoody (1989), which gives an account of Bass-Serre theory and some other of its applications with a very different flavour from the approach here. The results on ends of groups first appear (for finitely generated groups) in Stallings (1968), which also proves that groups of cohomological dimension 1 over $\mathbb{Z}$ are free. Swan (1969) extended this result on cohomological dimension to arbitrary groups; The book by Cohen (1972) gives a simpler account of these results. The structure of groups with a free subgroup of finite index was found by Karrass, Pietrowski, and Solitar (1973), Cohen (1973), and Scott (1974). Readers who know this material should not be surprised that they are proved easily from the Almost Stability Theorem, as the long proof of that theorem uses most of the ideas from the earlier proofs. The great advantage of the Almost Stability Theorem is that it offers the possibility of obtaining new results using that theorem, rather than having each time to give a new proof with only slight differences from the old ones (as happened, for instance, in extending the Stallings-Swan theorems to groups with a free subgroup of finite index).

Theorem 8.50 first appears in Dunwoody (1979), where he uses it to obtain Stallings' result on finitely generated groups with more than one end. Dunwoody regards this theorem as superseded by the Almost Stability Theorem, but I do not agree, as I have a fondness for this theorem. I had been endeavouring to find a proof of Stallings' theorem by obtaining a suitable tree on which the group acted. It was clear what the edges of the tree should be, but I could not discover what the vertices were to be. Dunwoody's insight was that it was unnecessary to find the vertex set explicitly; this kind of insight is an aspect of mathematics which I find very pleasing.

CHAPTER 9

For more about computability in free groups, see Boydron and Truffault (1974), Hoare (198), and Petresco (1962), among others. For instance, given finite subsets A and B of a free group, we know by Howson's Theorem 8.33 or 8.35 that there is a finite set C such that $\langle A\rangle \cap \langle B\rangle = \langle C\rangle$; it is possible to compute C, given A and B. There are also significant questions about the difficulty of solving problems in a free group; certain problems, for instance, are solvable in polynomial time, but are, in a sense which can be made precise, the most difficult problems solvable in polynomial time, as is shown by Avenhaus and Madlener (1981, 1982, 1984a,b).

Theorem 9.29 is due to Boone and Higman (1974).

The first account of the word problem is in Dehn (1912a,b; translated in the book Dehn (1980))

The first proof that there was a group with unsolvable word problem is due to Novikov (1955). This proof was simplifies by Boone (1959) and Britton (1958, 1963). The proof given here, taken from Aanderaa and Cohen (1980a) is the shortest.

The Embedding Theorem comes from Higman (1961). A proof by Valiev is given in Lyndon and Schupp's book. Other proofs occur in the books by Rotman (1973) and Shoenfield (1967). The proof here comes from Aanderaa and Cohen (1980b), which is again shorter than the other proofs, though it possibly gives less insight into why the theorem holds than do some other proofs (especially that in Shoenfield's book). The traditional proofs for several unsolvability results uses Turing machines. It seems that for group-theoretic decision problems the use of modular machines frequently simplifies the proofs considerably.

The various other unsolvability results proved comes from Baumslag, Boone, and Neumann (1959) and Rabin (1958); similar results were proved by Adian (1955, 1957, 1958).

Stilwell (1982) gives a survey of the word problem and related results, with further properties and more historical detail.

The results on one-relator groups were originally proved by Magnus (1930, 1932). Here we follow the approach of McCool and Schupp (1973).

BIBLIOGRAPHY

BOOKS

Adian, S.I., Boone, W.W., and Higman, G., eds. (1980) Word Problems II: the Oxford Book (North-Holland, Amsterdam)

Appel, K.I., Ratcliffe, J.G., and Schupp, P.E., eds. (1984) Contributions to Group Theory (Contemporary Mathematics 33, American Mathematical Society, Providence, Rhode Island)

Brown, R. (1968) Elements of Modern Topology (McGraw-Hill, New York; new edition, Ellis Horwood, Chichester, 1989)

Chandler, B. and Magnus, W. (1982) The History of Combinatorial Group Theory: a case study in the history of ideas (Springer, New York)

Cohen, D.E. (1972) Groups of Cohomological Dimension One (Lecture Notes in Mathematics 245, Springer, Berlin)

Cohen, D. E. (1987) Computability and Logic (Ellis Horwood, Chichester)

Dehn, M. (1987) Papers on Group Theory and Topology (translated and introduced by J. Stilwell) (Springer, New York)

Dicks, W. (1980) Groups, Trees and Projective Modules (Lecture Notes in Mathematics 790, Springer, Berlin)

Dicks, W. and Dunwoody, M.J. (1989) Groups Acting on Graphs (Cambridge University Press, Cambridge)

Gersten, S.M., ed. (1987) Essays in Group Theory (Springer, New York)

Gersten, S.M. and Stallings, J.R., eds. (1987) Combinatorial Group Theory and Topology (Annals of Mathematics Studies 111, Princeton University Press, Princeton)

Higgins, P.J. (1971) Notes on Categories and Groupoids (van Nostrand, London)

Johnson, D.L. (1976) Presentations of Groups (London Mathematical Society Lecture Notes 22, Cambridge University Press, Cambridge)

Lyndon, R.C. and Schupp, P.E. (1977) Combinatorial Group Theory (Springer, Berlin)

Magnus, W., Karrass, A., and Solitar, D. (1966) Combinatorial Group Theory (Wiley, New York)

Massey, W.S. (1967) Algebraic Topology: an introduction (Harcourt, Brace, and World, Inc., New York)

Nielsen, J. (1986) Collected Mathematical Papers (Birkhauser, Boston)

Reidemeister, K. (1932) Einfuhrung in die Kombinatorische Topologie (Vieweg, Braunschweig; reprinted, Chelsea, New York, 1950)

Rotman, J.J. (1973) The Theory of Groups (Allyn and Bacon, Boston)

Serre, J.-P. (1977) Arbres, Amalgames, SL_2 (Soc. Math. de France, series Asterisque 46)

Serre, J.-P. (1980) Trees (Springer, Berlin)

Shoenfield, J. (1967) Mathematical Logic (Addison-Wesley, Reading. Massachussets)

Stallings, J. R. (1971) Group Theory and Three-Dimensional Manifolds (Yale University Press, New Haven)

Stilwell, J. (1980) Classical Topology and Combinatorial Group Theory (Springer, New York)

PAPERS

Aanderaa, S. and Cohen, D.E. (1980a) Modular machines, the word problem for finitely presented groups, and Collins' Theorem, *in* Word Problems II: the Oxford Book (North-Holland, Amsterdam), 1-16

Aanderaa, S. and Cohen, D.E. (1980b) Modular machines and the Higman-Clapham-Valiev embedding theorem, *in* Word Problems II: the Oxford Book (North-Holland, Amsterdam), 17-28

Adian, S.I. (1955) Algorithmic unsolvability of problems of recognition of certain properties of groups, Dokl. Akad. Nauk SSSR (103), 533-535

Adian, S.I. (1957) The unsolvability of certain algorthmic problems in the theory of groups, Trudy Moskov. Mat. Obsc. (6), 231-298

Adian, S.I. (1958) On algorithmic problems in effectively complete classes of groups, Dokl. Akad. Nauk SSSR (123), 13-16

Althoen, S.C. (1974) A geometric realisation of a construction of Bass and Serre, J. Pure Appl. Alg. (5), 233-237

Avenhaus, J. and Madlener, K. (1981) *P*-complete problems in free groups, *in* Theoretical Computer Science: 5th GI conference, Karlsruhe (Springer Lecture Notes in Computer Science 104), 42-51

Avenhaus, J. and Madlener, K. (1982) The Nielsen reduction as key problem to polynomial reductions in free groups, *in* Computer Algebra: EUROCAM 82 (Springer Lecture Notes in Computer Science 144), 49-56

Avenhaus, J. and Madlener, K. (1984a) The Nielsen reduction and *P*-complete problems in free groups, Theoretical Computer Science (32), 61-76

Avenhaus, J. and Madlener, K. (1984b) On the complexity of intersection and conjugacy problems in free groups, Theoretical Computer Science (32), 279-295

Baer, R. and Levi, F. (1936) Freie Produkte und ihre Untergruppen, Compositio Math. (3), 391-398

Baumslag, B. and Tretkoff, M. (1978) Residually finite HNN extensions, Comm. Alg. (6), 179-194

Baumslag, G. (1963) On the residual finiteness of generalised free products of nilpotent groups, Trans. Amer. Math. Soc. (106), 193-209

Baumslag, G., Boone, W.W., and Neumann, B.H. (1959) Some unsolvable problems about elements and subgroups of groups, Math. Scand. (7), 191-201

Bergman, G.M. (1978) The diamond lemma for ring theory, Advances in Math. (29), 178-218

Bestvina, M. and Handel, M. (1989) Train tracks and automorphisms of free groups (preprint)

Book, R.V. (1988) Dehn's algorithm and the complexity of word problems, American Math. Monthly (95), 919-925

Boone, W.W. (1959) The word problem, Ann. of Math. (70), 207-265

Boone, W.W. and Higman, G. (1974) An algebraic characterisation of the solvability of the word problem, J. Austral. Math. Soc. (18), 41-53

Boydron, Y. and Truffault, B. (1974) Problemes de l'ordre generalisee pour les groupes libres, C.R. Acad. Sc. Paris (279), 843-845

Britton, J.L. (1958) The word problem for groups, Proc. London Math. Soc. (third series, 8), 493-506

Britton, J.L. (1963) The word problem, Ann. of Math. (77), 16-32

Brown, R. (1967) Groupoids and van Kampen's theorem, Proc. London Math. Soc. (third series, 8), 385-401

Burns, R.G. (1969) A note on free groups, Proc. Amer. Math. Soc. (23), 14-17

Burns, R.G. (1971a) On the intersection of finitely generated subgroups of a free group, Math. Zeit. (118), 121-130

Burns, R.G. (1971b) On finitely generated subgroups of free products, J. Austral. Math. Soc. (12), 358-364

Burns, R.G. (1972) On the finitely generated subgroups of an amalgamated product of two groups, Trans. Amer. Math. Soc. (169), 293-306

Burns, R.G. (1973) Finitely generated subgroups of HNN groups, Canadian. J. Math. (25), 1103-1112

Burns, R.G. and Chau, T.C. (1984) Another proof of the Grushko-Neumann-Wagner theorem for free products, *in* Contributions to Group Theory (op. cit.), 116-133

Chiswell, I.M. (1976a) Abstract length functions in groups, Math. Proc. Camb. Phil Soc. (80), 451-463

Chiswell, I.M. (1976b) The Grushko-Neumann theorem, Proc. London Math. Soc. (third series, 33), 385-400

Chiswell, I.M. (1979a) Embedding theorems for groups with an integer-valued length function, Math. Proc. Camb. Phil. Soc. (85), 417-429

Chiswell, I.M. (1979b) The Bass-Serre theorem revisited, J. Pure Appl. Alg. (15), 117-123

Chiswell, I.M. (1981) Length functions and free products with amalgamations of groups, Proc. London Math. Soc. (third series, 42), 42-58

Cohen, D.E. (1963) A topological proof in group theory, Proc. Camb. Phil. Soc. (59), 277-282

Cohen, D.E. (1973) Groups with free subgroups of finite index, *in* Conference on Group Theory (Springer Lecture Notes in Mathematics 319, Springer, Berlin)

Cohen, D.E. (1974) Subgroups of HNN groups, J. Austral. Math. Soc. (17), 394-405

Cohen, D.E. (1976) Finitely generated subgroups of amalgamated free products and HNN groups, J. Austral. Math. Soc. (22), 274-281

Cohen, D.E. (1977) Residual finiteness and Britton's Lemma, J. London Math. Soc. (second series, 16), 232-234

Cohen, M.M. and Lustig, M. (1988) On the dynamics and the fixed subgroup of a free group automorphism (preprint)

Collins, D.J. and Zieschang, H. (to appear) Combinatorial group theory and fundamental groups (Encyclopaedia of Mathematics, USSR)

Cooper, D. (1987) Automorphisms of free groups have finitely generated fixed point set, J. Alg. (111), 453-456

Culler, M. and Vogtmann, K. (1986) Moduli of graphs and outer automorphisms of free groups, Invent. Math. (84), 91-119

Dehn, M. (1912a) Transformation der Kurve auf zweiseitigen Flachen, Math. Ann. (72), 413-421

Dehn, M. (1912b) Uber unendliche diskontinuierliche Gruppen, Math. Ann. (71), 116-144

Dey, I.M.S. (1965) Schreier systems in free products, Proc. Glasgow Math. Assoc. (7), 61-79

Dunwoody, M.J. (1979) Accessibility and groups of cohomological dimension one, Proc. London Math. Soc. (third series, 38), 193-215

Federer, H. and Jonsson, B. (1950) Some properties of free groups, Trans. Amer. Math. Soc. (68), 1-27

Fischer, J. (1975) The subgroups of a tree product of groups, Trans. Amer. Math. Soc. (210), 27-50

Gersten, S.M. (1984) On Whitehead's algorithm, Bull. Amer. Math. Soc. (10), 281-284

Gersten, S.M. (1987) Fixed points of automorphisms of free groups, Advances in Math. (64), 51-85

Goldstein, R.Z. and Turner, E.C. (1984) Automorphisms of free groups and their fixed points, Invent. Math. (78), 1-12

Goldstein, R.Z. and Turner, E.C. (1986) Fixed subgroups of homomorphisms of free groups, Bull. London Math. Soc. (18), 468-470

Greendlinger, L. and Greendlinger, M.D. (1984) On three of Lyndon's results about maps, in Contributions to Group Theory (op. cit.), 212-213

Greendlinger, M.D. (1960) Dehn's algorithm for the word problem, Comm. Pure Appl. Math. (13), 67-83

Griffiths, H.B. (1954) The fundamental group of two spaces with a common point, Quart. J. Math. (4), 175-190

Griffiths, H.B. (1956) Infinite products of semigroups and local connectivity, Proc. London Math. Soc. (third series, 6), 455-485

Grushko, I.A. (1940) Uber die Basen einen freien Produktes von Gruppen, Sbornik (8), 169-182

Hall, M. Jr. (1949a) Coset representation in free groups, Trans. Amer. Math. Soc. (67), 421-432

Hall, M. Jr. (1949b) Subgroups of finite index in free groups, Canadian J. Math. (1), 187-190

Hall, M. Jr. and Rado, T. (1948) On Schreier systems in free groups, Trans. Amer. Math. Soc. (64), 386-408

Herrlich, F. (1988) Graphs of groups with isomorphic fundamental groups, Arch. Math. (51), 232-237

Higgins, P.J. (1964) Presentations of groupoids, with applications to groups, Proc. Camb. Phil. Soc. (60), 7-20

Higgins, P.J. (1966) Grushko's theorem, J. Alg. (4), 365-372

Higgins, P.J. and Lyndon, R.C. (1974) Equivalence of elements under automorphisms of a free group, J. London Math. Soc. (Second series, 8), 254-258

Higman, G. (1951) A finitely generated infinite simple group, J. London Math. Soc. (26), 61-64

Higman, G. (1961) Subgroups of finitely presented groups, Proc. Royal Soc. London, Series A (262), 455-475

Higman, G., Neumann, B.H., and Neumann, H. (1949) Embedding theorems for groups, J. London Math. Soc. (26), 247-254

Hoare, A.H.M. (1976) On length functions and Nielsen methods in free groups, J. London Math. Soc. (second series, 14), 188-192

Hoare, A.H.M. (1979) Coinitial graphs and Whitehead automorphisms, Canadian J. Math. (31), 112-123

Hoare, A.H.M. (1981) Nielsen methods in groups with a length function, Math. Scand. (48), 153-164

Hoare, A.H.M. (1984) Length functions and algorithms in free groups, *in* Contributions to Group Theory (op. cit.), 247-264

Holt, D.F. (1981) Uncountable locally finite groups have one end, Bull. London Math. Soc. (13), 557-560

Horadam, K.J. (1981) The word problem and related results for graph product group, Proc. Amer. Math. Soc. (820, 157-164

Howie, J. (1981) On pairs of 2-complexes and systems of equations over groups, J. Reine Angew. Math. (324), 165-174

Howson, A. G. (1954) On the intersection of finitely generated free groups, J. London Math. Soc. (29), 428-434

Imrich, W. (1977) Subgroup theorems and graphs, *in* Combinatorial Mathematics V (Lecture Notes in Mathematics 622, Springer, Berlin)

Juhasz, A. (1986) Small cancellation theory with a weakened small cancellation condition: (I) the basic theory, Israel J. Math. (55), 65-93

Juhasz, A. (1987a) Small cancellation theory with a weakened small cancellation condition: (II) the word problem. Israel J. Math. (58), 19-36

Juhasz, A. (1987b) Small cancellation theory with a weakened small cancellation condition: (III) the conjugacy problem, Israel J. Math. (58), 37-53

Karrass, A. and Solitar, D. (1970) The subgroups of a free product of two groups with an amalgamated subgroup, Trans. Amer. Math. Soc. (149), 227-255

Karrass, A. and Solitar, D. (1971) Subgroups of HNN groups and groups with one defining relation, Canadian J. Math. (23), 527-543

Karrass, A., Pietrowski, A., and Solitar, D. (1973) Finite and infinite cyclic extensions of free groups, J. Austral. Math. Soc. (16),

Karrass, A., Pietrowski, A., and Solitar, D. (1974) An improved subgroup theorem for HNN groups with some applications, Canadian J. Math. (26), 214-224

Kurosh, A.G. (1934) Die Untergruppen der freien Produkte von beliebigen Gruppen, Math. Ann. (109), 647-660

Lyndon, R.C. (1963) Length functions in groups, Math. Scand. (12), 209-234

Lyndon, R.C. (1965) Grushko's Theorem, Proc. Amer. Math. Soc. (16), 822-826

MacLane, S. (1958) A proof of the subgroup theorem for free products, Mathematika (5), 13-19

Magnus, W. (1930) Uber diskontinuierliche Gruppen mit einer definierenden Relation (Der Freiheitssatz), J. Reine Angew. Math. (163), 141-165

Magnus, W. (1932) Das Identitatsproblem fur Gruppen mit einer definierenden Relation, Math. Ann. (106), 295-307

McCool, J. (1974) A presentation for the automorphism group of a free group, J. London Math. Soc. (second series, 8) 259-266

McCool, J. (1975a) On Nielsen's presentation of the automorphism group of a free group, J. London Math. Soc. (second series, 10), 265-270

McCool, J. (1975b) Some finitely generated subgroups of the automorphism group of a free group, J. Algebra (35), 205-213

McCool, J. and Schupp, P.E. (1973) On one relator groups and HNN extensions, J. Austral. Math. Soc. (16), 249-256

Miller, C.F. III and Schupp, P.E. (1973) The geometry of HNN extensions, Comm. Pure Appl. Math. 26 (787-802)

Morgan, J.W. and Morrison, I. (1986) A van Kampen theorem for weak joins, Proc. London Math. Soc. (third series, 53), 562-576

Neumann, B.H. (1943) On the number of generators of a free product, J. London Math. Soc. (18), 12-20

Newman, M.H.A. (1942) On theories with a combinatorial definition of equivalence, Ann. of Math. (43), 223-243

Nickolas, P. (1985) Intersections of finitely generated free groups, Bull. Austral. Math. Soc. (31), 339-348

Nielsen, J. (1921) Om Regning med ikke kommutative Faktorer og dens Anvendlese i Gruppeteorien, Mat. Tidskrift B , 77-94

Novikov, P.S. (1955) On the algorithmic unsolvability of the word problem in group theory, Trudy Mat. Inst. Steklov 44 (translated in American Math. Soc. Translations (9))

Perraud, J. (1978) Sur les conditions de petite simplification qui permettent d'utiliser l'algorithme de Dehn, Math. Zeit. (163), 133-143

Petresco, J. (1962) Algorithmes de decision et de construction dans les groupes libres, Math. Zeit. (79), 32-43

Rabin, M.O. (1958) Recursive unsolvability of group theoretic problems,, Ann. of Math. (67), 172-194

Rabin, M.O. (1960) Computable algebra, general theory and theory of computable fields, Trans. Amer. Math. Soc. (95), 341-360

Schupp. P.E. (1971) Small cancellation theory over free products with amalgamations, Math. Ann. (193), 255-264

Schupp, P.E. (1984) Some aspects of Lyndon's work on group theory, *in* Contributions to Group Theory (op. cit.), 33-43

Scott, G.P. (1974) An embedding theorem for groups with a free subgroup of finite index, Bull. London Math. Soc. (6), 304-306

Scott, G.P. and Wall, C.T.C. (1979) Topological methods in group theory, *in* Homological Group Theory (London Mathematical Society Lecture Notes 36, Cambridge University Press, Cambridge)

Seifert, H. (1931) Konstruktion dreidimensionale geschlossener Raume, Ber. Sachs. Akad. Wiss. (83), 26-66

Stallings, J. R. (1965) A topological proof of Grushko's theorem on free products, Math. Zeit. (90), 1-8

Stallings, J.R. (1968) On torsion-free groups with infinitely many ends, Ann. of Math. (88) 312-334

Stallings, J.R. (1987) Foldings of G-trees (preprint)

Swan, R.G. (1969) Groups of cohomological dimension one, J. Alg. (12) 585-610

Stilwell, J. (1982) The word problem and the isomorphism problem for groups, Bull. Amer. Math. Soc. (new series, 6), 33-56

Tretkoff, M. (1973a) A topological proof of the residual finiteness of certain amalgamated free products, Comm. Pure Appl. Math. (26), 855-859

Tretkoff, M. (1973b) The residual finiteness of certain amalgamated free products, Math. Zeit. (132), 179-182

Tretkoff, M. (1975) Covering space proofs in combinatorial group theory, Comm. Alg. (3), 429-457

Tretkoff, M. (1980) A topological approach to the theory of groups acting on trees, J. Pure Appl. Alg. (16), 323-333

van der Waerden, B.L. (1948) Free products of groups, Amer. J. Math. (70), 527-528 van Kampen, E.R. (1933) On the connection between the fundamental groups of some related spaces, Amer. J. Math. (55), 261-267

Wagner, D.H. (1957) On free products of groups, Trans. Amer. Math. Soc. (84), 352-378

Weir, A.J. (1956) The Reidemeister-Schreier and Kurosh subgroup theorems, Mathematika (3), 47-55

Whitehead, J.H.C. (1936a) On certain sets of elements in a free group, Proc. London Math. Soc. (second series, 41), 48-56

Whitehead, J.H.C. (1936b) On equivalent sets of elements in a free group, Ann. of Math. (37), 782-800

INDEX